MICROFLUIDICS AND BIOMEMS APPLICATIONS

MICROSYSTEMS

Volume 10

OTHER BOOKS IN THE SERIES:

Methodology for the Modelling and Simulation of Microsystems
Bartlomiej F. Romanowicz
Hardbound, ISBN 0-7923-8306-0, October 1998
Microcantilevers for Atomic Force Microscope Data Storage
Benjamin W. Chui
Hardbound, ISBN 0-7923-8358, October 1998
Bringing Scanning Probe Microscopy Up to Speed
Stephen C. Minne, Scott R. Manalis, Calvin F. Quate
Hardbound, ISBN 0-7923-8466-0, February 1999
Micromachined Ultrasound-Based Proximity Sensors
Mark R. Hornung, Oliver Brand
Hardbound, ISBN 0-7923-8508-X, April 1999
Microfabrication in Tissue Engineering and Bioartificial Organs
Sangeeta Bhatia
Hardbound, ISBN 0-7923-8566-7, August 1999
Microscale Heat Conduction in Integrated Circuits and Their Constituent Films
Y. Sungtaek Ju, Kenneth E. Goodson
Hardbound, ISBN 0-7923-8591-8, August 1999
Scanning Probe Lithography
Hyongsok T. Soh, Kathryn Wilder Guarini, Calvin F. Quate
Hardbound, ISBN 0-7923-7361-8, June 2001
Optical MicroScanners and Microspectrometers using Thermal Bimoph Actuators
Gerhard Lommel, Sandra Schweizer, Philippe Renaud
Hardbound, ISBN 0-7923-7655-2, January 2002

Microfluidics and BioMEMS Applications

edited by

Francis E.H. Tay

The National University of Singapore

KLUWER ACADEMIC PUBLISHERS

BOSTON / DORDRECHT / LONDON

A C.I.P. Catalogue record for this book is available from the Library of Congress.

ISBN 1-4020-7237-6

Published by Kluwer Academic Publishers,
P.O. Box 17, 3300 AA Dordrecht, The Netherlands.

Sold and distributed in North, Central and South America
by Kluwer Academic Publishers,
101 Philip Drive, Norwell, MA 02061, U.S.A.

In all other countries, sold and distributed
by Kluwer Academic Publishers,
P.O. Box 322, 3300 AH Dordrecht, The Netherlands.

Printed on acid-free paper

Printed in the Netherlands.

TABLE OF CONTENTS

LIST OF CONTRIBUTORS

PART I *by Francis E. H. Tay and W. O. Choong*

PART II

Chapter 6
Development of Integrated Microfluidic Devices for Genetic Analysis
by Robin H. Liu and Piotr Grodzinski; Microfluidics Laboratory, PSRL, Motorola Labs, USA.

Chapter 7
Microfluidic Devices on Printed Circuit Board
by Stefan Richter, Nam-Trung Nguyen, Ansgar Wego, and Lienhard Pagel; University of Rostock, Germany; Nanyang Technological University, Singapore.

Chapter 8
Nano and Micro Channel Flows of Bio-Molecular Suspension
by Fan Xijun, Nhan Phan-Thien, Ng Teng Yong, Wu Xuhong, and Xu Diao; Institute of High Performance Computing, Singapore; Bioengineering Division, Natinal University of Singapore, Singapre.

Chapter 9
Transport of Liquid in Rectangular Micro-Channels by Electroosmotic Pumping
by Chun Yang; School of Mechanical and Production Engineering, Nanyang Technological University, Singapore.

Chapter 10
A Development of Slip Model and Slip-Corrected Reynolds Equation for Gas Lubrication in Magnetic Storage Device

by Eddie Yin-Kwee Ng, Ningyu Liu, and Xiaohai Mao; Nanyang Technological University, Singapore.

Chapter 11
Short Notes on Particle Image Velocimetry for Micro/Nano Fluidic Measurements
by Chee Yen Lim and Francis E. H. Tay; Institute of Materials Research and Engineering, Singapore.

FOREWORD

The past 10-15 years have shown an explosive growth in research on and development of BIOMEMS, Micro Total Analysis Systems (μTAS), or Lab-on-Chip devices. The field emerged as a promising marriage between the well-established area of MEMS and that of bio-analytical chemistry, and was in particular driven by the immense efforts put in the Human Genome Project and related, mostly US-based, funding. The field was originally constituted by researchers from micromechanical and analytical chemistry, as rapid amino acid and DNA separations carried out in micromachined channels in quartz were the most powerful demonstration of the potential of the Lab-on-Chip concept.

More recently the field has enormously widened, comprising areas such as protein analysis, genomics and proteomics, cell analysis, DNA diagnostics, environmental assays, drug discovery, chemistry on chips, and biochemical arrays. A very important common aspect in all of these applications, however, remains the handling and control of extremely small amounts of fluids (mostly liquids), also called microfluidics. After the first demonstration of a micromachined piezo-actuated pump in our institute in the late '90's (van Lintel et al.) it became clear that fluid flow control on a small scale was possible. Nowadays, the area of microfluidics is considered crucial for development of Lab-on-Chip applications, as it enables careful handling and treatment of minuscule amounts of precious biochemical samples.

Unfortunately up to now no textbooks exist that could serve to introduce students in this multidisciplinary field. In this book a two-track approach has been chosen: in the first part a case study, the development of a micropump, is systematically worked out. This will be a great help for those researchers that focus on efficiently developing particular, well-defined microfluidic devices using all modern techniques and tools currently available. The second part of the book forms a very useful complement to the first part, as it presents a variety of examples of microfluidics theory, phenomena, techniques and concepts authored by international leading experts.

Here you find a book that will undoubtedly help to develop skills needed for the design and realization of microfabricated fluid devices whereas it provides at the same time an introduction to the world of microfluidics in all its aspects. All people working in the area of microfabricated fluid devices, should consider this book as an indispensable and very valuable tool.

Albert van den Berg
Enschede, July 10th, 2002.

PREFACE

The idea of writing this book comes as a result of frequent enquiries about the possibility of publishing a book on documenting MEMS problems in its golden era. A first round survey of the literature showed that abundant researches on almost every aspect of MEMS have been carried out in recent years, so that any of the books can no longer give a comprehensive and balanced picture of the field. As a result, it was felt that a direct collection of peer-reviewed papers was hardly adequate; instead a new book based on substantially new concept was prepared, which is now placing before the reader. In planning this book, it soon became apparent that even if only the most important developments and relevant problems which took place from recent publications were documented, the book would seem impracticably thick. It was, therefore, necessary to restrict its scope to a narrower field. These subjects can be treated more appropriately if only the central ideas of MEMS development are tackled. The fact would be that, even after this limitation, the book would have been much larger than that in our expectation, giving some implication about the extent of the researches that have been carried out in MEMS related fields in recent times. We have envisaged that by demonstrating a comprehensive development of a micro pump, which is the heart of most lab-on-a-chip or BioMEMS devices for delivering fluid flows, we can illustrate the overall picture in the research and development of these tiny devices. We have also incorporated several selected articles that we think are useful in microfluidics and BioMEMS applications, hovering around the idea that MEMS devices are mostly associated with microfluidic flows; this is especially true for biomedical related devices, which have gained popularity and attention for their various miniaturization effects. These miniaturization effects include strict tolerance of high-yield, batched processes in fabrication, quick mechanical response due to small inertia, maximum saving on chemicals and reagents, minimum waste on samples, etc.

We have aimed at providing, within the skeleton we just outlined, a reasonably self-contained picture of our present knowledge and compilation that are separated into two parts in the book. In the first part, we have attempted to present the approach of tackling a research problem in MEMS device development from the fundamentals, including a complete literature survey, modelling and simulation, process development and fabrication, and verification and testing. We have intentionally left out packaging and commercialisation stages for their industrial and commercial complexities. The second part of the book incorporates some valuable articles that are related to microfluidics and its applications, mostly of biomedical use or the related advanced modellings. The purpose of their incorporation as a separate part in the book serves the purpose of complementing the subject discussed in Part I, whose central idea is tackling a MEMS development problem. A specific device – the micro pump, deemed to be the most appropriate and relevant being the heart for any biomedical related and lab-on-a-chip products, has been exploited for this demonstration purpose. Flowchart I below depicts the concept of the presentation of the book, showing how the second part of article collection complement the first part of the book, which weighs equally with its counterpart in terms of length and importance:

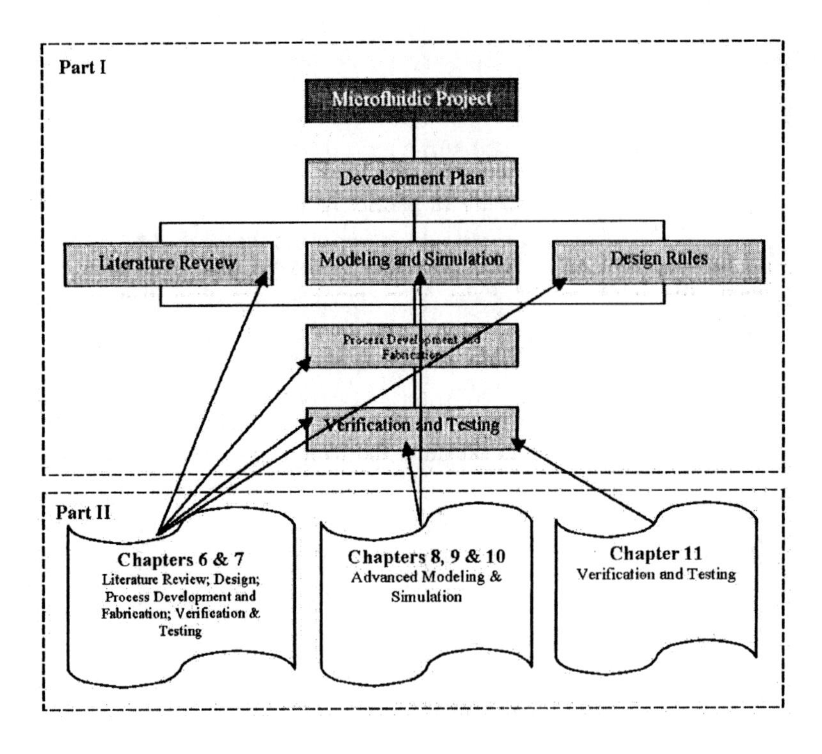

Flowchart I. Complementary relationship between Part I and Part II

As stated above, this book is divided into two parts. Part I consists of five chapters that deal with conception, design, fabrication and testing of a MEMS device. Chapter 1 presents a detailed literature review for a micro-pump, which is always a primary and mandatory step in performing any research irrespective of their nature. It categorizes micro pump into mechanical and non-mechanical categories (without moving parts, see Flowchart II). The development trend for micro pump and its respective pumping principles are also studied and reported. The last portion of this chapter discusses the motivation of this entire part of the book.

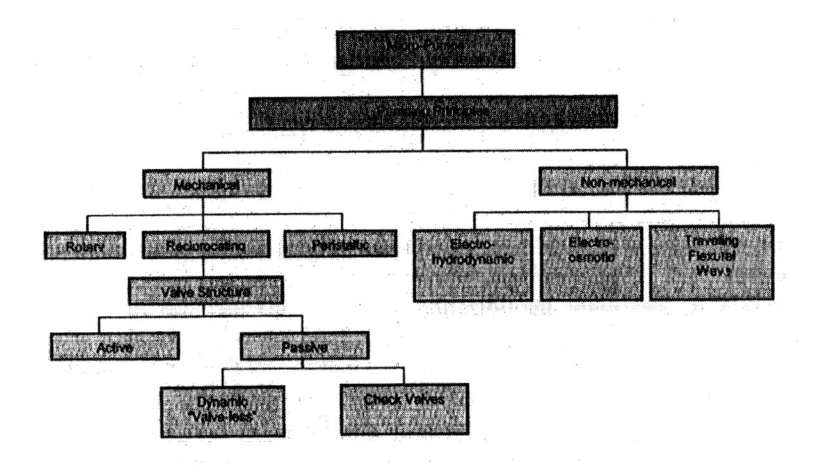

Flowchart II. Micro pumps categories by their pumping principles

Chapter 2 presents the design rules which are applicable to other development projects, reporting some mandatory steps in preliminary designing and consideration for a micro pump. In this chapter, compression ratio, valve switching criteria, self-priming capability and bubble tolerance are discussed. While the approach is limited to designing a micro pump, it is the author wish that the reader can realize the general rules of thumb in designing similar devices by induction.

Chapter 3 is mainly concerned with modelling and simulation, which is often necessary before any fabrication of real prototype. It covers some governing equations based on Navier-Stokes equations, individual modellings of actuator, inlet valve and exit valve, and construction of a complete model. Again, by the demonstration of micro pump, the message behind the text is clear, and realization of the hidden application idea is quite clear in applying to other MEMS development projects.

Chapter 4 deals with process development and fabrication of a micro pump, which involve mainly photolithography, thin film deposition, and wet etching. The entire process flow is also shown from silicon and glass processes, to wafer bonding.

The last chapter in Part I, Chapter 5, contains the experimental verification and preliminary functional test for the micro pump developed based on previous chapters, which also includes some recommendations as final comments.

Part II comprises six chapters, which are contributions from various prominent authors in the world. All but the last chapter were selected from contributions made to the International MEMS Workshop 2001, from 4th to 6th July 2001 at National University of Singapore, Singapore. Chapter 6, a contribution by Robin H. Liu and Piotr Grodzinski that bears the title "Development of integrated microfluidic devices for genetic analysis," discusses the design, fabrication, and testing of plastic microfluidic devices for on-chip genetic sample preparation and DNA microarray detection. Genetic analysis and microfluidics theory were included in its introduction, from which some beginners may find them very useful for quick references. Devices such as micromixers, microvalves, cell capture devices, on-chip micro-PCR, and other hybridization enhancements are also discussed in details.

Chapter 7 "Microfluidic devices on printed circuit board," is contributed by Stefan Richter *et al.*, revealing a new approach for fabrication of microfluidic devices based on printed circuit board (PCB) technology. Descriptions on the basic process steps in PCB fabrication considering the special needs of fluidic components to make it easier to understand the technological approach. Similar to Chapter 6, design, fabrication and characteristics of a number of sensors and actuators are also presented. Their Results of active components, including various types of pumps and sensors prove the feasibility of this new fabrication concept.

Chapter 8 "Nano and micro channel flows of bio-molecular suspension," contributed by Xijun Fan *et al.*, describes two particle methods, the molecular dynamics (MD) and dissipative particle dynamics (DPD), for simulating flow problems in micro and nano scales. Their simulation principles, governing equations and numerical implementations are presented in this paper, using molecular models that are used in polymer rheology to model bio-macromolecules. Results for both MD and DPD simulations are also presented and discussed.

Chapter 9 "Transport of liquid in rectangular microchannels by electroosmotic pumping," contributed by Chun Yang, provides a theoretical framework for the phenomenon of electroosmotic flow in microchannels, in which the author analyses the characteristics of transient and steady-state electroosmotic flows in rectangular microchannels. An analytical solution to two-dimensional Poisson-Boltzmann equation that governs the electrical double-layer filed near the solid-liquid interface is developed here based on Debye-Hückel approximation and Greens' function formula. Effects of various parameters on the electroosmotic velocity distributions in this chapter provide insights into the electroosmotic phenomenon.

Chapter 10 "A development of slip model and slip-corrected Reynolds equation for gas lubrication in magnetic storage device," contributed by Eddie Yin-Kwee Ng *et al.*, reports the implementation of direct simulation of Monte Carlo (DSMC) for ultra thin film lubrication environment. The authors find that the slip velocity is affected not only by shear stress, but also by the density. Through this study, a new slip velocity model – stress-density ratio (SDR) model particularly suitable to the flows with large Knudsen numbers is developed, and a slip-corrected Reynolds equation is subsequently derived and solved numerically.

Chapter 11 "Short notes on particle image velocimetry for micro/nano fluidic measurements," contributed by Chee Yen Lim and Francis Eng Hock Tay, discusses the application of a mature fluid visualization and measurement technique to measure fluidic flows in micro/nano scales. Several limiting issues, such as particle size, light diffraction limit, and the instrumentation were discussed. Some potential research directions were also included. Table I below summarises the central topics and key ideas of each chapter for both parts in the book:

Table I. Summary of chapters

Ch.	Title/Authors	Central Topic	Methodology	Key Ideas	Key words
1	Literature Review	Micro pump development and categorization	Complete search on the literature	Gathering, understanding and analysing facts	Micro pumps, mechanical pumps, non-mechanical pumps
2	Design Rules	Design considerations for piezoelectric actuation micro pump	Breakdown of consideration factors and design criteria	Determining criteria for dead volume, valve switching, self-priming capability and bubble tolerence	Piezoelectric actuation, compression ratio, volume stroke, dead volume, finite element model, ABACUS, ANSYS
3	Modelling & Simulation	Performing simulations of components and the entire	Computational, by commercial software	Development and simulations of inlet and outlet valves, then	Frequency mode shapes, FEM simulations

Ch.	Title/Authors	Central Topic	Methodology	Key Ideas	Key words
		system before fabrication		the entire system	
4	Process Development & Fabrication	Prototype building	Process and fabrication	Building the pump by thin-film deposition, etching, and wafer bonding	Bulk silicon micromachining, photolithography, thin film deposition, wet etching, wafer bonding
5	Verification & Testing	Verifying the fabricated pump	Experimental	Pump functionality test and characterization	Functional tests, FEM model verifications
6	Robin H. Liu and Piotr Grodzinski	Device development, and system integration for on-chip functionalities	Process development and system integration	Demonstrating plastic micro-fabrication, micromoulding, etc. for bio-technology applications	Micro-mixers/ valves, cell capture device, integrated micro PCR device, and microchannel DNA hybridazation arrays
7	Stefan Richter *et al.*	Fabrication of microfluidic devices by PCB technology	Design, fabrication, and characterization; numerical simulations	Development of micro pumps, pH, flow & pressure sensors on PCB	PCB, sensors, actuators, micro pumps
8	Xijun Fan *et al.*	Advanced modelling on bio-molecular suspension	Computational, using molecular dynamics (MD) and dissipative particle dynamics (DPD)	Implementing molecular models for simple and suspended liquids in nano-channel	Biomacromolecular suspensions, MD, DPD, nano-/micro-channel
9	Chun Yang	Advanced modelling on electroosmotic pumping	Computational, based on basic transport equations	Analysing characteristics of transient/ steady state electroosmotic flow in rectangular micro-channel	Electroosmotic flow, BioMEMS pumping, rectangular micro-channel, Greens' function approach

Ch.	Title/Authors	Central Topic	Methodology	Key Ideas	Key words
10	Eddie Yin-Kwee Ng *et al.*	Slip consideration in film lubrication	Computational, by slip-corrected Reynolds equation	Development of Stress-Density ratio (SDR) model, and comparisons to various models	SDR model, DSMC, Reynolds equation, slip coefficient, film lubrication
11	Chee Yen Lim and Francis E. H. Tay	Discussion on modern flow visualization technique app-lied to micro-/nano-fluidic measurements	Experimental in nature, by particle image velocimetry (PIV)	Flow visualization and characterization in micro-/nano-scales by nano-particles	CCD cameras, pulsed laser, microscope, nano-particles, cross-correlation, fast Fourier transform (FFT)

Though we have produced a book which in its methods of presentation and general approach would seem micro pump orientated in Part I, it would be evident from its later part that the present book is neither a documentary about developing a micro pump nor entirely a compilation of prestigious articles. Flowchart I above illustrates how the second part of the book complements the first part by exhibiting its related applications in literature surveying, designing, fabrication, and testing. Some advanced numerical modellings and simulations are also included for this purpose. We hope the reader can benefit from a series of related chapters and gain proficiency in tackling other research projects with similar nature and capacity. The references that are given, which we hope to cover the most important and relevant papers, are to equip the reader with a starter kit, or to help them attain some bearings in the literature; an omission of any particular research work should not be interpreted as due to our lack of regard for its merit. We have not included in our consideration about attempting the task of referring to all the relevant publications, which will be extremely difficult if not impossible. Lastly, we have, not always been able to use the most elegant notation but we hope that we have succeeded, at least, in avoiding the use in any one section of the same symbol for different quantities.

Francis E. H. Tay
Editor

ACKNOWLEDGEMENTS

It is a pleasure to thank many friends and colleagues for advice and help. In the first place we wish to record our gratitude to Mr Andojo Ongkodjojo Ong and Mr Yen Peng Kong for useful advice and assistance in early stages of this project, as well as their support for coordinating the article contributions from various authors through communications. We are also greatly indebted to Mr Chee Yen Lim, who helped in final stages of coordination work and in preparing camera-ready draft, including the index of the book.

PART I

Chapter 1

LITERATURE REVIEW FOR MICROPUMPS

1.1 Origins of Micropump Research

A survey through the literature reveals that one of the very first documented reports of the concept of a micropump (or miniaturized pump to be exact) can be found in [3] in 1975. In the patent, the proposed miniaturized pump, which was designed for implantation into the human body, consists of a variable volume chamber bounded by two apposed piezoelectric disk benders, forming a bellows, connected to a solenoid valve (Fig.1). The pump is driven by a rectangular wave pulse generator coupled with a transformer for high voltage amplification. As illustrated in Fig. 1-1(a), in response to a pulse of one polarity (positive), the solenoid valve is activated to retract the plunger against the spring to open the outlet and close the inlet. At the same time, the variable volume chamber begins to expel some of its contents because the voltage induced in the secondary of the transformer causes the disk benders to flex inward. With reference to Fig. 1-1(b), the current flow from the pulse generator stops suddenly at the end of the rectangular wave pulse. This deactivates the solenoid and allows the spring to return the armature to its extended position thus opening the inlet and closing the outlet. In addition, this sudden cessation of current flow through the primary of the transformer induces a large voltage in the secondary which now has a polarity opposite to that of the preceding half cycle. This causes the disk benders to flex in the opposite direction (outward) so that fluid is drawn into the variable volume chamber from the reservoir.

The pump was designed to deliver down to slightly under 0.2μl of fluid per pump stroke and the total delivery is a function of the number of pulses from the controller. It was capable of developing a maximum pressure head of about 80mmHg at an input voltage of 35V. The overall dimensions of the pump were not available but the chamber was estimated to be about 20mm thick and 35mm in diameter.

Several other patents based on conventional fabrication methods such as milling and plastic moulding were also filed in the seventies through to the nineties [4-9]. Even though the idea of a micropump was considerably

F.E.H. Tay (ed.), Microfluidics and BioMEMS Applications, 3–140.

different in terms of the fabrication method and general design, an interesting point was the realisation of the practical potential of such micropumps in the field of biotechnology.

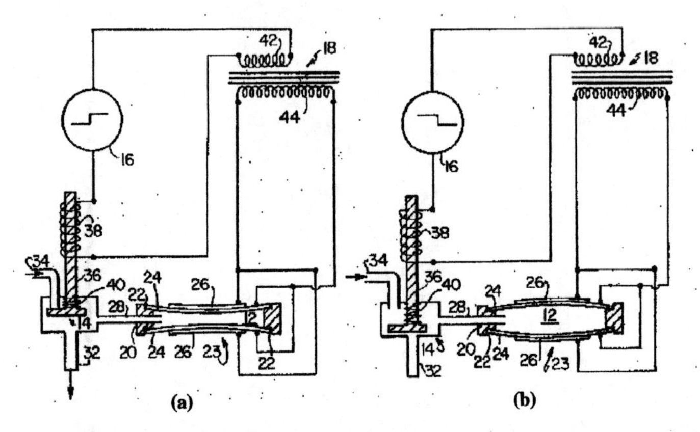

Figure 1-1. Schematic of a micropump powered by piezoelectric disc benders. (a) Pumping stroke and (b) intake stroke [3]

It was not until 1984 when the first patent of a micropump based on integrated chip (IC) microfabrication technologies was filed [10] by J.G. Smits. The micropump was a peristaltic pump with three inline active valves which were actuated by piezoelectric disks, arranged in different configurations (Fig. 1-2). In fact, Smits was the first to initiate research on micropumps based on microvalves in the early eighties at Stanford University.

Conventional peristaltic pumps typically force the fluid along by waves of contraction produced mechanically by means of flexible tubing. This principle is derived from the phenomenon of peristalsis in biology which refers to successive waves of involuntary contraction passing along the walls of hollow muscular structures (such as the oesophagus or intestine) that force the contents onward.

The pump proposed in the above-mentioned patent was based on the use of an inlet valve and an outlet valve to be opened and closed selectively, and a displacing member. The valves and displacing member were realised by positioning either piezoelectric mono-morphs or bimorphs on a diaphragm to form piezoelectrically controlled elements. The valves and displacing member are connected in series in a channel and were functionally controllable for obtaining a peristaltical displacing of the fluid to be pumped through the channel. Two designs were proposed, one being oblong (Fig. 1-2(a)) and the other triangular in shape (Fig. 1-2(b)). Both pumps can be realised by means of micromachining of silicon and glass wafers and then anodically bonding them together.

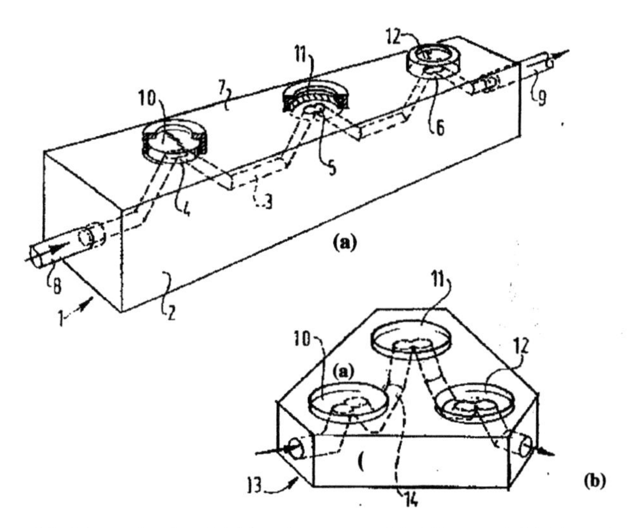

Figure 1-2. Schematic of peristaltic micropumps. (a) Oblong and (b) triangular design [10]

Following this pioneering work by Smits, many patents on micropumps especially those based on IC fabrication technologies surfaced in the next twenty years [11-16]. The patents were not confined to the prior mentioned types of micropumps (i.e. positive displacement) but other innovative types such as ultrasonic, electroosmotic, electrohydrodynamic and rotary micropumps. It was also during this period when there was a marked increase in academic activity and interest in the area of micropumps and microfluidics research as a whole. In the following sections, a survey of these micropumps will be presented in detail.

1.2 Mechanical Micropumps

Micropumps can generally be classified into two groups: mechanical and non-mechanical (without moving parts) (Fig. 1-2). At least three kinds of mechanical micropumps have been developed: peristaltic, reciprocating and rotary pumps. For non-mechanical micropumps, there are electrohydrodynamic, electroosmotic and ultrasonic micropumps among others. As to date, the reciprocating displacement micropump has the most popular choice for researchers (Fig. 1-3 and 1-4). A breakdown of some of the mechanical micropumps developed thus far and their performances is given in Table 1-1.

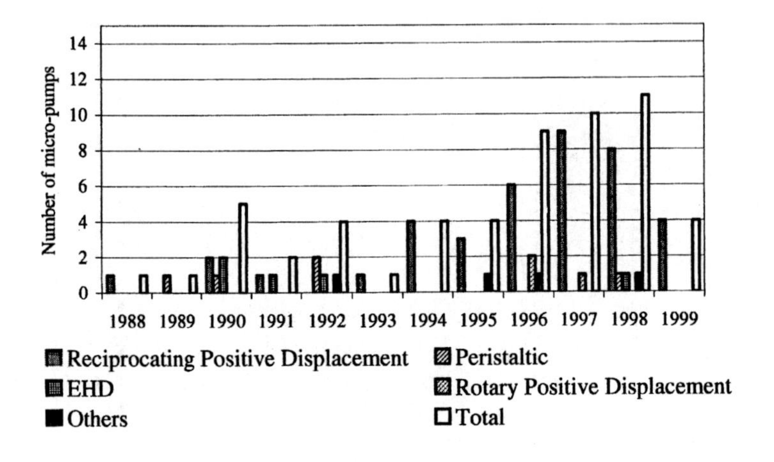

Figure 1-3. Trend in micropump research: pumping principle

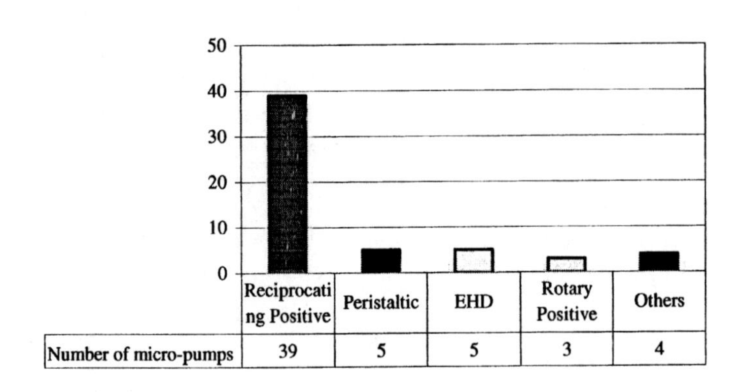

Figure 1-4. Breakdown of micropump research: pumping principle

Table 1-1. Characteristics of mechanical micropumps

Pumping Principle	Actuating Principle	Date	Author	Maximum Flow rate (μl/min)*	Maximum Pressure
Reciprocating displacement	Piezoelectric disc	1988	H. T. G. Van Lintel et al.	<10	2 mH_2O
Peristaltic	Piezoelectric disc	1989	Jan G. Smits	3	0.6 mH_2O

Reciprocating displacement	Thermo-pneumatic	1990	F. C. M. Van De Pol	34	0.5 mH_2O
Reciprocating displacement	Piezoelectric stack	1990	Shuichi Shoji et al.	40	1 mH_2O
Reciprocating displacement	Electrostatic	1991	Jack W. Judy et al.	12 to 640 nl per cycle	
Peristaltic with 4 valves	Laser light driven (microcells)	1992	Hideo Mizoguchi et al.	5.4 (30nl per cycle at 3Hz)	0.03 mH_2O
Reciprocating displacement	Electrostatic	1992	R. Zengerle et al.	70	0.25 mH_2O
Reciprocating displacement	Piezoelectric disc	1993	Erik Stemme et al.	16 ml/min	2 mH_2O
Reciprocating displacement	Thermo-pneumatic	1994	B. Bustgens et al.	44	3.8 mH2O
Reciprocating displacement	External pneumatic drive	1994	R. Rapp et al.	83 (gas)	0.47 mH_2O (gas) and 240 mmH_2O
Rotary positive displacement	Driven by external motor	1996	Thomas Weisener et al.	Max. 100 ml/min at 6000rpm 8 ml/min at 20000rpm	Max. 500mH_20 at 6000rpm Max. 150mH_20 at 20000rpm
Reciprocating displacement	Piezoelectric disc (bimorph and cantilever type)	1996	M. Stehr et al.	1600□l/min in forward direction and 1200□l/min in reverse direction for liquids. 8000□l/min in both directions for gas.	1.7 mH_2O
Reciprocating displacement	Piezoelectric disc	1996	Anders Olsson et al.	2.3 ml/min	7.6 mH_2O
Rotary positive displacement	Miniaturized electromagnetic asynchron motors or stepping motors	1996	J. Dopper et al.	1 ml/min (oil) and 200 ml/min (glycerin/water)	12 mH_2O (oil) and 0.7 mH_2O (water)
Reciprocating displacement	Piezoelectric disc	1996	J. Dopper et al.	300	10 mH_2O
Reciprocating displacement	Electromagnetic	1996	P Dario et al.	13 ml/s	0.55 mH_2O
Reciprocating displacement	Piezoelectric disc	1997	Anders Olsson et al.	2.3 ml/min	7.6 mH_2O
Reciprocating displacement	SMA	1997	W. L. Benard	50	
Reciprocating displacement	Piezoelectric disc	1997	Ingo Ederer et al.	48 g/h of RME fuel	
Reciprocating displacement	Piezoelectric (thick film)	1997	M Koch et al.	120	0.2 mH_2O
Reciprocating displacement	Piezoelectric disc	1997	Torsten Gerlach	7.8 ml/min (gas)	0.28 mH_2O (gas)
Rotary positive displacement	Electromagnetic	1997	Andrew S. Dewa et al.	350	0.14 mH_2O
Reciprocating displacement	SMA	1998	William L. Bernard et al.	50	0.053 mH_2O
Reciprocating displacement	Piezoelectric disc	1998	R. Linnemann	1 to 1.4 ml/min	1 mH_2O
Reciprocating displacement	Piezoelectric disc	1998	K. P. Kamper et al.	400	21 mH_2O
Peristaltic	Pneumatic	1998	Norbert Schwesinger et	800	

			al.		
Reciprocating displacement	Piezoelectric disc	1999	Didier Maillefer et al.	100 ml/h	2 mH_2O
Reciprocating displacement	Piezoelectric stack	1999	Jung-Ho Park et al.	80 mm^3/s	32 mH_2O

Note: * flow rate in mm/min unless units otherwise stated

1.2.1 Reciprocating Micropumps

This type of micropump generally comprises of a pressure chamber bounded by a flexible diaphragm driven by an actuator and passive or dynamic microvalves. The check valves and the actuator play very important roles in the flow rate and the maximum output pressure. Many types of actuators such as piezoelectric, pneumatic, electrostatic and thermo-pneumatic have been used. The maximum pressure head realizable by the micropump depends directly on the available force of the actuator used (Table 1-2).

Table 1-2. Comparison of different actuation means

ACTUATION METHODS	PRESSURE	DISPLACEMENT	RESPONSE TIME
External			
Disk type piezoelectric	Small	**Medium**	Fast
Stack type piezoelectric	**Very large**	**Very small**	Fast
Pneumatic	Small	Large	Slow
Shape Memory Alloy	Large	Large	Slow
Electromagnetic	Small	**Large**	Fast
Micromachinable			
Electrostatic	Small	Very small	Very fast
Thermo-pneumatic	Large	Medium	Medium

Note:

Pressure:

Very large: P > 100kgf/cm2

Large: 100kgf/cm2 > P > 1

Medium: 1 > P > 0.5

Small: 0.5 > P

Displacement:

Large: d > 100 mm

Medium: 100 > d > 30

Small: 30 > d > 10

Very small: $10 > d$

Response time:

Very fast: $t < 0.1ms$

Fast: $1 > t > 0.1ms$

Medium: $1s > t > 1ms$

Slow: $t > 1s$

Source: [17]

1.2.1.1 Piezoelectric Actuation

Even though Smits was the first to toy with the concept of a silicon micropump, it was Van Lintel et al who first published their work on a silicon based micropump [18]. In the paper, two pumps based on the micromachining of silicon were described. The pumps, which were of the reciprocating displacement type, comprise one or two pump chambers, a thin glass pump membrane actuated by a piezoelectric disc and passive silicon check valves to direct the flow (Fig. 1-5). The deflection of the pump chamber will cause the pump chamber pressure to either increase or decrease depending on the direction of deflection. This in turn creates a differential pressure across the valves that either opens or closes them. This sequential upwards and downwards deflection of the membrane results in a pumping effect. In this piece of work, the chambers, channels and valves were realized in a 2" silicon wafer by wet chemical etching. The silicon wafer was then bonded, top and bottom, to two borosilicate glass wafers by anodic bonding. The piezoelectric disc was attached by means of cyano-acrylate adhesive. This was a significant piece of work because it was the first reported work on a successfully fabricated micropump using micromachining technologies.

Shortly following that, M. Esashi et al also reported on a silicon micropump in 1989 [19]. A normally closed microvalve and micropump were fabricated on a silicon wafer by micromachining techniques. The developed micropump was a diaphragm-type pump that consisted of two polysilicon one-way valves and a diaphragm driven by a small piezoelectric actuator. The maximum pumping flow rate and pressure were 20 µl/min and 780 mmH_2O/cm^2 respectively. Both of the two earliest developed micropumps are fundamentally of the reciprocating positive displacement type. This type of micropump is the most researched upon and popular choice of researchers because of its relative simplicity and self-priming characteristic in addition to its capability of developing a high pressure head and high flow rate. However, such pumps suffer from some major drawbacks, namely, the flow is of a periodic nature and they are highly susceptible to blockage especially for the case of integrated passive valves. Another major disadvantage of these piezoelectrically driven pumps is the fact that high supply voltages (more than 100V) are needed. In addition, the application of piezoelectric discs is not compatible with integrated fabrication.

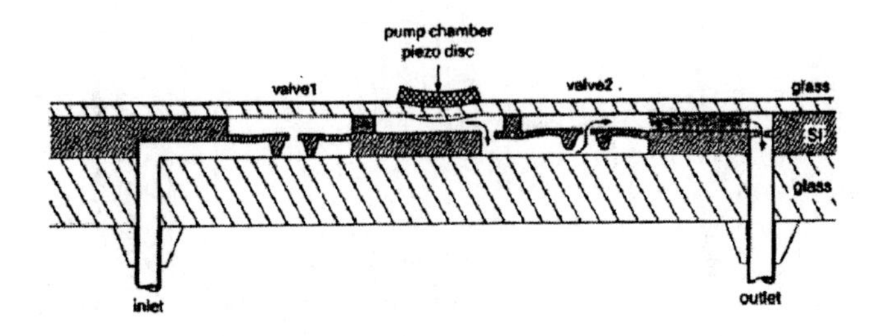

Figure 1-5. Schematic of the first reciprocating displacement micropump [18]

Nevertheless, reciprocating micropumps, especially those based on piezoelectric actuation, have grown to be the dominant type of micropump and tremendous improvements have been made [20-22]. In 1996, one of the very first commercial micropumps was developed by IMT of Germany [23]. The pump, made predominantly from polycarbonate, was fabricated using a combination of silicon micromachining and conventional thermoplastic replication technologies.

1.2.1.2 Thermo-pneumatic Actuation

Until 1990, all the micropump prototypes developed made use of piezoelectric bimorph or monomorph discs for the actuation. Driven by the aim to realize a pump by merely applying microengineering techniques like thin-film technology, photolithography techniques and silicon micromachining, researchers have to look into "micromachinable" actuators (Fig. 1-3). The first piece of work on the utilization of micromachinable actuators in micropumps was carried out by F.C.M. Van De Pol et al in 1990 [24]. This actuation principle was adopted from Zdelblick & Angell who used it in a micro miniature valve, using a gas/liquid system and resistive heating [25]. The micropump is basically a reciprocating displacement pump with passive valves. The actuator comprises of a cavity filled with air, a square silicon pump membrane and a built-in aluminium meander, which served as a heater resistor (Fig. 1-6). A thin silicon sheet suspended by four small silicon beams, which serve as thermal insulators, supports the resistor. Aluminium current leads connect the meander to the bond pads, running through narrow channels that form a restriction to gas flow. The flexible membrane can displace liquid in the pump chamber. The application of an electric voltage to the heater resistor will cause a temperature rise to the air inside the cavity and a related pressure increase inducing a downward deflection of the pump membrane. As the pump membrane deflects downwards, the pressure in the pump chamber increases. This causes a pressure difference across the two valves which results in the opening and closing of the inlet and outlet valves respectively. The converse is when the pump membrane deflects upwards.

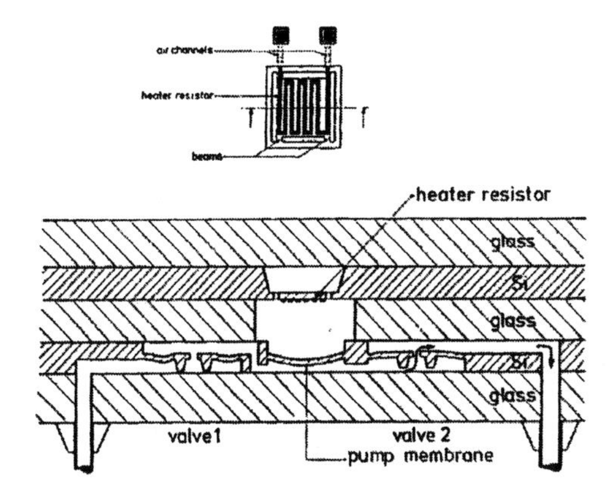

Figure 1-6. Schematic of the first thermo-pneumatic micropump [25]

A maximum yield and built-up pressure of 34 ml/min and 0.05 Bar at 6 V respectively were reported. Two drawbacks that the author finds about thermo-pneumatic pumps are the slow response time (i.e. small flow rate) and the relatively complex structure. Moreover, from a fabrication point of view, it is not easy to obtain a hermetic seal for the cavity and to carry out the multi-layer bonding process.

1.2.1.3 Electrostatic Actuation

As researchers began to venture into micromachinable actuators, the first micromachined membrane pump based on electrostatic actuation was developed in 1991 [26]. Other than the fact that the micropump was first of its kind in term of electrostatic actuation, it was also the first surface micromachined micropump as compared to previous bulk micromachined pumps. No bulk silicon etchants (e.g. KOH) or wafer-bonding techniques were used in its fabrication. Instead, selective deposition and etching of sacrificial layers were used to define the structures. This is significant because it was a truly IC compatible process which means that a hybrid system consisting of both the micropump and the driving circuit can be fabricated in one process. Each pump consisted of an active inlet valve, a pumping membrane, and an active outlet valve. All parts were encapsulated by silicon nitride and were actuated by electrostatic forces. Actuation voltages of approximately 50 V were required for observable valve closure and membrane deflections. However, no pumping action was reported.

Nevertheless, the first working electrostatic micropump was realized by R. Zengerle et al in the following year [27]. The micropump was made up of 4 silicon layers, which formed two passive cantilever passive valves, a pump membrane and a counter-electrode for electrostatic actuation [Fig. 1-7].

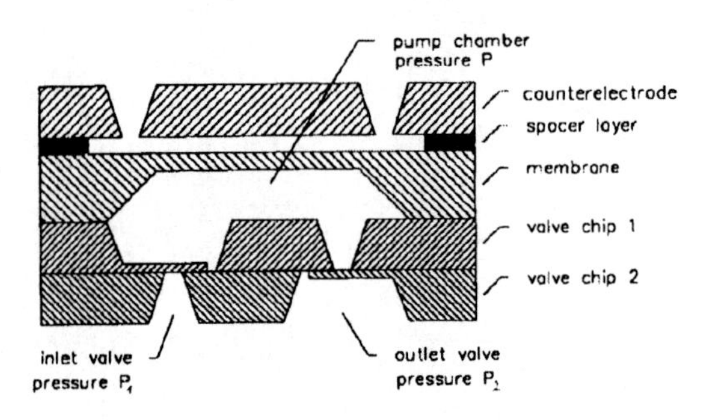

Figure 1-7. Schematic of an electrostatically actuated micropump [27]

Direct electrostatic forces were used for the deflection of the pump membrane, which had an area of 4 x 4 mm^2 and a thickness of 25 μm. Upon application of an electrical voltage, the membrane would deflect outward, leading to a fluid flow through the inlet valve into the pump chamber. The volume stroke of the membrane was between 10 to 50 nl. When the supply voltage was shut off, the membrane would bend back and the fluid is forced through the outlet valve. The liquid to be pumped was not subjected to any electrical field. The separation between the movable membrane and the electrically isolated stator was 4 μm. The passive valves were basically cantilevers measuring 1 x 1 mm^2 with thickness varying from 10 to 20 μm. Pumping was achieved for the first time at actuation frequencies in the range of 1 to 100 Hz. At a frequency of 25 Hz, a flow rate of 70 μl/min at zero backpressure was realized. In addition, a maximum pressure head of 2500 Pa was developed.

1.2.1.4 Pneumatic Actuation

A micropump based on pneumatic actuation was also developed by Zengerle et al [27] in 1992. In the paper, they reported on a variation to their electrostatic micropump based on pneumatic actuation. However, this type of actuation is not very popular with researchers. This may be attributed to the relative complexity in terms of design and fabrication. Nevertheless, in 1994, an external pneumatically micropump was fabricated using the LIGA process and thermoplastic microreplication [28-29]. The micropump is made of a gold structure with a titanium diaphragm as shown in Fig. 1-8. The check valve consists of a titanium membrane and a polyimide membrane. A maximum flow rate of 80 μl/min and a maximum pressure of 0.47 mH_2O were obtained for pumping of air at a driving frequency of 5 Hz.

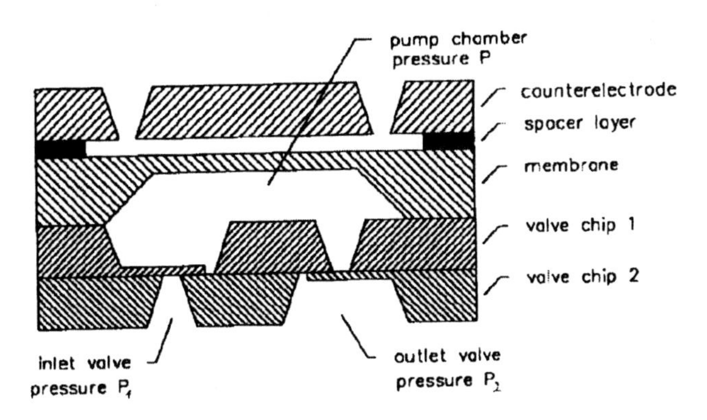

Figure 1-8. Schematic of a pneumatic micropump fabricated by LIGA process [29]

An interesting point of this micropump is the utilization of thermoplastic microreplication. It provides a way of producing microstructure components in high volumes at low cost compared with the more commonly used but relatively more expensive materials (e.g. single crystalline silicon) and fabrication methods (e.g. silicon etching). Microreplication also enables the use of different materials like polymers, metals and ceramics. Most of the time, these materials are more suitable in terms of mechanical characteristics. For example, the use of a polymeric pump membrane will generally result in a bigger displacement than for a silicon membrane for a given input. These factors combined to greatly increase the potential use of microstructures in industry. In fact, the author believes that polymers will eventually take over silicon as the primary material in microfluidics devices in the future and the current trend suggests such a direction too.

1.2.1.5 Shape Memory Alloy (SMA) Actuation

As the name implies, SMA actuated micropumps make use of the shape-memory effect (SME) in SMA materials (such as titanium nickel) as an actuator as they are reportedly capable of both high forces and high strains relative to other microactuator technologies. The SME involves a phase transformation between two solid phases, namely, the austenite (high-temperature) and martensite (low-temperature) phases. In SMA materials, the martensite is much more ductile than the austenite, and this low-temperature state can undergo significant deformation through the selective migration of variant boundaries in the multi-variant grain structures. When heated to the austenitic start temperature (i.e. the beginning of the "reverse" phase transformation), the material starts to form single variant austenite. If the material is not mechanically constrained, it will reassume the pre-deformed shape, which it retains if cooled back to the martensite phase. If the material is mechanically constrained, the material will exert a large force while assuming the pre-deformed shape. This is also commonly known as the one-way SME.

One of the main advocates of SMA micropumps is W. L. Bernard et al [30-31]. They developed a reciprocating micropump based on a pair of SMA actuators (Fig. 1-9). A pump rate of 50 μl/min was achieved at a driving frequency of 0.9 Hz and a power input of 0.54 Watts. This performance is comparable to other reported micropumps and effectively demonstrated the feasibility of SMA actuation.

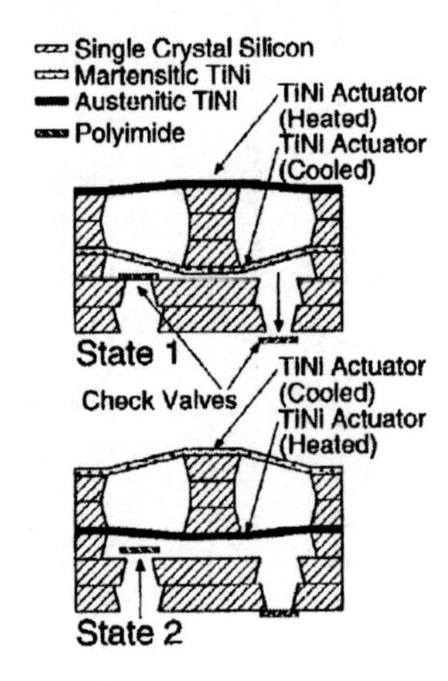

Figure 1-9. Schematic of a SMA actuated micropump [31]

1.2.2 Peristaltic Micropumps

As to date, there are two types of peristaltic micropumps reported in the literature. Except in the earlier years, peristaltic micropumps are not popular with researchers. The author believes that this could be attributed to the fact that the control of such pumps is generally much more involved and the structures are relatively complicated in design.

1.2.2.1 Piezoelectric Actuation

In 1989, Smits published on a peristaltic micropump [32] as proposed in his earlier patent. The pump has a pumping rate of 3 μl/min at zero backpressure and a maximum pressure head of 60 cmH_2O (Fig. 1-10). Even though its performance was modest by today's standards, a number of interesting observations were made. First of all, the pumping rate was found to vary linearly proportional to the backpressure. This backpressure dependent flow rate is in fact one the major shortcomings of micropumps. Another observation that Smits made was that the pumping rate increased linearly with the driving frequency until a threshold value (15Hz in this case) where it started to fall off. At frequencies 50Hz and higher, no pumping effect could be observed anymore. He attributed this reduction to zero of the pumping rate to the viscosity of the pumping fluid, which at higher frequencies resists being pushed through the narrow channels. This observation effectively means that there is a cap to the maximum pumping rate for each micropump design. In the following year, Smits reported on a marked improvement in his micropump which realized a flow rate of 100 μl/min at zero backpressure [33].

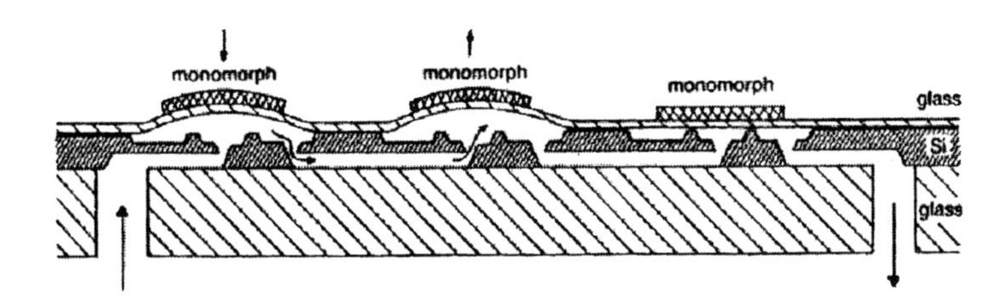

Figure 1-10. Schematic of the first peristaltic micropump [33]

1.2.2.2 Thermo-pneumatic Actuation

A peristaltic micropump based on thermo-pneumatic was also reported in the literature [34]. Three active pressure chambers with flexible membranes were used to provide the peristaltic pumping motion (Fig. 1-11). Microresistive heaters were used to change the temperature of the pressure chambers. In another development [35], the pressure chambers were filled with light-heated working fluid and heated up using laser light produced from an array of microcells.

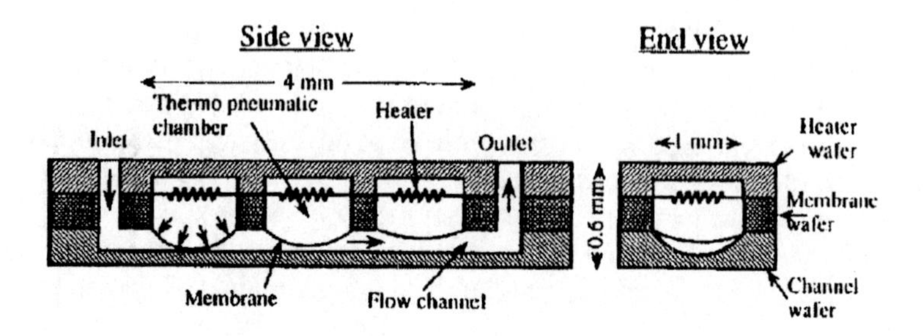

Figure 1-11. Schematic of a thermo-pneumatic actuated peristaltic micropump [34]

1.2.3 Reciprocating Micropumps With Dynamic Microvalves

This type of micropump uses diffuser/nozzle as the flow directing elements in place of passive check valves. Consequently, they are often known as valve-less diffuser micropumps. Valve-less reciprocating pumps have flow channels at the inlet and the outlet that are designed to have different flow resistances in the forward and the reverse directions. This eliminates wear and fatigue in the check-valves and reduces the risk of valve clogging. In addition, this type of micropump is relatively easier to fabricate. The idea to use such channels in pumps was first mentioned in 1989 [36].

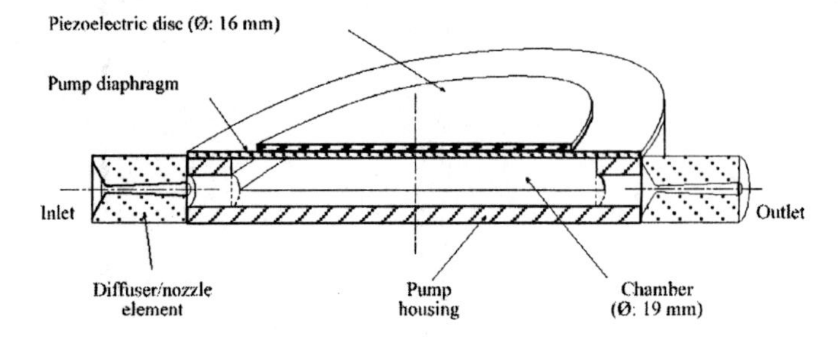

Figure 1-12. Cross-sectional view of the single-chamber metal (brass) pump consisting of a circular housing with an oscillating top diaphragm and two conical diffuser elements [37]

The first valve-less reciprocating pump was the valve-less diffuser pump presented in 1993 [37] and illustrated in Fig. 1-12. In the diffuser pump, diffuser elements are used as the flow directing elements. The opening angles of the diffusers are small, normally less than 20°, and the diffuser direction is the positive flow direction.

In 1994, the first valve-less pump fabricated in silicon using anisotropic wet etching was presented [38-39] (Fig. 1-13). The sharp inlet and outlet together with the large opening angle defined by the <111>-planes in the silicon makes the converging wall direction the forward direction and the diverging wall direction the reverse direction. Consequently, the pump works in the opposite direction as compared with the earlier presented valve-less diffuser pump, which has its forward direction in the diverging wall direction. Considerable work has been carried out in this type of micropump but majority of them are on the theoretical modelling [40-41]. In addition, other fabrication means were also attempted [42].

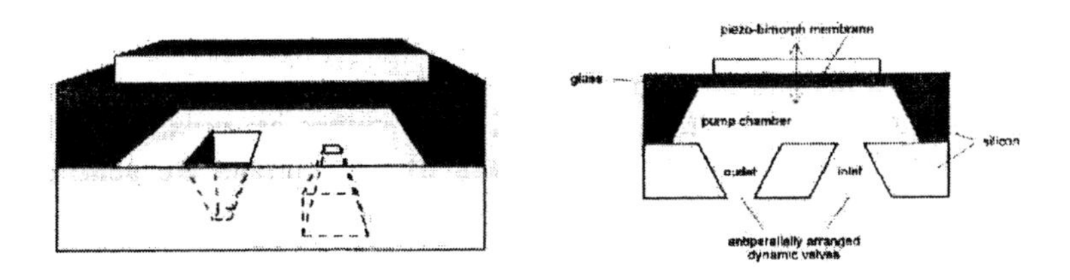

Figure 1-13. Schematic view of the first silicon dynamic micropump [38-39]

A third type of valve-less micropump was proposed in 1995 [43]. This pump uses the valvular conduit [44] as flow directing element (Fig. 1-14). Valvular conduits have been fabricated using micromachining technology and show a direction dependent flow behaviour, but no data for pump performance was found as to date.

Three of the biggest shortcomings of valve-less pumps are leakage (i.e. back-flow), susceptibility to cavitations and difficulty in priming. However, a valve-less pump with self-priming capability was reported very recently [45]. The pump utilizes a novel two-level pump chamber geometry and enables both gas and liquid pumping. It is also reportedly fully self-priming and insensitive to cavitations and gas bubbles in the liquid.

Source:

Figure 1-14. Micrograph of a Tesla valve. The geometry is designed to give a lower flow resistance in the positive flow direction than in the opposite, the negative, low direction [43]

1.3 Non-mechanical Micropumps

In contrast to mechanical micropumps, non-mechanical micropumps generally have neither moving parts nor valves. Consequently, advantages include simplicity in design and fabrication, longer life cycle and minimal risk of blockage. However, such pumps are generally inferior in performance and are restricted to certain types of pumping fluids (e.g. electrohydrodynamic micropumps are limited to liquids of low conductivity and dielectric liquids such as organic solutions). In addition, the actuation means are such that they will interfere with the pumping liquids. Examples of such pumps include electrohydrodynamic, electroosmotic, ultrasonic and magnetohydrodynamic pumps. Non-mechanical pumps are usually feasible only a microscale as the forces generated by such means are generally unable to move larger quantity of fluids unless one is looking at input in the order of some tens of kilovolts.

1.3.1 Electrohydrodynamic (EHD) Micropumps

The fundamental phenomenon that allows the transduction of electrical to mechanical energy in an EHD pump is an electric field acting on induced charges in a fluid. In EHD pumping, fluid forces are generated by the interaction of electric fields with the charges they induce in the fluid. The fluid must be of low conductivity and dielectric in nature in order for EHD pumping to occur. This idea of using electric forces acting directly on a fluid to achieve pumping is not a new one. In fact, EHD pumps were first proposed and built back in the 1960's although not by IC technologies [46-47].

Two types of EHD micropumps, namely, DC-charge injection and travelling wave drive micropumps have been presented in the literature. They will be briefly described in the following sections.

1.3.1.1 DC-charge injection EHD micropump

The EHD injection micropump was first proposed and realized by A. Ritcher & H. Sandmaier in 1990 [48]. The fluid force is the Coulomb force on ions injected from one or both electrodes into the fluid by electrochemical reactions. EHD pumps require 2 permeable electrodes in direct contact with the fluid to be pumped. A pressure gradient will develop between the electrodes and this leads to fluid motion between the emitter and collector. The simplest arrangement consists of two electrically isolated grids (Fig. 1-15) which is the configuration adopted by Ritcher (Fig. 1-16).

A flow rate of about 15 ml/min and a pressure head of around 0.25 mH_2O were reported using an active area of 3 x 3 mm^2 and driving at around 800V.

It was proposed that a reduction of the grid distance could allow the driving voltage to be reduced and stacking of devices to achieve higher pressure heads was plausible. However, aqueous solution cannot be pumped due to their high ionic conductivity and hence the use in medical or biological systems is restricted.

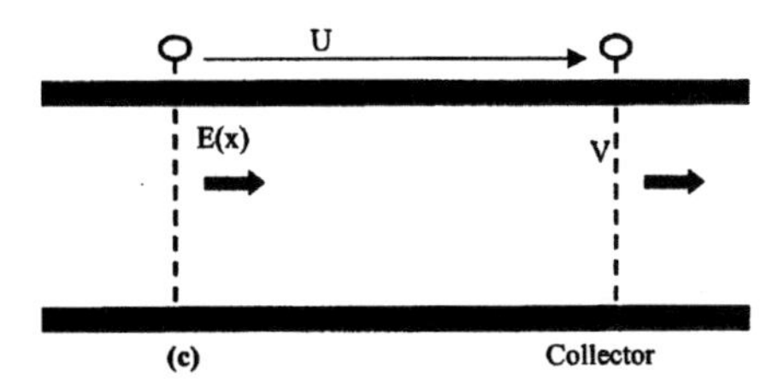

Figure 1-15. Basic geometry of EHD injection pumps [48]

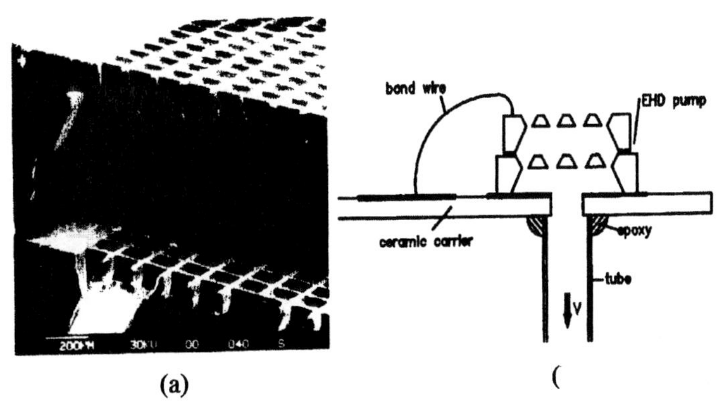

Figure 1-16. (a) Cross section of the grid used in the EHD pump by Ritcher & Sandmaier and (b) schematic view of the mounted EHD pump [48]

1.3.1.2 Travelling wave voltage EHD micropump

One of the very first working EHD pump based on travelling wave-induced electroconvection was reported by G. Fuhr et al in 1992 [49]. Travelling-wave induced electroconvection requires a gradient in natural electrical conductivity or permittivity. Fluids normally do not exhibit such inhomogenities but by imposing a temperature gradient, an appropriate gradient of conductivity is plausible. It has been demonstrated that waves of electric fields travelling perpendicular to the temperature and conductivity gradient, induce charges in the liquid bulk [50]. These charges are slightly displaced and interact with the travelling field due to charge relaxation processes. As a result, volume forces act and depending on the sign of the temperature gradient, a fluid transport occurs in the same (or opposite) direction of the travelling wave.

In the pump by G. Fuhr et al, a structure as shown by Fig. 1-17 was adopted. The electrode array is formed on the substrate and the flow channel is formed across the electrodes. Travelling waves are produced by 90°-phase shifted rectangular voltages of high frequency (100 kHz to 30 MHz) and low voltage (20 to 50 V), which are applied along the channel direction. Flow in the range of 0.05 to 5 μl/min was obtained.

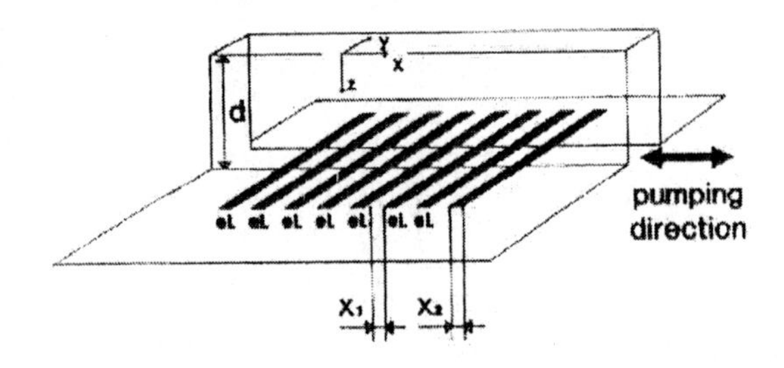

Figure 1-17. Schematic view of the structure of a travelling wave type EHD micropump [49]

1.3.2 Electroosmotic Micropumps

Electroosmotic micropumps make use of the electrokinetic phenomenon of electroosmosis for pumping electrically conductive solutions. They are used in micro chemical analysis systems and DNA chips. This phenomenon is aptly described by S.L. Zeng et al [51]:

"When coming into contact with electrolyte solutions, silica capillary walls and (silica) particle surfaces spontaneously develop an electric double layer caused by electrochemical reactions at the surface. The typical reaction for glass is the deprotonation of acidic silanol groups at the surface to give a negatively charged wall. Counter ions from the bulk liquid shield this wall charge and form a layer of diffuse ions that span a distance of about 10 nm from the wall... When an external electric field is applied parallel to the wall, ion drag in this electrical double causes a net motion of bulk liquid along the wall that is termed electoosmotic flow."

The process is illustrated in Fig. 1-18. By micromachining techniques, it is possible to fabricate a complex manifold of channels in a planar substrate as a network of capillaries for electroosmotic pumping. The capillaries are typically of a diameter less than 100 μm. Figure 1-19 shows an example of such a network of capillaries and the typical layout of an electroosmotic pump.

Electroosmotic pumps have three major advantages, namely, no mechanically moving parts are required, pumping is not localized at a certain point but over the entire length of the capillary channel and flow can be controlled by switching the voltages, without the need for valves. Generally, high pressure heads are associated with such pumps, reportedly up to 10 Bar at an input voltage of 2 kV [51]. However, such pumps are restricted to pumping electrolytes and moderate flow rates in the order of less than 10 μl/min. In addition, back-flow will take place when not pumping, as there are no valves to check back-flow and they consume a considerable amount of energy to generate the necessary electric field strength for most applications (typical input voltages of 20 kV in chemical analysis systems).

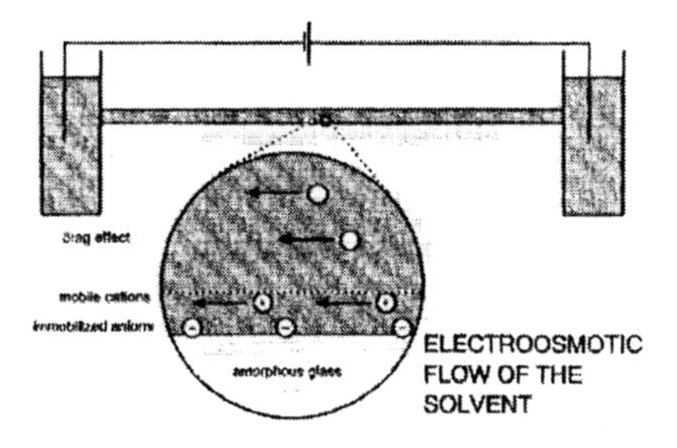

Figure 1-18. Schematic illustrating electroosmosis in a capillary [52]

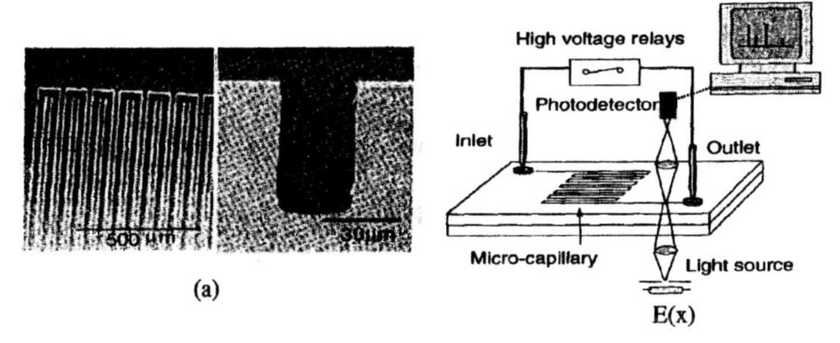

Figure 1-19. (a) SEM images of microcapillary network and (b) schematic of the proposed microosmotic pump [53]

1.3.3 Ultrasonic Micropumps

Ultrasonically driven or flexural plate wave (FPW) micropumps make use of the phenomenon of acoustic streaming, whereby a finite-amplitude acoustic field is used to set the host fluid in motion. The acoustic field is set up by FPWs, generated by an array of piezoelectric actuators, which propagate along a thin plate that constitutes one wall of a flow channel.

One of the first ultrasonic micropump was developed by R.M. Moroney et al [54-55]. The use of flexural plate waves in microfabricated membranes for fluid transport is derived from the use of flexural ultrasonic wave in rotary motors. The structure for ultrasonic pumping is very simple and typically comprises of a cap and a thin membrane to define an encapsulated microchannel (Fig. 1-20). There is effectively no dead volume associated with such micropumps, as the channel itself is the pump. In addition, these pumps do not require any valves and the operating voltage is in the order of 10 volts. However, such pumps suffer from back-flow when not pumping and highly backpressure dependent flow.

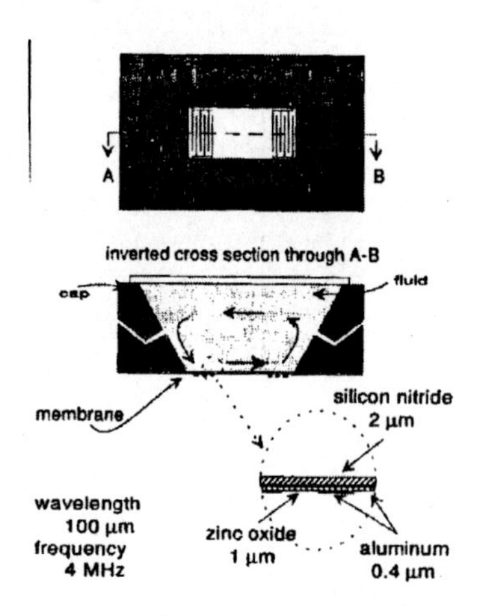

Figure 1-20. Schematic a flow channel with acoustic pumping [54-55]]

1.3.4 Magnetohydrodynamic (MHD) Micropumps

The concept of a MHD micropump is relatively new and one of the very first developed MHD micropumps is by J. Jang & S.S. Lee [56] in 1999. MHD refers to the flow of electrically conducting liquids in electric and magnetic fields. Lorentz force is the pumping source of conductive, aqueous solutions in the MHD micropump. The typical structure of a MHD pump is relatively simple, comprising of microchannels with two walls bounded by electrodes to generate the electric field while the other two walls bounded by permanent magnets of opposite polarity for generating the magnetic field (Fig. 1-21). Conducting fluid in the microchannels of the MHD micropump is driven by Lorentz force in the direction perpendicular to both magnetic and electric fields (Fig. 1-21(a)). This type of micropumps is similar to EHD micropumps with the main advantage being that MHD micropumps can be used to pump fluids with higher conductivity. This greatly widens the latter's utilization in medical and biological applications. In the MHD pump by J. Jang & S.S. Lee (Fig. 1-22), a pressure head of 18 mmH2O at an input current of 38 mA and a maximum flow rate of 63 µl/min at 1.8 mA were obtained.

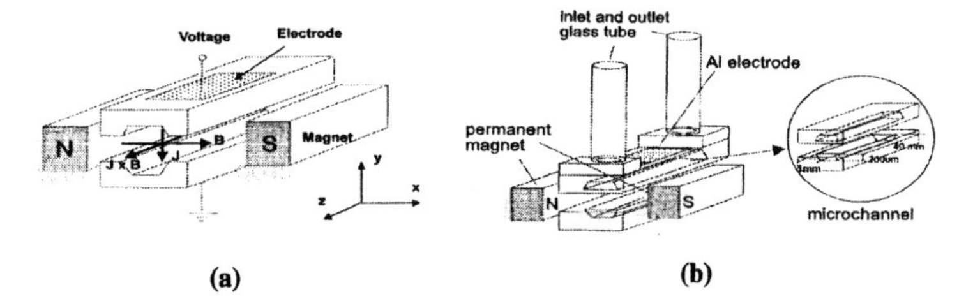

Figure 1-21. Schematic of the a) actuation principle of the MHD micropump using Lorentz force as the driving force and (b) a possible structure of a MHD pump [56]

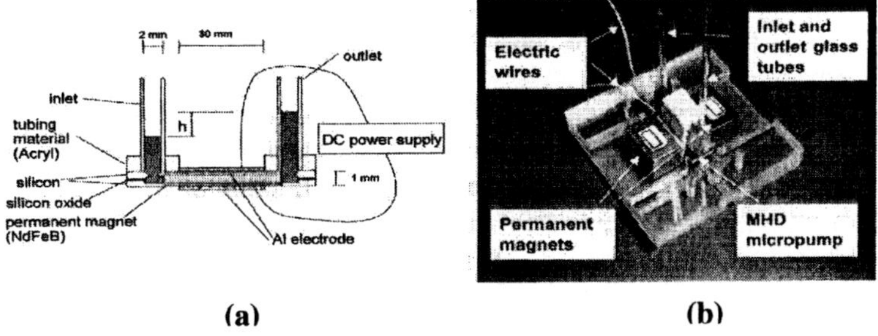

Figure 1-22. Schematic of the structure and (b) photograph of the MHD pump by J. Jang & S.S. Lee [56]

Compared with other types of non-mechanical micropumps, the MHD micropump has several advantages, such as simple fabrication process, bi-directional pumping ability, and its suitability in pumping of conductive fluids.

1.4 Motivation

After a survey through the literature and the research works that have been carried out thus far, the author found that there are no established design rules for the designing of micropumps. Therefore, in this project, the author aims to formulate the design rules for the designing of micropumps.

As the development of micropumps is highly application driven, the micropump under investigation is for the purpose of integration into a drug delivery system, more specifically, insulin delivery. The author has chosen the reciprocating displacement micropump because such pumps are capable of delivering relatively high pressure heads and flow rates. They are also simple in design and hence the fabrication process will not be overly complicated. Piezoelectric actuation is chosen because piezoelectric materials are very established materials and the control of such actuators is relatively easier as compared to the other means. Nevertheless, it must be stressed the rules developed in this project are not confined to this application and type of micropump.

In addition, the author aims to develop a fabrication process for the practical realisation of micropumps. In the design of micropumps, it is often an inter-play between theoretical modelling and technological limitations in terms of fabrication constraints. The micropump designer must always bear in mind and be fully aware of fabrication limitations. Consequently, he must not only design a particular structure but also have a process sequence that will enable the practical realisation of his design. Hence, the development of a realistic and feasible process for the fabrication of the proposed micropumps constitutes a major part of this project

In the third part of this project, the author aims to establish an experimentation set-up for the characterization of micropumps. Performance measures such as pressure head, flow rate and backpressure dependency are of interest. As to date, researchers have not established a standardized method and set-up for testing micropumps. In fact, some of them still use relatively primitive and inaccurate methods for evaluating their micropumps. This issue will be also addressed in this project.

Chapter 2

DESIGN RULES FOR MICROPUMPS

In the very first step in the development cycle of micropumps, the designer must have an idea of the final application of his proposed micropump, from which he will then choose the most appropriate type of micropump. For example, if the micropump is to be used in micro total analysis systems (μTAS) or electrophoresis systems, electroosmotic pumps will be most suitable. For this project, the end application is for an insulin delivering system. The specifications are self-priming capability, minimum pressure head of 15 mmHg (i.e. tissue pressure) and stable flow rate of less than 1 μl/min. The reciprocating displacement micropump with piezoelectric actuation is identified as a most suitable candidate for the intended specifications for reasons as mentioned in earlier chapters.

In this chapter, the designs rules for the design of a reciprocating type micropump based on piezoelectric actuation will be presented in detail.

2.1 Preliminary Design

In this project, a reciprocating displacement micropump as illustrated in Fig. 2-1 is developed. The pump consists of a glass-silicon-glass structure with the silicon layer forming the valves. The top glass wafer forms the actuating membrane (Pyrex™ 7740) while holes are drilled in the bottom wafer to form the inlet and outlet.

Diaphragm valves are used with silicon oxide patterned on it to form the valve-seats. Details of the fabrication process will be given in Chapter 5. The dimensional parameters are determined by the design rules that are discussed in the following sections. Prior to that, a brief introduction to piezoelectric material and its application in microactuation is given in the following section.

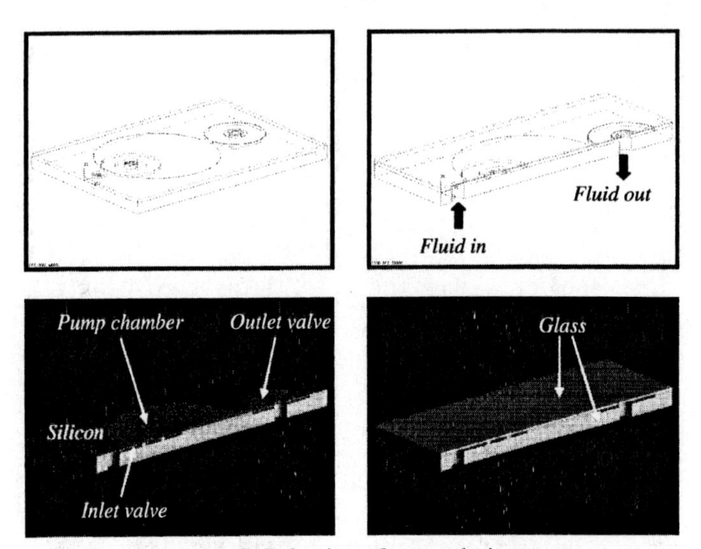

Figure 2-1. CAD drawings of proposed micropump

2.1.1 Piezoelectric Actuation

Piezoelectric excitation has been widely used for micromechanical devices such as gyroscopes and microvalves. The effect was discovered by Jacques and Pierre Curie in 1880. They discovered that if special crystals (e.g. quartz, tourmaline and Rochelle salt) were subjected to mechanical tension, they became electrically polarized and this polarization was proportional to the extension. They also discovered that the opposite was true, that is, if an electrical field was applied across, the material deformed proportionally to the electrical field. This is known as the inverse piezoelectric effect. Piezoelectricity involves the interaction between the electrical and mechanical behaviour of the medium. Piezoelectricity has since become the general term to describe the property exhibited by certain crystals of becoming electrically polarized when stress is applied to them [57]:

In the absence of mechanical stresses, the strain S (i.e. $l/\Delta l$) experienced by a piezoelectric ceramic medium when subjected to an external electric field E is given by (Fig. 2-2):

$$S = dE \quad (2\text{-}1a)$$

$$S = gD \quad (2\text{-}1b)$$

$$g = \frac{d}{\varepsilon^T} \tag{2-1c}$$

where d, g and ε^T are material constants respectively known as the piezoelectric charge constant, piezoelectric voltage constant and permittivity at constant stress, and D is the electric displacement.

The strain experienced by an elastic medium subjected to a tensile stress T is in accordance with Hooke's Law:

$$S = s^E T \tag{2-2}$$

where s^E is the compliance of the medium at a constant electrical field. Generally, however, the response of a stressed piezoelectric medium is a complex interaction between both electrical and mechanical parameters. To a good approximation, the total strain S experienced by the medium is:

$$S = s^E T + dE \tag{2-3}$$

$$S = s^D T + gD \tag{2-4}$$

in which s^E and s^D are respectively the specific compliances at constant electric field and constant electric displacement (note that in SI units, d is expressed in C/N, or its equivalent m/V, and g is expressed in Vm/N, or its equivalent m^2/C).

Alternatively, the constitutive equations can be expressed as:

$$T = c^E S - eE \tag{2-5}$$

$$D = eS + \varepsilon^S E \tag{2-6}$$

where c^E, e and ε^S are the elasticity (evaluated at constant electric field), piezoelectric and dielectric (evaluated constant mechanical strain) material constants respectively.

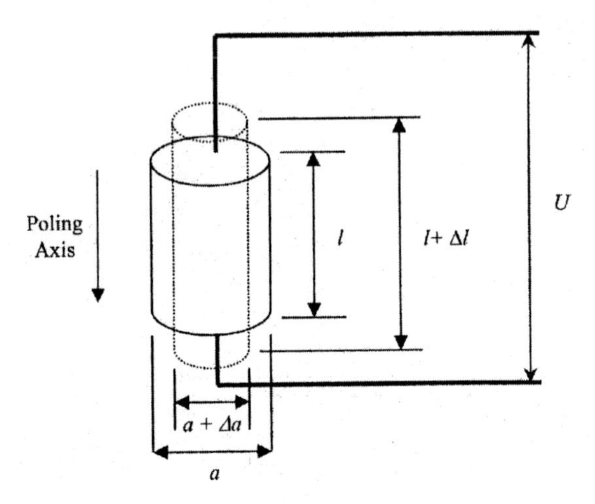

Figure 2-2. Deformation of a piezoelectric device under the application of an electric field

In general, Eqs. (2-3) to (2-6) collectively constitute the *Linear Theory of Piezoelectricity* whereby the linear equations of elasticity are coupled to the charge equation of electrostatics by means of piezoelectric constants. All the elastic, piezoelectric and dielectric coefficients are taken as constants independent of the magnitude and frequency of applied stresses and electric fields. Real materials involve mechanical and electrical dissipation. In addition, they may show strong nonlinear behavior, hysteresis effects, temporal instability (aging) and a variety of magneto-mechano-electric interactions [58].

Examples of piezoelectric materials are quartz, $LiTaO_3$, PZT and ZnO. Non-piezoelectric materials such as silicon can nevertheless be excited by depositing a thin film of piezoelectric material (such as PZT or ZnO) onto the latter. Another solution is to mount a piezoelectric disk on the non-piezoelectric material. This eliminates the problem of having to grow the film thick enough such that high voltages can be applied without dielectric breakdown (sparks/short circuits across the film). The piezoelectric effect can be used and is commonly used to bend a diaphragm, for example, in a micropump or microvalve. The principle is illustrated in Fig. 2-3 where a piezoelectric disk is glued onto a diaphragm. When a voltage is applied across the piezoelectric disc, a strain results which creates moments at the interface that force the diaphragm to bend. This configuration is used in the proposed micropump.

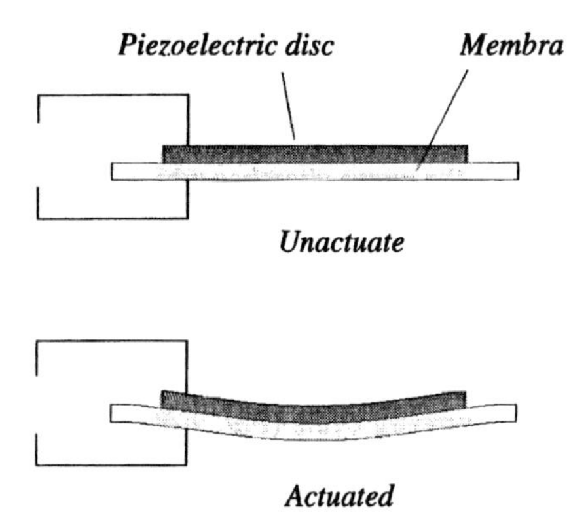

Figure 2-3. The bending of a bimorph consisting of a piezoelectric disc glued onto a membrane (e.g. brass and silicon)

2.2 Compression Ratio

The basic structure of a reciprocating membrane micropump with two passive check valves can be simplified as illustrated in Fig. 2-4. An oscillating membrane induces periodic volumetric changes in the pump chamber and sequentially causes an over/under pressure to be generated in the chamber. In the following sections, the combined structure of the piezoelectric disc and the membrane will be termed as the actuator unit.

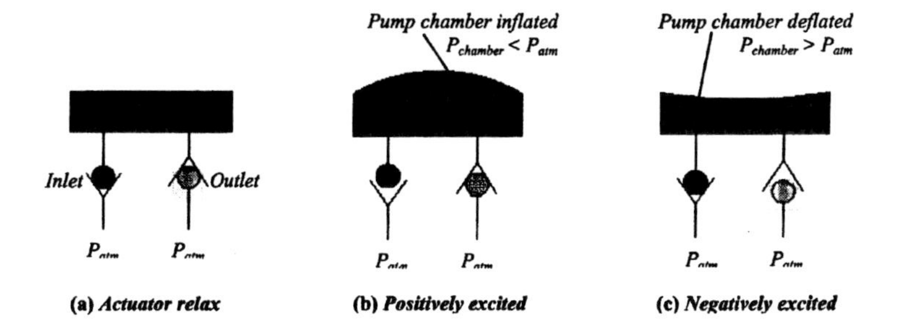

Figure 2-4. Working principle of a reciprocating micropump

The operation of the pump under zero backpressure is illustrated in Fig. 2-4. When the actuator unit is not actuated by any input, both check valves are closed (Fig. 2-4(a)) and there is no flow. In the supply cycle (Fig. 2-

4(b)), the input voltage is of such a polarity that it causes a strain in the piezoelectric disc. However, the elongation of the piezoelectric disc is constrained by the membrane and as such, moments at the interface are induced. These moments causes the actuator unit to deflect outwards thus enlarging the volume of the pump chamber which in turn results in an under-pressure to be generated in the pump chamber. The pressure difference ($\Delta P_{iv} = P_{atm} - P_{chamber}$) across the inlet valve causes it to open. Similarly, the pressure difference across the outlet valve causes it to close. A flow of the pumping medium into the pump chamber is induced by the pressure differential.

Conversely, in the pumping cycle (Fig. 2-4(c)), the actuator unit deflects inward and an over-pressure is generated in the pump chamber. The inlet valve is forced to close while the outlet valve is forced to open. The pumping medium in the pump chamber is then expelled from the pump chamber.

With reference to Fig. 2-4(b), denoting the dead volume of the pump chamber and the stroke of the actuator diaphragm as V_o and ΔV respectively, the compression ratio, ε, is defined as [59]:

$$\varepsilon = \frac{\Delta V}{V_o} \qquad (2\text{-}7)$$

The compression ratio of a micropump can be increased by either increasing the stroke of the actuator unit or reducing the dead volume of the pump chamber. A larger compression ratio is desirable for a micropump as will be explained in later sections. However, due to the small stroke of many of the microactuators and the inevitable large dead volume, the compression ratio of a typical micropump is usually small. The method for determining the compression ratio will be discussed in the following sections.

2.2.1 Determination of Volume Stroke

In order to model the actuator unit, one can use either analytical [18, 60-61] or numerical means (i.e. finite element method) [62]. In this project, both analytical and Finite Element Method (FEM) models of the actuator unit are developed by the author. Experiments are also carried to verify these models. These will be described in the following sections.

In the proposed micropump, commercial piezoceramic discs (PXE5™) from Philips™ are used. The discs are 10 mm in diameter and 200 μm in thickness. Further material properties can be found in Appendix B.

2.2.1.1 Analytical Model

When a voltage U is applied to across the piezoelectric disc, the electric field E_3 will cause the disc to undergo a free strain of (Fig. 2-6):

$$S_f = d_{31} E_3 \tag{2-8}$$

$$E_3 = \frac{U}{t_p} \tag{2-9}$$

where S_f is the free strain; d_{31} is the piezoelectric constant and t_p is the thickness of the piezoelectric disc.

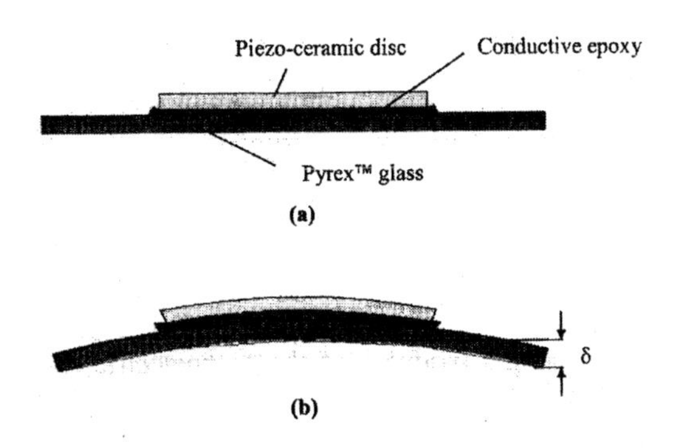

Figure 2-5. Schematic of the actuator unit in (a) un-actuated state and (b) excited state

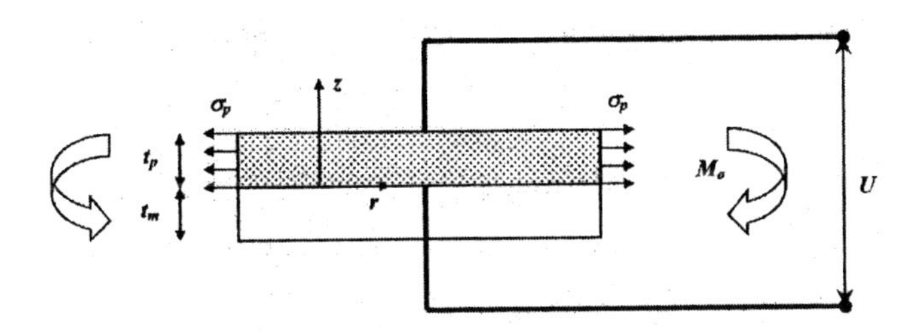

Figure 2-6. Schematic of the stress state in the bimorph

The free strain refers to the strain that the piezoelectric material will undergo under the application of an input voltage without any external mechanical stress acting on it. This assumption is usually made in most of the analytical models because of the complexity involved in modelling the coupling stress that the membrane in turn induces back on the piezoelectric disc as a result of the strain induced by the electric field.

Now, a stress σ_p is induced in the piezoelectric disc and a bending moment M_o (per unit length) is set up in the piezoelectric disc/membrane double layer:

$$\sigma_p = c_p^E S_f \tag{2-10}$$

$$\begin{aligned} M_o &= \frac{\sigma_p t_p t_m}{2} \\ &= \frac{c_p^E d_{31} U}{t_p} \frac{t_p t_m}{2} \\ &= \frac{c_p^E d_{31} t_m}{2} U \end{aligned} \tag{2-11}$$

where c_p^E is the Young's modulus of the piezoelectric disc evaluated at constant electric field.

As illustrated in Fig. 2-7, the actuator unit can be divided into 2 parts, namely, an inner bi-laminar plate under the condition of pure bending produced by the uniformly distributed moments M_o and the outer portion, an annular plate under a line force (Q_o) along its inner diameter.

The governing partial differential equation for the bending of plates (in polar coordinates) that gives the deflection surface of a laterally loaded plate is given by [63]:

$$\begin{aligned} \nabla^4 w &= \left(\frac{\partial^2}{\partial r^2} + \frac{1}{r}\frac{\partial}{\partial r} + \frac{1}{r^2}\frac{\partial^2}{\partial \theta^2} \right)\left(\frac{\partial^2 w}{\partial r^2} + \frac{1}{r}\frac{\partial w}{\partial r} + \frac{1}{r^2}\frac{\partial^2 w}{\partial \theta^2} \right) \\ &= \frac{q}{D} \end{aligned} \tag{2-12}$$

where w is the lateral displacement; r is the radial distance in polar coordinate; θ is the angular coordinate; q is the intensity of the lateral load over the plate and D is the flexural rigidity given by

$$D = \frac{Eh^3}{12(1-\upsilon^2)}$$

with E being the Young's modulus, h the thickness and υ the Poisson's ratio of the material.

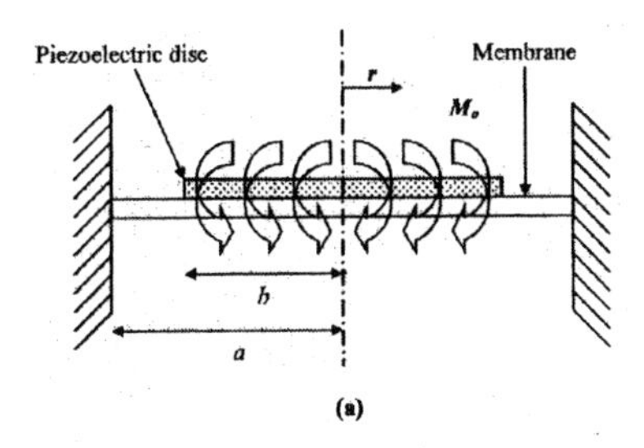

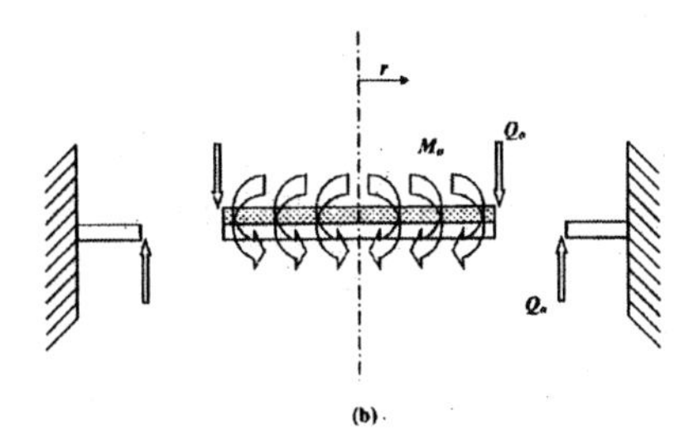

Figure 2-7. Schematic view of the loading in the actuator unit. (a) Actual loading and (b) equivalent loading

By the substitution of the appropriate boundary and load conditions into Eq. (2-12) and the utilization of the principle of superimposition, the equation governing the deflection of the inner bi-laminar plate can be formulated as [64]:

$$w_I = \frac{M_o}{2D_e(1+\upsilon_e)}\left(b^2 - r^2\right) + \frac{Q_o a^3}{D_m}\left(\frac{C_3 L_2}{C_4} - L_1\right)$$
$$= L_3 \frac{t_m}{4}\left(b^2 - r^2\right)\cdot U + \frac{t_m ab}{2}\frac{C_4 L_3}{L_2}\left(\frac{C_3 L_2}{C_4} - L_1\right)\cdot U \tag{2-13}$$

Similarly, the equation governing the deflection of the outer annular portion can be shown to be given by:

$$w_{II} = w_b + \theta_b r F_1 - Q_o \frac{r^3}{D_m} F_2 = \frac{t_m ab}{2}\frac{C_4 L_3}{L_2}\left(\frac{C_3 L_2}{C_4} - L_1\right)\cdot U$$
$$+ L_3 \frac{t_m b}{2}\cdot F_1 \cdot U \cdot r - \frac{t_m b}{2a^2}\frac{C_4 L_3}{L_2}\cdot F_2 \cdot U \cdot r^3 \tag{2-14}$$

The terms given in Eqs. (2-13) and (2-14) and the detail formulation of the governing equations is given in Appendix A.

Equations (2-13) and (2-14) completely defines the deflection surface of the actuator unit. From these two equations, stresses, moments and deflection at any point in the actuator can be determined. More importantly, by integration of the two equations, the volumetric displacement, V_{total}, of the actuator unit can be obtained, that is:

$$V_I = \int_0^b \int_0^{2\pi} w_I \cdot r \cdot d\theta \cdot dr \quad \text{and}$$

$$V_{II} = \int_b^a \int_0^{2\pi} w_{II} \cdot r \cdot d\theta \cdot dr \quad \text{which gives}$$

$$V_{total} = \int_0^b \int_0^{2\pi} w_I \cdot r \cdot d\theta \cdot dr + \int_b^a \int_0^{2\pi} w_{II} \cdot r \cdot d\theta \cdot dr \tag{2-15}$$

In this analytical model, several simplifications are made:

1) The epoxy glue layer has been ignored and perfect bonding between the piezoceramic disc and the membrane is assumed.

2) The coupling stress that the membrane induces back onto the piezoceramic disc is ignored (i.e. free strain is assumed for the piezoceramic disc).
3) A perfect clamped boundary condition is assumed.
4) Anisotropic nature (mechanical and dielectric properties) of the piezoceramic disc is ignored.

Nevertheless, this analytical model can be extended to include the epoxy layer by means of the *Classical Laminate Theory*, that is, by considering the inner portion as a three-layer laminate. The anisotropic nature of the piezoceramic material can also be factored into the analysis. In addition, other geometries such as square or rectangular plates can also be analysed. However, this will result in a very complicated model that requires numerical methods (such as Ritz or Levy's method) to solve. The model, the author, will then be too involved to be used as a first estimate of the performance of actuator unit. Another use for this actuator model is for optimisation purposes as the governing equations are all parameterised.

2.2.1.2 Finite Element Model

In order to be able to model the epoxy layer, the anisotropic nature of the piezoceramic material, the coupling effect and more complicated geometries, the Finite Element Method (FEM) is used. A commercial software, ANSYS™, is utilized for this purpose. The following sections will give a brief description of the modelling of piezoelectric structures using the FEM, with focus on the mechanical-electrical coupling (coupled-field piezoelectric analysis) and the anisotropic material properties.

Background to the Finite Element Method for Piezoelectric Media

The Finite Element Method was first used for the analysis of vibrations in elastic structures at around 1950 [65]. The method is based on variational methods and early uses of the method were for the analysis of elastic structures in the 1940s, for electromagnetic resonators in the 1950s and for piezoelectric media in the 1960s. Since then, FEM has become very popular among people from both the research and industrial fields because of its versatility in handling a myriad of problems. In addition, FEM software has also mushroomed throughout the years, the more prominent ones being Abaqus™, Patran™, Nastran™ and ANSYS™.

The variational principle for a piezoelectric medium was first formulated by EerNisse in 1967 [66], but without a complete derivation. EerNisse used variational methods for the analysis of piezoelectric disks with various diameter/thickness (D/T) ratios. At the same time, a corresponding variational principle was derived from Hamilton's principle by Tiersten, with applications to piezoelectric plate vibrations [67]. The FEM was applied to piezoelectric media by several researchers in the late 1960s and early 1970s [68-69] but Allik & Hughes [70] were the first to formulate the FEM for a three-dimensional piezoelectric medium in 1968 using the same variational principle as developed by EerNisse. In the subsequent development of the finite element formulation by J.T. Oden & E. Kelley [69], appropriate forms of the law of conservation of energy for an element were formulated instead of using variational principles. Inspired by J.T. Oden & E. Kelley, Schmidt [71] in 1973 formulated the FEM for three-dimensional piezoelectric bodies using the Galerkin method.

The first computational examples of applications of the FEM to the modelling of piezoelectric media were given by Allik [72], where the FEM was used to analyse a complicated three-dimensional transducer. In the years to follow, the FEM became increasingly popular for the modelling of piezoelectric media, but the method was not widely used until the 1980s. In 1986, piezoelectric elements were included in the commercial finite element program, ANSYS™ [73], which has since been used by many groups for finite element modelling of piezoelectric media [74] and by MEMS researchers in the 1990s [61, 75]. Subsequently, piezoelectric finite elements have also been included in other commercially available finite element programs such as ABAQUS™ [76].

Piezoelectric Modelling with ANSYS™ – an Electro-structural Coupled Field Analysis

In a nutshell, piezoelectricity is the coupling of structural and electric fields and it is a natural property of materials such as quartz and ceramics. Applying a voltage to a piezoelectric material creates a displacement and conversely, vibrating a piezoelectric material generates a voltage. They are basically two types of coupling, strong and weak, distinguished by the finite element formulation techniques used to develop the matrix equations. These are illustrated here with two types of degrees of freedom ($\{X_1\}$, $\{X_2\}$):

1) Strong (simultaneous, full, left) coupling – where the matrix is of the form:

$$\begin{bmatrix} [K_{11}] & [K_{12}] \\ [K_{21}] & [K_{22}] \end{bmatrix} \begin{Bmatrix} \{X_1\} \\ \{X_2\} \end{Bmatrix} = \begin{Bmatrix} \{F_1\} \\ \{F_2\} \end{Bmatrix} \tag{2-16}$$

where $[K_{ij}]$, i, j = 1, 2, are the stiffness submatrices and $\{F_i\}$, i = 1, 2, are the force vectors.

The coupling effect is accounted for by the presence of the off-diagonal submatrices $[K_{12}]$ and $[K_{21}]$. This method provides for the coupled response in the solution after one iteration. An example is the piezoelectric problem.

2) Weak (sequential, right) coupling – where the coupling in the matrix equation in the most general form:

$$\begin{bmatrix} [K_{11}(\{X_1\},\{X_2\})] & [0] \\ [0] & [K_{22}(\{X_1\},\{X_2\})] \end{bmatrix} \begin{Bmatrix} \{X_1\} \\ \{X_2\} \end{Bmatrix} = \begin{Bmatrix} \{F_1(\{X_1\},\{X_2\})\} \\ \{F_2(\{X_1\},\{X_2\})\} \end{Bmatrix} \tag{2-17}$$

The coupling effect is accounted for in the dependency of $[K_{11}]$ and $\{F_1\}$ on $\{X_2\}$ as well as $[K_{22}]$ and $\{F_2\}$ on $\{X_1\}$. The coupling between the FE equations is in such a way that the FE matrices or the generalized force vector of one system of matrix equations is a function of the unknown quantities in the other system of matrix equations and vice versa. At least two iterations are required to achieve a coupled response. An example is the magneto-mechanical problem.

In ANSYS™, variational principles are used to develop the finite element equations that incorporate the piezoelectric effect [70]. The coupling is of the strong form being that the entire system of finite element equations is solved in one iteration where the coupling effect is accounted by means of off-diagonal submatrices in the stiffness matrix. Details of the formulation of the finite element equations can be found in [77]. Bearing in mind the concept behind the formulation, the next most important thing that the user needs to know is the material definition and convention. These will be discussed in the following.

The electromechanical constitutive equations used in ANSYS™ are:

$$\begin{Bmatrix} [T] \\ [D] \end{Bmatrix} = \begin{bmatrix} [c] & [e] \\ [e]^T & -[\varepsilon] \end{bmatrix} \begin{Bmatrix} [S] \\ -[E] \end{Bmatrix} \tag{2-18}$$

where [T] is the stress vector; [D] is the electric flux density vector; [S] is the strain vector; [E] is the electric field vector; [c] is the elasticity matrix (6 x 6) evaluated at constant electric field; [e] is the piezoelectric matrix (6 x 3) and [ε] is the dielectric matrix (3 x 1) evaluated at constant mechanical strain.

In most of the literature, Eq. (2-18) is expressed as a system of two equations in tensor form. This allows a more systematic way of defining and expressing the anisotropic properties of piezoelectric material. The constitutive relations for a piezoelectric media in Einstein's summation convention are [58]:

$$T_{ij} = c^{E}_{ijkl} S_{kl} - e_{kij} E_k \tag{2-19}$$

$$D_i = e_{ikl} S_{kl} + \varepsilon^{S}_{ik} E_k \tag{2-20}$$

where i, j, k, l = 1, 2, 3; T_{ij} are components of the mechanical stress tensor [N/m^2]; S_{kl} are components of the mechanical strain tensor; D_i are components of the electric flux density [C/m^2]; E_k are components of the electric field vector [V/m]; c^{E}_{ijkl} are components of the elastic stiffness constant tensor evaluated at constant electric field [N/m^2] and e_{ikl} are components of the piezoelectric constant tensor [C/m^2]; ε^{S}_{ik} are components of the dielectric const. tensor evaluated at constant mechanical strain [F/m].

For the loss-less case, the constant tensors c^{E}_{ijkl} , e_{ikl} and ε^{S}_{ik} are real. Complex values can be introduced to represent elastic, piezoelectric and dielectric losses. These constitutive relations may be given in several forms, but this form is especially suitable for finite element formulation because displacement (i.e. strain) is the primary degree of freedom in the FE solution. However, the form that is most widely used by piezoelectric material manufacturers is the "strain" form as given by:

$$S_{ij} = s^{E}_{ijkl} T_{kl} - d_{kij} E_k \tag{2-21}$$

$$D_i = d_{ikl} T_{kl} + \varepsilon^{T}_{ik} E_k \tag{2-22}$$

where i, j, k, l = 1, 2, 3; T_{kl} are components of the mechanical stress tensor [N/m^2]; S_{ij} are components of the mechanical strain tensor; D_i are components of the electric flux density [C/m^2]; E_k are components of the electric field vector [V/m]; s^{E}_{ijkl} are components of the compliance tensor evaluated at constant electric field [m^2/N]; d_{ikl} are components of the piezoelectric constant tensor [C/N] and ε^{T}_{ik} are components of the dielectric const. tensor evaluated at constant mechanical stress [F/m].

In order to express the elastic and piezoelectric tensors in the form of a matrix array, a compressed matrix notation is usually introduced in place of the tensor notation. This matrix notation consists of replacing ij or kl by p or q, where i, j, k, l take the values 1, 2, 3 and p, q take the values 1, 2, 3, 4, 5,6 according to Table 2-1, namely:

$$c_{ijkl}^{E} \equiv c_{pq}^{E},\ e_{ikl} \equiv e_{ip},\ T_{ij} \equiv T_{p} \tag{2-23}$$

Hence the constitutive equations reduce to:

$$T_{p} = c_{pq}^{E} S_{q} - e_{kp} E_{k} \tag{2-24}$$

$$D_{i} = e_{iq} S_{q} + \varepsilon_{ik}^{S} E_{k} \tag{2-25}$$

where $S_{ij} = S_{p}$ when i = j, p = 1, 2, 3 and $2S_{ij} = S_{p}$ when i = j, p = 4, 5, 6.

Table 2-1. Matrix notation

ij or kl	p or q
11	1
22	2
33	3
23 or 32	4
31 or 13	5
12 or 21	6

Now, as the piezoelectric properties are given in terms of their d_{ikl}, s_{ijkl}^{E} and ε_{ij}^{T} values (see Appendix B for PXE5 material properties as provided by the manufacturer), respective transformations have to be carried out to obtain the respective e_{ikl}, c_{ijkl}^{E} and ε_{ij}^{S} values. These transformations are governed by:

$$\begin{aligned} e_{ip} &= d_{iq} c_{qp}^{E} \\ \varepsilon_{ij}^{T} &= \varepsilon_{ij}^{S} + d_{iq} e_{jq} \end{aligned} \tag{2-26}$$

Values for c_{pq}^{E} are obtained from s_{pq}^{E} by means of matrix inversion. For this purpose, the author wrote a MATLAB® script, which together with the results, is given in Appendix C. These material property definitions are then input into ANSYS™ to run the simulations.

FEM Simulation Results

First of all, a comparison study is carried out to study the differences between the use of two-dimensional and three-dimensional models to simulate the piezoceramic actuator unit. Both the 2D and 3D models comprise of three layers, namely, the glass membrane (200μm thick), epoxy (100μm thick) and piezoelectric disc (200μm thick. The models are fully clamped at the extended edges and voltages are applied across the piezoceramic layer (Fig. 2-8). Results of the comparison are given in Fig. 2-9.

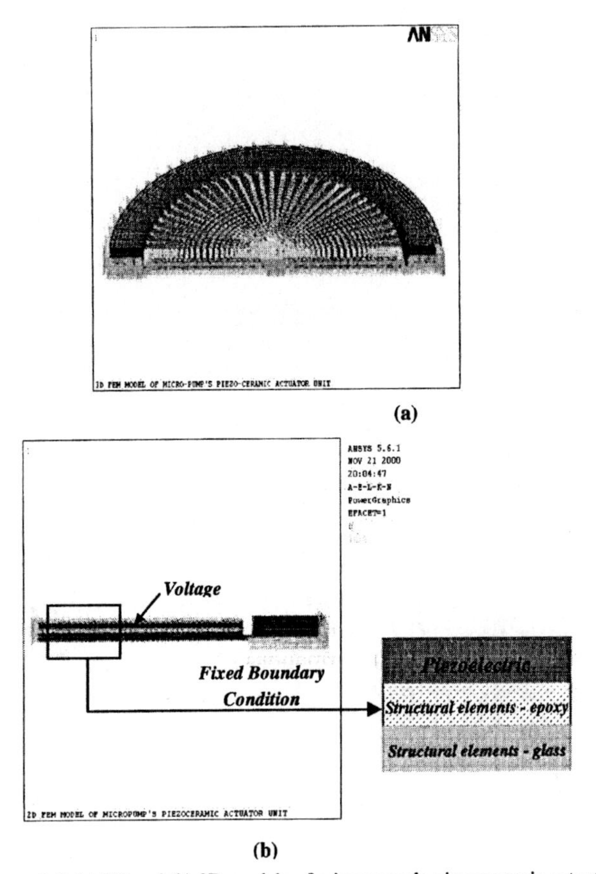

Figure 2-8. (a) 3D and (b) 2D models of micropump's piezoceramic actuator unit

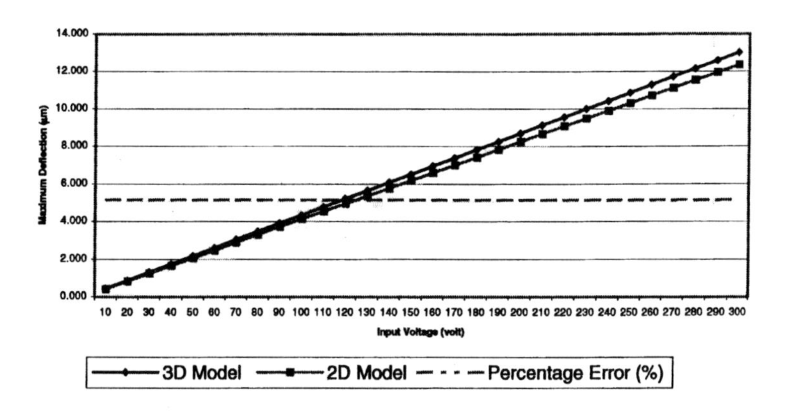

Figure 2-9. Comparison of results obtained using 3D and 2D piezoceramic actuator model

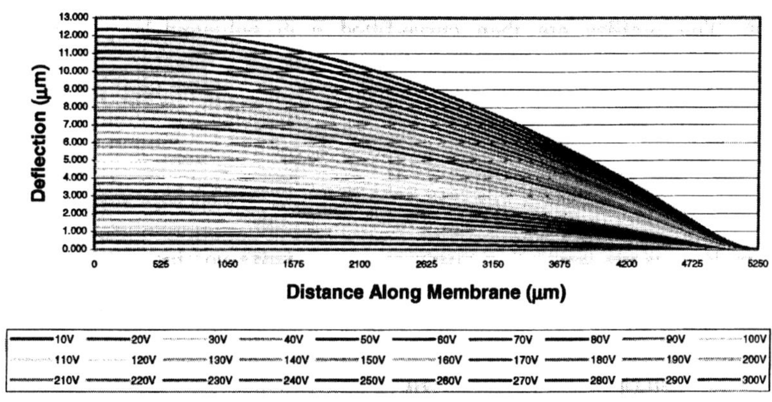

Figure 2-10. Deflection profile of piezoceramic actuator at various input voltage

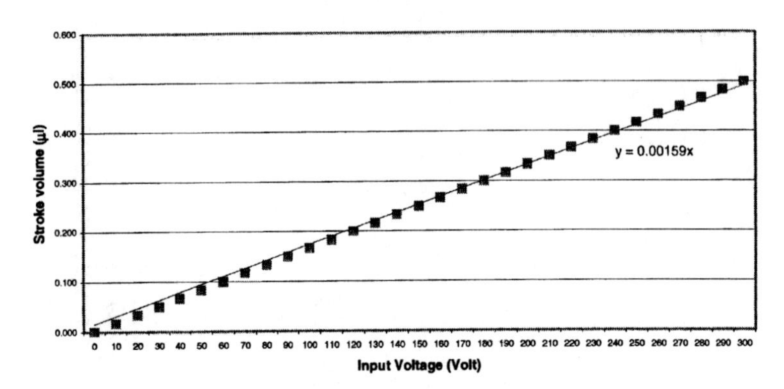

Figure 2-11. Stroke volume of piezoceramic actuator

From Fig. 2-9, it can be seen that the percentage difference is only about 5% and hence it is sufficiently accurate to use an axisymmetric 2D model. This will greatly reduce the computation time.

In order to determine the volumetric deflection (i.e. stroke volume), the deflection profiles at different input voltages have to be obtained first (Fig. 2-10). The profiles are then curve-fitted with polynomial curves and integrated to obtain the volume swept by the profile using MATLAB®. A linear relationship between the volumetric displacement and the input voltage (Fig. 2-11) is obtained:

$$V_{total} = 0.00159U \tag{2-27}$$

where V_{total} is the volumetric displacement or stroke volume (µl) and U is the input voltage.

2.2.2 Determination of Dead Volume

The dead volume is obtained by determining the total volume between the inlet and outlet valve that encompasses the pump chamber. In the author's design, the dead volume comprises of four parts, namely, the inlet channel (V_{inlet}), the pump chamber (V_{pump}), the outlet channel (V_{outlet}) and the outlet valve chamber (V_{valve}) (Fig. 2-12).

The dead volume, V_o, of the proposed micropump is:

$$V_o = V_{inlet} + V_{pump} + V_{outlet} + V_{valve}$$
$$\approx 1.145 \mu l$$

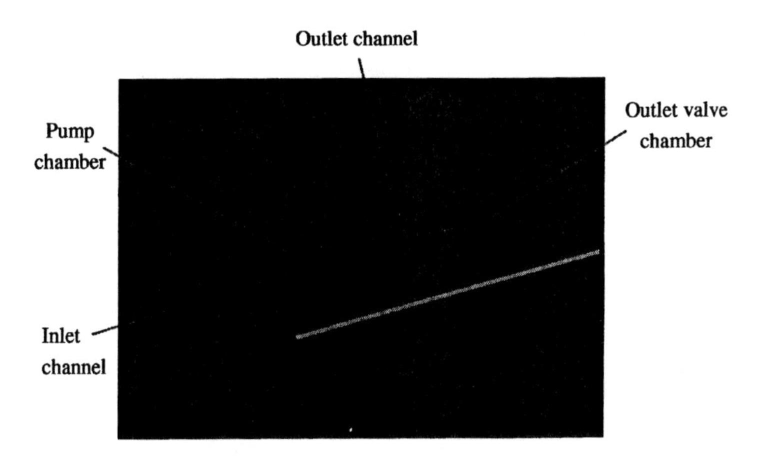

Figure 2-12. Dead volume of the micropump

2.2.3 Determination of Compression Ratio

The compression ratio of the proposed micropump, based on the FEM model, is given by (Eq. 2-7):

$$\varepsilon = \frac{V_{total}}{V_o} = \frac{0.00159U}{1.145} \tag{2-28}$$

The importance of the compression ratio will be further elaborated on in the next few sections.

2.3 Criterion for Switching of Valve

During the supply and pumping cycle, a differential pressure, $|\Delta P|$, is generated in the pump chamber. It can be assumed that a critical pressure difference $|\Delta P_{crit}|$ is required to switch the valves and this forms the basic operation criterion for the proper functioning of the valves:

$$|\Delta P| > |\Delta P_{crit}| \tag{2-29}$$

The criterion given by Eq. (2-29) determines whether the pressure generated in the pump chamber by the actuator is sufficient in opening and closing of the check valves. This is the foremost design rule for the designer.

In order to determine this critical pressure, the surface energies at the water-air and silicon-water interfaces (also known as interfacial energy) have to be taken into consideration [59]. In an extension to the model in [59], the author has included the effect of a pre-tension on the critical pressure. These will be presented in detail in the following sections.

2.3.1 Surface Energy Consideration

Prior to the filling of the micropumps, a thin layer of water is usually trapped between the valve and the valve-seat (Fig. 2-13(a)). In the proposed micropump, silicon oxide is grown on the silicon surface to create hydrophilic oxide-water interfaces. This will help in the priming of the micropump has the risk of bubbles being trapped in the pump is minimized.

With reference to Fig. 2-13, assume that a thin layer of water with thickness z_o between the valve-seat and if the valve diaphragm and the valve diaphragm moves an infinitesimal distance dz from z_o to z_o + dz, the surface area, A, at the water-air interface is increased by:

$$dA = 2 \times 4bdz = 8bdz$$

where 4b is the perimeter of valve seat (b = 1700 μm).

The corresponding change in surface energy brought about by this movement is:

$$dW_{surf1} = \sigma_{a-w} 8bdz \tag{2-30}$$

where $\sigma_{a\text{-}w}$ is the surface energy between water and air ($\sigma_{a\text{-}w}$ = 75 mNm^{-1}).

Now, since the layer water has no contact with a water reservoir, so the lateral dimension has to correspondingly shrink by 2dx during the lifting of the valve (Fig. 2-13(b)). The energy that is needed to reduce the wetted area between the silicon oxide and water is:

$$dW_{surf2} = 2 \times \sigma_{o-w} \times 8bdx = 16\sigma_{o-w} bdx \tag{2-31}$$

where $\sigma_{o\text{-}w}$ is the surface energy between water and air ($\sigma_{o\text{-}w}$ = -35.82 mNm^{-1}).

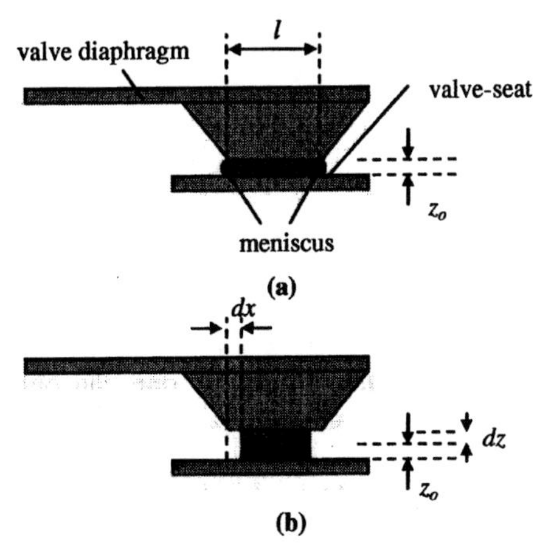

Figure 2-13. Schematic view of a wetted valve during filling. (a) Initial thin layer of water prior to filling and (b) after the valve is lifted by infinitesimal distance dz

These two contributions will effectively cause a change in the total surface energy given by:

$$dW_{surf\_total} = 8b\left(\sigma_{a-w}dz - 2\sigma_{o-w}dx\right) \tag{2-32}$$

Applying the principle of conservation of mass for the water layer and assuming that the fluid is incompressible:

$$z_o l 4b = \left(z_o + dz\right)\left(l - 2dx\right)4b$$

where l is the length of the valve seat (l = 300 μm).

Neglecting the small term, dxdz, the relationship between dx and dz can be obtained:

$$dx = \frac{l}{2z_o}dz \tag{2-33}$$

In order to overcome this change in the surface energy, an external work, dW_{ext1}, given by the pressure difference ΔP_{ext1} and the area of the membrane valve over which this differential pressure acts, has to be applied:

$$dW_{ext1} = \Delta P_{ext1} \times \left(\pi R_v^2 - b^2\right) dz \tag{2-34}$$

where R_v is the radius of the valve (R_v = 2500 μm).

Balancing the two energy terms (i.e. Eqs. (2-32) and (2-34)) and making use of Eq. (2-49):

$$\Delta P_{crit1} = \frac{8b\left[\sigma_{a-w} - \sigma_{o-w}\left(l/z_o\right)\right]}{\left(\pi R_v^2 - b^2\right)} \tag{2-35}$$

Now, the height of the liquid layer, z_o, can in the extreme case be estimated by the roughness of the wetted surface ($z_o \approx 0.1 - 0.5$ μm). Substituting the necessary values into Eq. 2-51, one can obtain a critical pressure (ΔP_{crit1}) of 16.5 kPa based on surface energy considerations (i.e. neglecting elastomechanics of the valve membrane). This value is of the same order of magnitude as that obtained experimentally in [59] for a cantilever valve.

From Eq. (2-35), some interesting observations can be made, the most important being the fact that the critical pressure is largely influenced by the l/z_o ratio. Hence, to design a valve with a small critical pressure, the length of the valve-seat should be minimized. It must also be noted that this model is based on the worst-case scenario, which will give an upper bound for the critical pressure. A similar analysis can be found in [59] which give the case whereby the valve is completely dried prior to filling, in which case the critical pressure can be reduced by as much as a factor of 10.

2.3.2 Elastomechanical Consideration

As above-mentioned, the critical pressure obtained based purely on surface energy consideration effectively neglects the elastomechanics of the valve. Now, in most of the valve designs, the valve is subjected to a pre-tension so as to improve the leak tightness of the valves in the backward direction for small back pressures and to enable blocking of the valves in the forward direction for forward pressures below a critical value as pre-determined by the designer. In order to determine this pre-tension, an axisymmetric FEM model of the valve is built and simulated (Fig. 2-14).

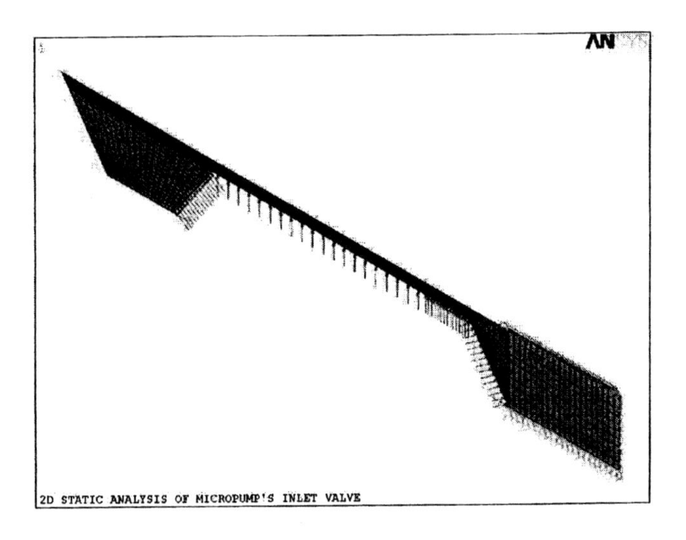

Figure 2-14. Geometry and boundary conditions of FEM model for inlet valve

The results for the simulation of the valve are depicted in Figures 2-14 to 2-17. The pre-stress in the inlet valve is due to the pre-deflection of 0.85 μm (due to 56% of the 1.5 μm of oxide that protrudes out from the silicon substrate). Hence the pressure to open the valve is equivalent to the pressure needed to lift the valve by 0.85 μm. From the simulation, this pressure (i.e. ΔP_{crit2}) is found to be 2 kPa.

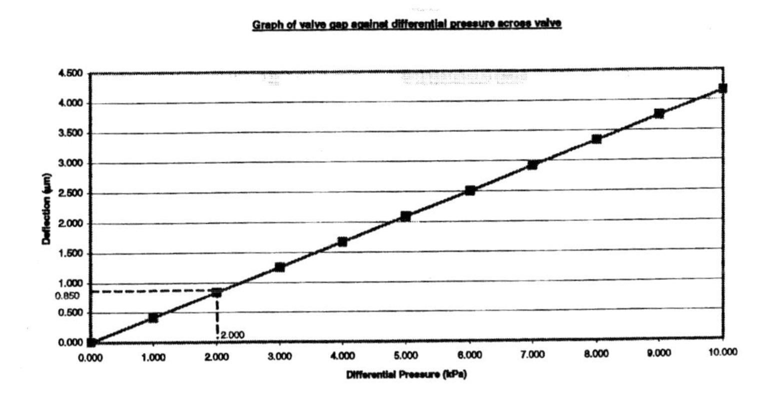

Figure 2-15. Variation of valve gap with differential pressure across valve

The total pressure, ΔP_{crit}, needed to open the valve can be found as a summation of ΔP_{crit1} and ΔP_{crit2} which gives a value of 18.5 kPa. Hence, the piezoelectric actuator of the micropump must be able to generate a chamber pressure above this valve for pumping to be possible. This is the foremost design rule in the design of a reciprocating displacement of micropump.

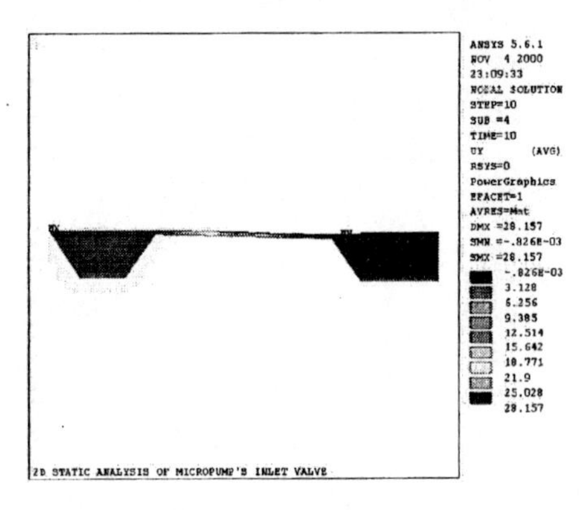

Figure 2-16. Deflection of valve at 100 kPa differential pressure across valve

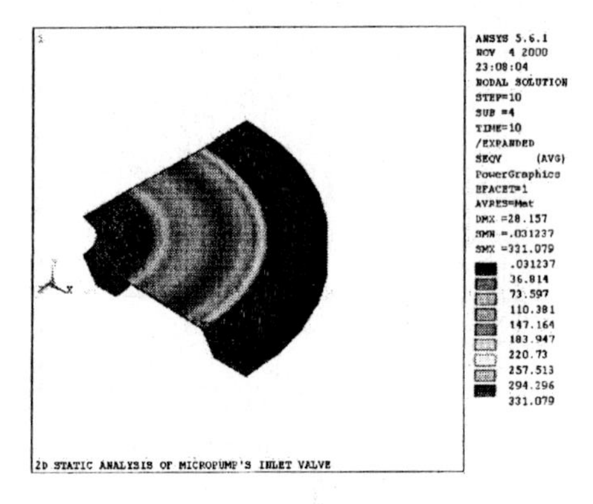

Figure 2-17. Von Mises stress state of valve at 100 kPa differential pressure across valve

2.4 Criterion for Self-priming Capability

To formulate the design rule for self-priming capability, the worst-case scenario has to be assumed: the whole pump chamber is assumed to be filled with gas and the valves are wetted. In the analysis, an ideal gas undergoing adiabatic state changes with an adiabatic coefficient of γ is assumed. Suppose the volume of the gas in the chamber is V_o under atmospheric pressure P_a and the stroke volume of the actuator unit generated in the pump chamber is ΔV, then the corresponding pressure differential generated, ΔP, is given by:

$$P_a V_o^\gamma = (P_a + \Delta P)(V_o + \Delta V)^\gamma \tag{2-36}$$

Substituting Eqs. (2-7) and (2-36) into the condition given by Eq. (2-29), the criterion for the minimum compression ratio of a self-priming pump is:

$$\varepsilon_{self-priming} > \left(\frac{P_a}{P_a + |\Delta P_{crit}|} \right)^{1/\gamma} - 1 \tag{2-37}$$

At low frequencies, there is an isothermal behaviour (i.e. $\gamma = 1$), hence:

$$\varepsilon_{self-priming} > \frac{|\Delta P_{crit}|}{P_a + |\Delta P_{crit}|} \tag{2-38}$$

As the critical pressure ΔP_{crit} is normally small as compared to the atmospheric pressure, Eq. (2-38) can be simplified into:

$$\varepsilon_{self-priming} > \frac{|\Delta P_{crit}|}{P_a} \tag{2-39}$$

Equation (2-39) is the design rule governing whether a micropump is self-priming or not. For the author's proposed micropump, under a 150 V input voltage, the compression ratio (Eq. (2-28)) is given by:

$$\varepsilon = \frac{0.00159 \times 150}{1.145} \approx 0.208$$

In addition: $\dfrac{|\Delta P_{crit}|}{P_a} = \dfrac{18500}{100000} = 0.185$

Clearly, in this case: $(\varepsilon = 0.208) > \left(\frac{|\Delta P_{crit}|}{P_a} = 0.185 \right)$

Therefore, one can conclude that the proposed micropump should be self-priming in nature. Incidentally, the criterion given by Eq. (2-39) is also the governing criterion for a micropump to be able to pump gases.

2.5 Criterion for Bubble Tolerance

The analysis for bubble tolerance is somewhat similar to that for self-priming. In fact, a pump that is self-priming is in theory bubble tolerance as the case for determining self-priming is equivalent to the case whereby the gas bubble fills the entire pump. However, such an analysis may be too stringent at times, hence the author has formulated a design rule for bubble tolerance based on the amount of bubble trapped in the pump.

To develop the model for bubble tolerance analysis, the case for a completely liquid filled micropump is first looked into. Assuming a liquid of compressibility K_{liq}, when the volume of the pump chamber (Vo) changes by ΔV, the pressure peak ΔP occurring in the pump chamber can be calculated by:

$$\Delta V + V_o = V_o\left(1 - K_{liq}\Delta P\right) \tag{2-40}$$

Substituting Eqs. (2-7) and (2-40) into the condition given by Eq. (2-29), the criterion for the minimum compression ratio, ε_{liq}, for a micropump to pump liquid is:

$$\varepsilon_{lq} > K_{liq}\left|\Delta P_{crit}\right| \tag{2-41}$$

Most liquids have a small compressibility (for example, water has a compressibility factor of K_{liq} = 5 x 10^{-9} m^2/N), hence the design rule for liquid pumps can be fulfilled with relative ease. However, this consideration is only valid when the pump chamber is completely filled with liquid and no gas is trapped in the pump at any one time.

Equations (2-39) and (2-41) give the criteria for the two extreme cases, namely, a pump completely filled with gas and a pump completely filled with liquid respectively. To formulate the design rule for a pump which has a certain amount of gas trapped in it, a relative gas content, α_{gas}, is introduced by the author and this is given by:

$$\alpha_{gas} = \frac{V_{gas}}{V_o}$$

where V_{gas} is the volume of gas trapped within the pump evaluated at atmospheric pressure and V_o is the dead volume of the micropump.

Hence, the criterion for bubble tolerance can be given by:

$$\varepsilon_{bubble-tolerance} > (1-\alpha_{gas})K_{liq}|\Delta P_{crit}| + \alpha_{gas}\frac{|\Delta P_{crit}|}{P_a} \tag{2-42}$$

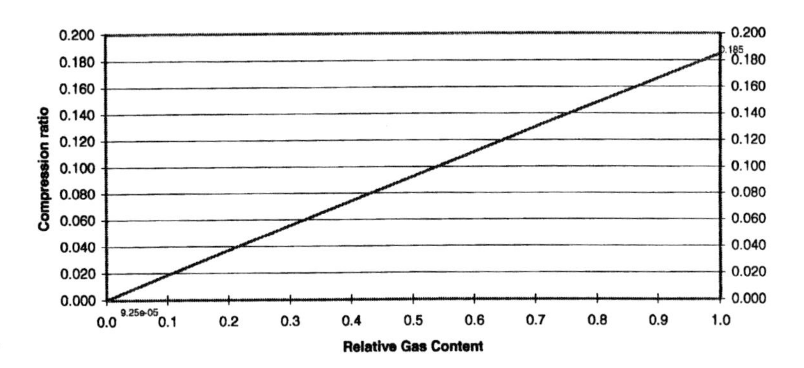

Figure 2-18. Criterion for bubble tolerance for proposed micropump

Figure 2-18 gives a plot of the criterion given by Eq. (2-42) for various relative gas ratios for the proposed micropump. As the relative gas content has to be pre-determined, Eq. (2-42) is not very useful for the design phase. Nevertheless, such a plot can be used as a gauge for the degree of bubble tolerance for a particular pump design once its compression ratio has been determined. In addition, it can be used as a guide for the amount of gas that a particular micropump design can take before pumping will fail.

Chapter 3

MODELLING AND SIMULATION

After formulation of the designs rules for a reciprocating displacement micropump based on piezoelectric actuation, the modelling of the micropump will be presented in this chapter. In the full system modelling of the micropump, quasi-static conditions have been assumed in consideration of the low operating frequencies, which effectively render the inertia effects to be insignificant. The modelling technique described in this chapter is used to improve the pump design and possibly shorten the time required from the design to fabrication of marketable devices.

3.1 Background of Microfluidics Systems Modelling

Modelling of microfluidics devices and systems often involve elastomechanics, fluid transmission, energy transformation among electrical, magnetic and thermal energies, as well as energy coupling effects. Take for instance the case of micropumps whereby it is commonplace to have strong interactions between fluid and structural elements which may also be coupled electrically or thermally to describe the actuation mechanism. The equations governing the behaviour of these systems often have no analytical solutions, particularly if they include non-linear effects such as large deflection and stress stiffening.

Previous researches were mainly focused on functional element scale modelling such as the modelling of microvalves and actuating membrane, most of which involving the use of the FEM. For example, the FEM has been used to analysed microvalves by Ulrich & Zengerle [78] and Koch et al [79]. Non-linear membrane deformation under electrostatic actuation was analysed using differential equations by Francais et al [80] and using FEM by Cozma & Puers [81] and Gong et al [82].

In recent years, researchers have also direct their effort towards full system modelling. One of the early works in this area includes Zengerle & Ritcher [83] who reported a full system theoretical model for diaphragm micropumps based on a set of governing differential equations which were solved with a dedicated software, PUSI, developed by the authors. In addition, the bond graph method, a mathematical method based on energy and power relationships in lumped elements, was adopted by Van der Pol et al. [84] to model a thermo-pneumatic micropump.

Regarding full system modelling and simulation, one method is to develop differential equations to describe the micropump's working procedure after all individual elements have been functionally construed. Another method applies mechanical–electrical equivalent network representation [85-86] to build system transfer functions, and then determine the micropump's state or process parameters by computer-aided electrical circuit analysis software such as SPICE and AMS. In this method, the equivalent network is built by first sub-dividing the complete device structure into lumped elements. The behaviour and characteristics of each element is then analysed analytically or numerically. Following that, each of these elements is then described on the basis of analogies between relevant physical parameters of the dominating phenomenon and electrical parameters. For example, when considering fluid flow, flow rate and pressure are analogous to electrical current and voltage respectively. A complete fluidic device can then be modelled with an equivalent network by linking lumped elements in accordance to Kirchoff's laws adapted to mechanical and fluidic systems. Analysis is then carried out with an electrical simulation software.

In this project, the first method is used, that is, the micropump is first sub-divided into functional elements such as valves and actuating membrane and their characteristics analysed using the FEM. A system of differential equations governing the behaviour or the micropump is then determined and used to integrate all the elements together.

3.2 Governing Equation for Micropumps

The governing equation for a micropump is derived from the principle of *conservation of mass* which states that, for any defined control volume [87]:

Mass of fluid	–	Mass of fluid	=	Increase of mass of fluid
entering per unit time		leaving per unit time		in the control volume per unit time

Taking the pump chamber as the control volume (Fig. 3-1) and using the continuity equation:

$$\oint\!\!\oint_{CS} (n \cdot \rho v)\, dA = -\frac{\partial}{\partial t} \iiint_{CV} \rho\, d\forall \tag{3-1}$$

where *CS* denotes control surface; *CV* denotes control volume; *n* is the normal to the control surface; ρ is the density of the fluid and v is the fluid velocity.

Equation (3-1) physically means that the net efflux of mass through a control volume is equal to the rate of decrease of mass inside the control volume. The term on the LHS of Eq. (3-1) is equivalent to the net efflux of fluid out of the pump chamber and if the inertia of the check valves can be neglected, it is given by:

$$\oint\!\!\oint_{CS} (n \cdot \rho v) dA = \rho\left(\phi_{ov}(P_{ov}) - \phi_{iv}(P_{iv})\right) \tag{3-2}$$

where ϕ_{iv} is the steady state flow characteristic of the inlet valve; ϕ_{ov} is the steady state flow characteristic of the outlet valve; P_{iv} is the hydrostatic pressure differential across the inlet valve ($P_{iv} = P_{in} - P$) and P_{ov} is the hydrostatic pressure differential across the outlet valve ($P_{ov} = P - P_{out}$).

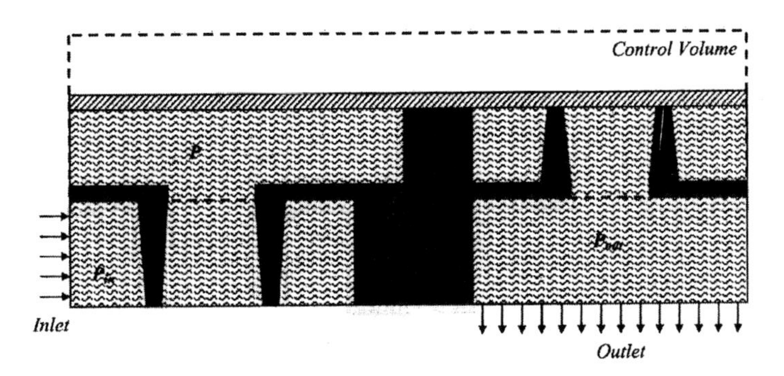

Figure 3-1. Control volume of a reciprocating micropump

The term on the RHS of Eq. (3-1) is equivalent to the rate of change of mass within the control volume and assuming that the pumping medium is incompressible, it is given by:

$$-\frac{\partial}{\partial t}\iiint_{CV} \rho dV = -\left(\rho\frac{\partial V_{chamber}}{\partial t} + \rho\frac{\partial V_o}{\partial t} - \rho\frac{\partial V_{iv}}{\partial t} + \rho\frac{\partial V_{ov}}{\partial t} - \frac{\partial(\rho_{gas}V_{gas})}{\partial t}\right) \tag{3-3}$$

where ρ is the density of the pumping medium; $V_{chamber}$ is the stroke volume of the actuator; V_o is the total pump chamber volume; V_{iv} is the volumetric displacement of the inlet valve; V_{ov} is the volumetric displacement of the outlet valve; ρ_{gas} is the density of any gas trapped within the control volume and V_{gas} is the volume of gas trapped within the control volume.

Equating Eqs. (3-2) and (3-3):

$$\rho(\phi_{ov}(P_{ov})-\phi_{iv}(P_{iv})) = -\left(\rho\frac{\partial V_{chamber}}{\partial t}+\rho\frac{\partial V_o}{\partial t}-\rho\frac{\partial V_{iv}}{\partial t}+\rho\frac{\partial V_{ov}}{\partial t}-\frac{\partial(\rho_{gas}V_{gas})}{\partial t}\right)$$

which gives:

$$\begin{aligned}\frac{\partial V_{chamber}}{\partial t} &= \phi_{iv}(P_{iv})-\phi_{ov}(P_{ov}) \\ &-\frac{\partial V_o}{\partial t}+\rho\frac{\partial V_{iv}}{\partial t}-\rho\frac{\partial V_{ov}}{\partial t}+\frac{1}{\rho}\frac{\partial(\rho_{gas}V_{gas})}{\partial t}\end{aligned} \tag{3-4}$$

Since $V_{chamber}$ is a function of the chamber pressure (P) and the input voltage (U):

$$\begin{aligned}\frac{\partial V_{chamber}}{\partial t} &= \left.\frac{\partial V_{chamber}}{\partial t}\right|_P+\left.\frac{\partial V_{chamber}}{\partial t}\right|_U \\ &= \left.\frac{\partial V_{chamber}}{\partial U}\right|_P\cdot\frac{dU}{dt}+\left.\frac{\partial V_{chamber}}{\partial P}\right|_U\cdot\frac{dP}{dt}\end{aligned} \tag{3-5}$$

Substituting Eq. (3-5) into Eq. (3-4) and manipulating the differential for the other terms, one will arrive at:

$$\frac{dP}{dt}=\frac{\phi_{iv}(P_{iv})-\phi_{ov}(P_{ov})-\left.\frac{\partial V_{chamber}}{\partial U}\right|_P\cdot\frac{dU}{dt}}{\left.\frac{\partial V_{chamber}}{\partial P}\right|_U+\frac{dV_o}{dP}-\frac{\partial V_{iv}}{\partial P}+\frac{dV_{ov}}{dP}-\frac{1}{\rho}\frac{d(\rho_{gas}V_{gas})}{dP}} \tag{3-6}$$

Equation (3-6) is the basic governing equation for evaluating the time dependent pressure, P, inside the pump chamber. Since Eq. (3-6) contains terms that are non-linear, solving it will require numerical methods. The assumptions made in deriving the governing equation are:

1) Valve leakage is neglected;
2) Inertia effects can be neglected (i.e. input can be transient but response is steady- state, that is, pseudo steady state);
3) Pressure distribution within the chamber is spatially uniform (i.e. local fluid redistributions inside the pump chamber occurs quasi-statically, that is, instantaneously on the time scale considered) and
4) The pumping medium is incompressible.

To further simplify Eq. (3-6), performance-reducing effects such as bubbles in the pump chamber and valve deformation are ignored. Assuming a rigid pump chamber, Eq. (3-6) will hence reduce to:

$$\frac{dP}{dt}=\frac{\phi_{iv}(P_{iv})-\phi_{ov}(P_{ov})-\left.\frac{\partial V_{chamber}}{\partial U}\right|_{P}\cdot\frac{dU}{dt}}{\left.\frac{\partial V_{chamber}}{\partial P}\right|_{U}} \qquad (3\text{-}7)$$

In this project, Eq. (3-7) is used to model the behaviour of the pump. Essentially, the model will give better performance than the actual pump. In the following sections, the FEM is used for determining the various terms (i.e. characteristics of the valves and actuating membrane) on the RHS of Eq. (3-7), which together with the results, will be presented in detail.

3.3 Modelling of the Actuator Unit

First of all, the natural frequencies and mode shapes of the piezoelectric actuator unit are determined using a 3D model. The results are tabulated in Table 3-1 and Fig. 3-2. As can be seen, the fundamental frequency is around 14 kHz, which is very much larger than the operating frequency (< 200 Hz). In the actual situation, with the relatively large damping of the fluid, the resonance frequency can be very much lesser. Nevertheless, at such a low operating frequency, the piezoelectric actuator unit can be assumed to follow a quasi-static behaviour [88-89].

Table 3-1. Natural frequencies of the micropump's piezoelectric actuator unit

Mode	*Frequency (Hz)**
1	*14709*
2	*35455*
3	*58884*

Mode	*Frequency (Hz)**
4	*67328*

**Note: Medium is vacuum.*

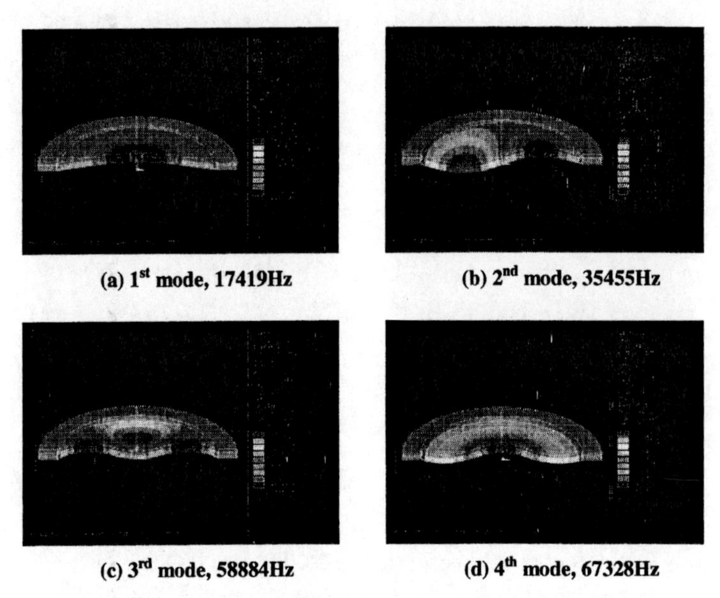

Figure 3-2. Mode shapes of micropump's piezoelectric actuator unit

In order to determine

$$\left.\frac{\partial V_{chamber}}{\partial U}\right|_P \text{ and } \left.\frac{\partial V_{chamber}}{\partial P}\right|_U,$$

the characteristics of the actuator unit at different input voltages and pressure differences (while keeping the other variable fixed) have to be evaluated respectively.

Using a similar model as that described in Chapter 2 except that a pressure is applied on the actuator to represent the chamber pressure (Fig. 3-3), the relationship between the stroke volume and the input voltage under various pressure differences across the actuator can be determined.

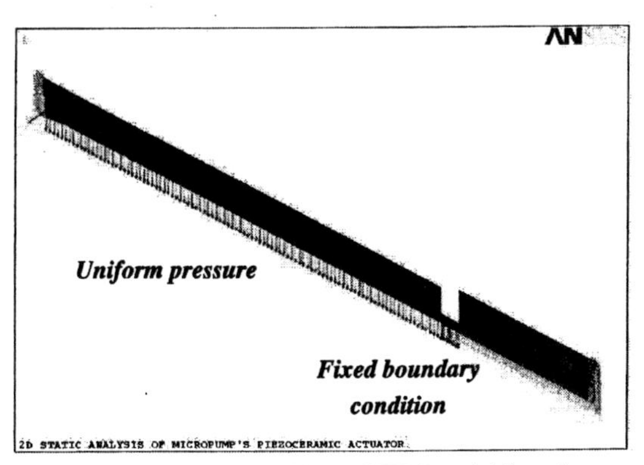

Figure 3-3. Geometry and boundary conditions of FEM model for micropump's piezoelectric actuator unit under chamber pressure

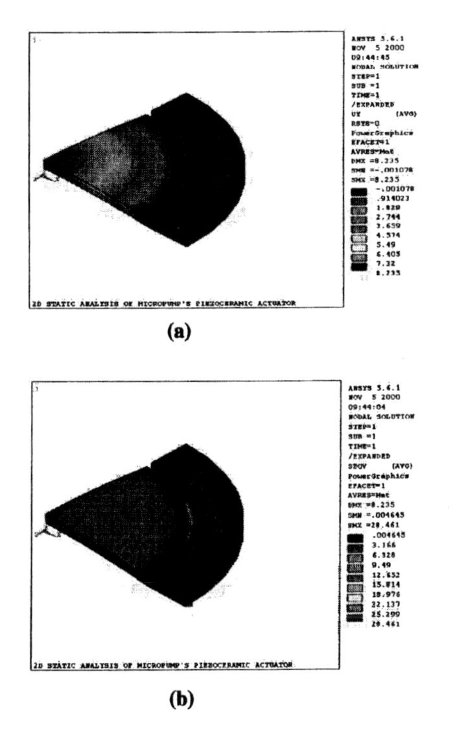

Figure 3-4. (a) Lateral deflection and (b) Von Mises stress state of micropump's piezoelectric actuator unit at 200 V input voltage and 0.1 kPa chamber pressure

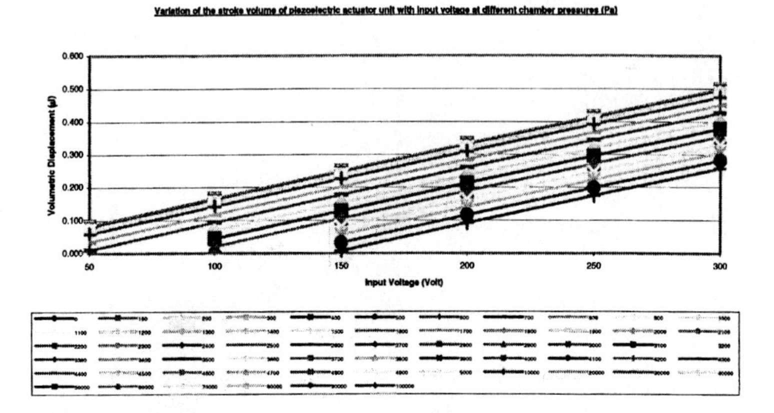

Figure 3-5. Variation of the stroke volume of micropump's piezoelectric actuator unit with input voltage at different chamber pressures

From the FEM simulations, the relationships between the stroke volume and input voltage at various chamber pressures are determined. It has been found that the stroke volume is linearly proportional to the input voltage while keeping the chamber pressure fixed. By polynomial curve fitting of the results using *MATLAB®*, the following relationship is obtained:

$$V_{chamber} = 0.001666U + \beta \tag{3-8}$$

where $V_{chamber}$ is the stroke volume (µl); U is the input voltage (Volt) and β is a constant dependent on the magnitude of the chamber pressure.

From Eq. (3-8):

$$\left.\frac{\partial V_{chamber}}{\partial U}\right|_{P} = 0.001666 \tag{3-9}$$

Likewise, from the FEM simulations, the relationships between the stroke volume and chamber pressure at various input voltages are determined. Once again, it has been found that the stroke volume is linearly proportional to the chamber pressure while keeping the input voltage fixed. By polynomial curve-fitting of the results using *MATLAB®*, the following relationship is obtained:

$$V_{chamber} = \alpha + 2.42329e\text{-}3P \tag{3-10}$$

where $V_{chamber}$ is the stroke volume (μl); P is the chamber pressure (kPa) and α is a constant dependent on the magnitude of the input voltage.

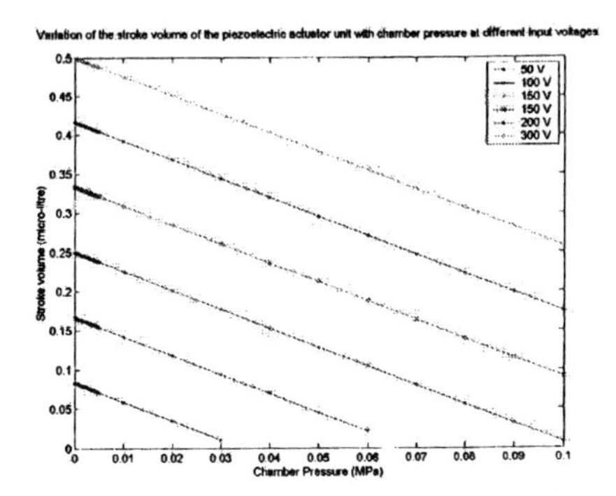

Figure 3-6. Variation of the stroke volume of micropump's piezoelectric actuator unit with chamber pressure at different input voltages

From Eq. (3-10):

$$\left.\frac{\partial V_{chamber}}{\partial P}\right|_{U} = 2.42329e\text{-}3 \qquad (3\text{-}11)$$

3.4 Modelling of the Inlet Valve

The modelling of the valves is a coupled field simulation involving fluid and structure interactions. In *ANSYS™*, fluid-structure interaction is a sequential field-coupling analysis involving iterations between two or more physics fields until a certain convergence criterion is met. This type of coupling belongs to the category of *weak* coupling (Chapter 2). The process flow for the approach used in the simulation of the valves is shown in Fig. 3-7. Essentially, the fluid region is first solved with the necessary initial conditions (i.e. for non steady-state analysis) and boundary conditions. Then the pressure distribution is read from the solution of the fluid region and applied as loads on the structural regime. The structural regime is then solved. In this case, the criterion for convergence is that when the maximum deflection of the valve membrane between two consecutive structural executions is less than the tolerance value, the iteration is terminated and solution is deemed complete. In the case whereby the difference in the maximum deflection between two consecutive structural executions is larger than the tolerance value, the fluid region is solved again but this time taking into account the change in the flow regime due to the deformation of the structural regime. This loop repeats until either the criterion is satisfied or the number of loops allowable is exceeded (set by the designer).

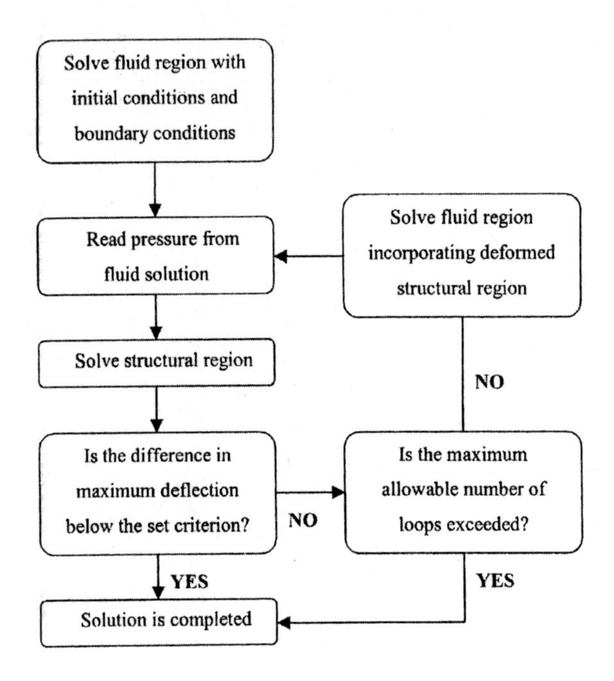

Figure 3-7. Flow chart of fluid-structure simulation of valves

The FEM model of the inlet valve is shown in Fig. 3-8, depicting both the fluid and structural regimes. In order to determine the flow characteristic of the inlet valve, the inlet valve is simulated at different pressure differences, from which the resulting flow rates are obtained (Fig. 3-9). The flow characteristic is given by:

$$\phi_{iv} = 0.0291(P_{iv})^{3.7031} \tag{3-12}$$

where ϕ_{iv} is the volume flow rate (μl/min) and P_{iv} is the pressure difference (kPa).

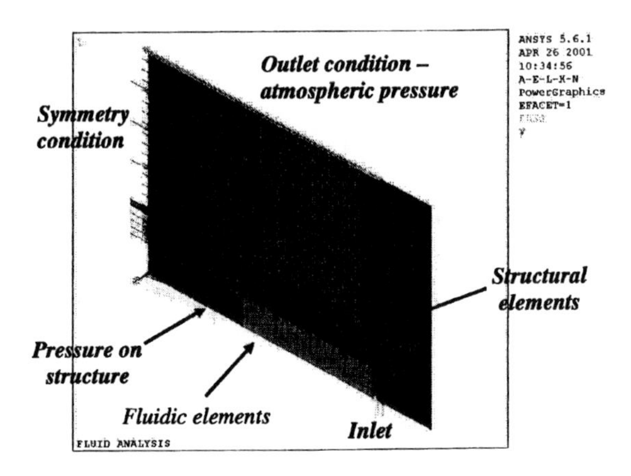

Figure 3-8. FEM model of inlet valve showing the fluid region (green) and structural region (purple)

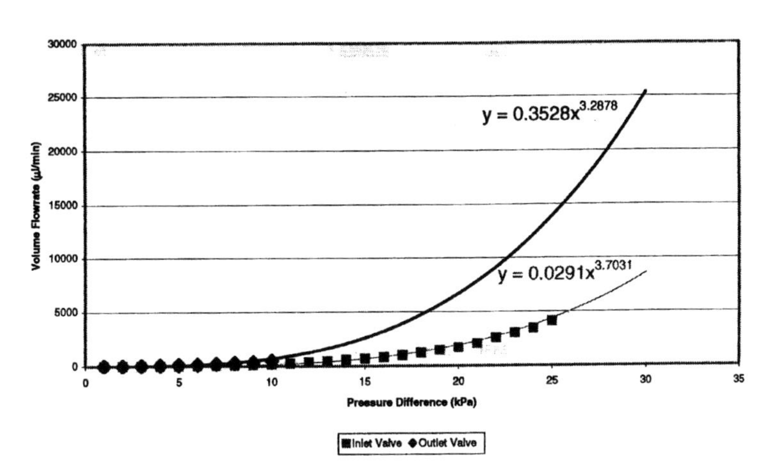

Figure 3-9. Flow characteristics of valves

As depicted in Fig. 3-10, a large proportion of the pressure drop occurs at the region near to the valve-seat. This is expected since the valve gap acts as a constriction to the flow and the entrance and exit losses are very significant. Figure 3-11 shows the velocity plot for the inlet valve. The flow velocity is accelerated at the valve-seat which results in dynamic losses.

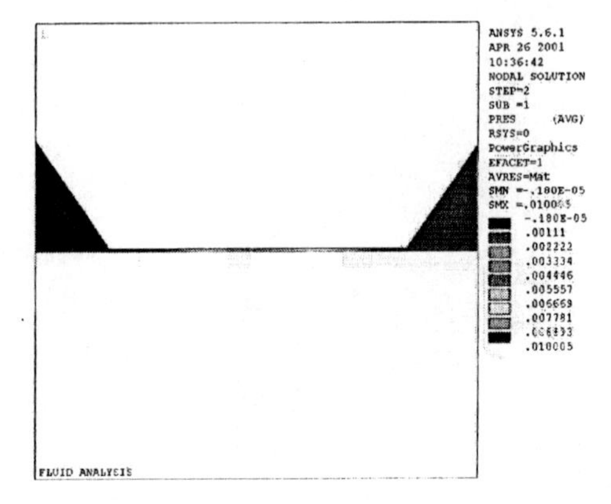

Figure 3-10. Pressure distribution at the valve-seat of inlet valve at a pressure difference of 0.01MPa

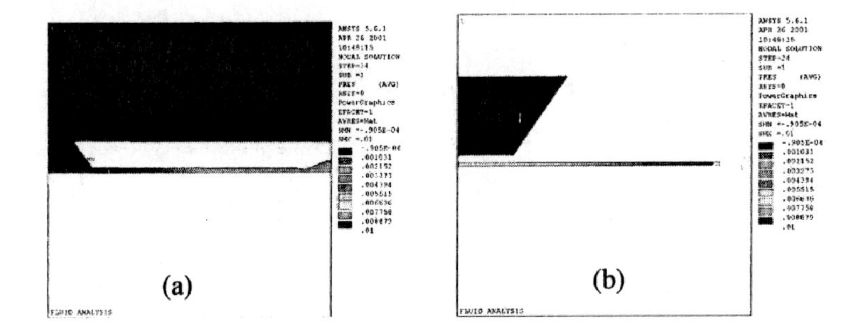

Figure 3-11. Vector plots showing flow in the (a) inlet valve and (b) at the valve-seat region at a pressure difference of 0.01MPa

3.5 Modelling of the Outlet Valve

The modelling of the outlet valve is similar to the inlet valve except for the inclusion of the outlet channel which contributes to pressure loss due to laminar friction losses. Once again, a large part of the pressure drop occurs at the valve-seat region. However, for the outlet valve, the valve gap is much larger than for the inlet valve and hence the constriction to the flow is not that significant. In addition, in this case, there is also considerable pressure drop in the outlet valve chamber region because of its small height (i.e. 10 μm). Therefore, in the design of the micropump, even though one desires to have as small a dead volume as possible, one must also bear in mind that the chamber and channel heights will cause restriction to the flow if they are too shallow.

The flow rate to pressure relationship is given by:

$$\varphi_{ov} = 0.3528(P_{ov})^{3.2878} \tag{3-13}$$

where ϕ_{ov} is the volume flow rate (μl/min) and P_{ov} is the pressure difference (kPa).

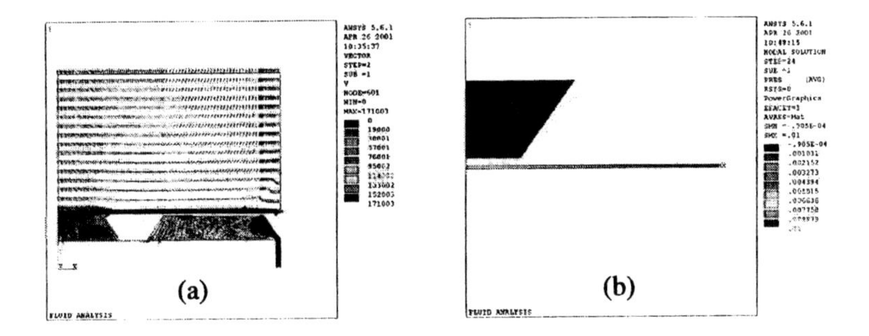

Figure 3-12. Pressure distribution at the (a) valve-seat region of inlet valve and (b) entrance (i.e. outlet channel) at a pressure difference of 0.01MPa

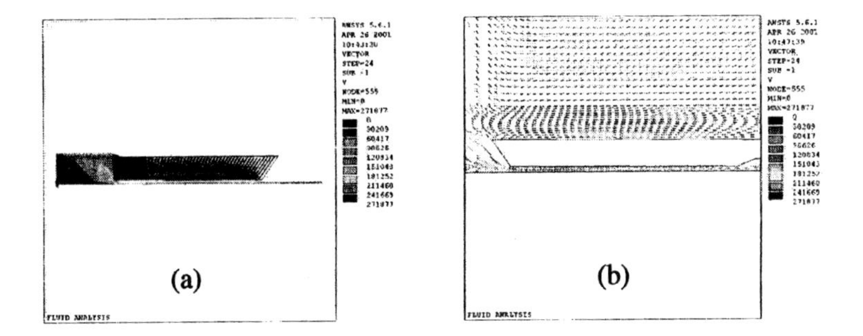

Figure 3-13. Vector plots showing flow in the (a) outlet valve and (b) at the valve-seat region at a pressure difference of 0.01MPa

3.6 System Model of Proposed Micropump

The terms

$\left.\frac{\partial V_{chamber}}{\partial U}\right|_P$, $\left.\frac{\partial V_{chamber}}{\partial P}\right|_U$, ϕ_{iv} and ϕ_{ov} are given in Eqs. (3-9), (3-11), (3-12) and (3-13) respectively. Substituting them into Eq. (3-7):

$$\frac{dP}{dt}=\frac{\frac{0.0291P_{iv}^{3.7031}}{60}-\frac{0.3528P_{ov}^{3.2878}}{60}-0.001666\cdot\frac{dU}{dt}}{2.42329e-3}$$
$$=\frac{\frac{0.0291(P_{in}-P)^{3.7031}}{60}-\frac{0.3528(P-P_{out})^{3.2878}}{60}-0.001666\cdot\frac{dU}{dt}}{2.42329e-3}$$

(3-14)

In order to examine the response of the pump, a harmonic driving voltage wave (i.e. $U(t)=U_o[1-\cos(2\pi ft)]$ is used for driving the piezoceramic actuator unit with f being the actuating frequency. Equation (3-14) then becomes:

$$\frac{dP}{dt}=\frac{\frac{0.0291(P_{in}-P)^{3.7031}}{60}-\frac{0.3528(P-P_{out})^{3.2878}}{60}-0.001666\cdot 2\pi fU_o\sin(2\pi ft)}{2.42329e-3} \qquad (3\text{-}15)$$

Since an analytical solution of Eq. 3-15 is not possible, it is solved numerically using the commercially available *Mathematica™* software package. The chamber pressure and average flow rate are determined with varying peak actuation voltages and driving frequencies. The variation of the chamber pressure with time for a peak-to-peak actuation voltage of 80V and a driving frequency of 10 Hz is shown in Fig. 3-14. In addition, the variations of the average flow rate with frequency and actuation voltage are depicted in Figs. 3-15 and 3-16.

From the simulation results, it is found that the pump rate varies approximately linearly with voltage for frequencies less than 300 Hz. In addition, the flow rate increases with the driving frequency at lower frequencies but becomes saturated at higher frequencies. This saturation frequency is dependent on the actuation voltage. Hence, to obtain high flow rates, it will be better to apply higher actuation voltages rather then applying high driving frequencies. On the other hand, too high a voltage may cause breakdown or de-polarization of the piezoelectric material. These two factors limit the maximum flow rate obtainable for a particular micropump design.

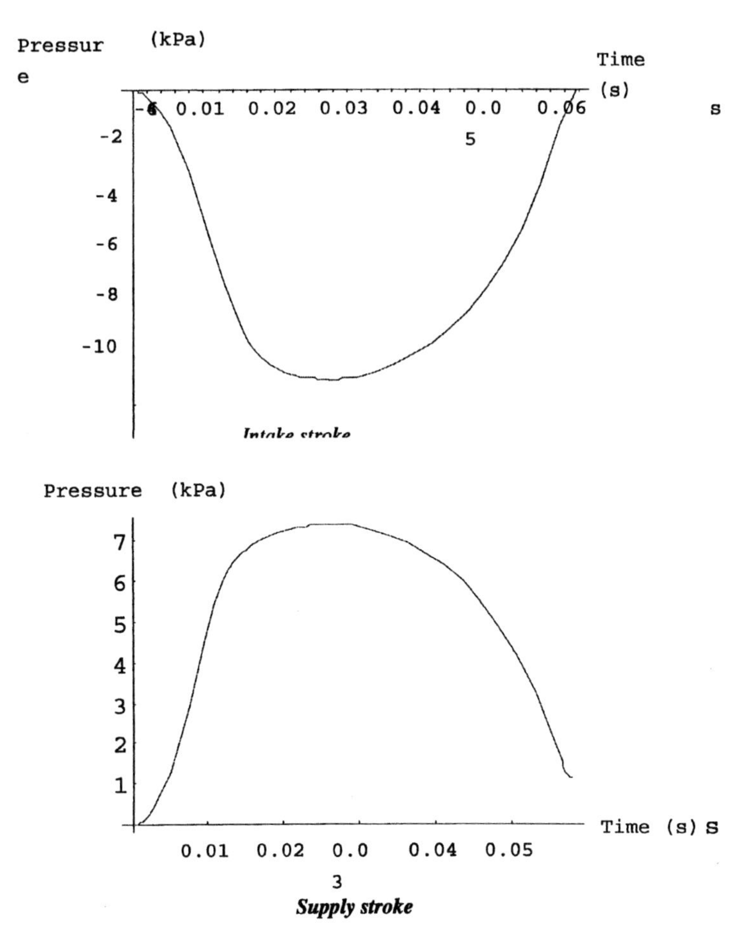

Figure 3-14. Variation of chamber pressure with time at V_{pp} = 80V and f = 10 Hz

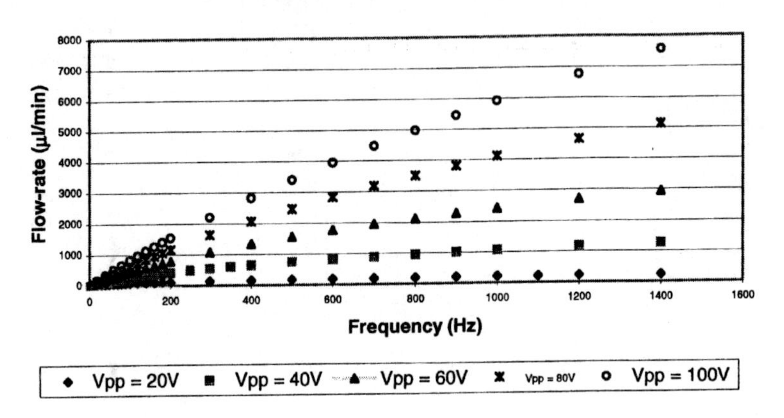

Figure 3-15. Relationship between flow rate and driving frequency at zero backpressure

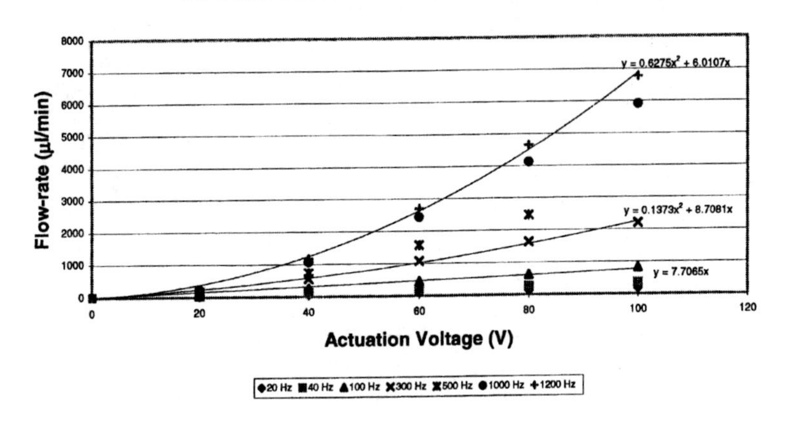

Figure 3-16. Relationship between flow rate and actuation voltage at zero backpressure for various driving frequencies

Chapter 4

PROCESS DEVELOPMENT AND FABRICATION

Micropumps have been fabricated using different techniques in metal (e.g. aluminium), silicon/glass and polymers (e.g. polycarbonate). Figure 4-1(a) shows the breakdown of materials used in micropumps based on a survey of the literature. The type of material used is in general closely related to the fabrication technique employed (Fig. 4-1(b)), each having its advantages and limitations. The metal pump has the advantage of fast prototype fabrication and readily available equipment. The fabrication of a pump in silicon can be a much more complicated and involved process. However, the advantages are that many pumps can be fabricated at the same time (i.e. batch fabrication) and consistent micrometer size features can be produced. The use of silicon and glass has also advantage over metals as pump materials as they are more resilient to aggressive media thus making them more suitable for applications such as chemical analysis systems. The use of polymers as pump material has the advantage that the material is inexpensive compared with more commonly used materials in MEMS devices. Based on one master (i.e. the mould), it is possible to fabricate thousands of pumps. The main disadvantage of using polymers at this point of time is the lack of maturity of the fabrication technique (usually by means of the *LIGA* process (a German acronym for *Lithographie*, *Galvanoformung*, *Abformung* meaning lithography, electroplating, and moulding respectively)) and the unavailability of the equipment needed to support this process. For details of these processes, the reader can refer to [90][91].

In general, the techniques used in micropumps fabrication can be categorized as conventional and micromachining technologies. The former constitutes well-known techniques such as milling, electro-discharge machining and thermoplastic replication while the latter consists of bulk and surface micromachining, *LIGA* and wafer bonding techniques among others. In this project, bulk silicon micromachining together with wafer bonding technique are used to fabricate the proposed micropump. The various processes associated with bulk silicon micromachining technology will be described in this chapter, followed by the process development and fabrication results of the proposed micropump.

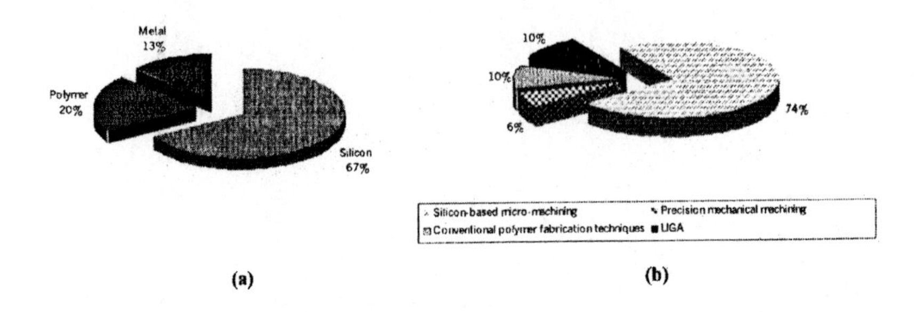

Figure 4-1. Breakdown of (a) materials and (b) fabrication techniques used in micropump research

4.1 Bulk Silicon Micromachining

In a nutshell, bulk micromachining is the selective and controlled removal of the silicon substrate by means such as photolithography, wet chemical etching and dry plasma etching. This is in contrast to the other popular silicon micromachining technique, namely, surface micromachining, in which case the structures are built by sequential additive (of thin films) and patterning processes.

There are four basic techniques associated with bulk silicon micromachining. These are photolithography (patterning), the deposition of thin films of materials, the removal of material by wet chemical etching and the removal of material by dry etching techniques. Another technique that is commonly utilised is the introduction of impurities into the silicon to change its properties (i.e. doping). However, doping is not utilized in this project. These processes will be described briefly in the following sections, with focus on the areas relevant to this project.

4.1.1 Photolithography

Photolithography in translated literally from Latin means "light-stone-writing". Essentially, it is an optical means for transferring patterns onto a substrate. Patterns are first transferred to an imagable photoresist layer which is a liquid film that can be spread out evenly onto a substrate (by means of a spin-coater), exposed with a desired pattern, and developed into a selectively placed layer for subsequent processing. Photolithography is a binary pattern, that is, there is no grey-scale, colour, nor depth to the image. The basic steps in the photolithography process are illustrated in Fig. 4-2 and the process parameters used in this project are given in Appendix D.

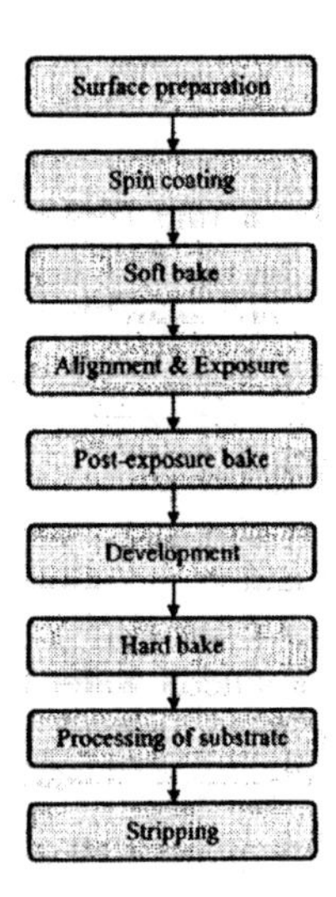

Figure 4-2. Process flow-chart of photolithography process

Surface preparation comprises of cleaning of the wafer, dehydration to remove water from the wafer surface and wafer priming which involves the utilization of primers (such as hexamethyldisilazane (HMDS)) to form bonds with the surface and produce a polar surface (CH_3-groups) to enhance the adhesion of the resist to the wafer surface. The resist is then spin coated onto the wafer to obtain an uniform layer of resist. Soft-baking or pre-baking evaporates the solvents within the resist and densify the resist after spinning. Alignment and exposure is then carried out to transfer the pattern of the photomask onto the resist by means of a mask aligner and typically a UV lamp. A positive resist is used in this project, in which case the radiation from the exposure lamp induces a chemical reaction in the exposed areas of the photoresist, altering the solubility of the resist in the developer (i.e. exposed areas become more soluble). Post-exposure bake is then carried out to bring to a halt the reactions initiated during the exposure. Development involves the selective dissolution of the exposed resist in the developer to transform the latent resist image during exposure into a relief image which will serve as a mask for subsequent processes. Hard-baking stabilizes and hardens the developed photoresist prior to further processing steps that the resist will mask. The wafer is then processed (e.g. etching) with the resist as the mask. Finally, the resist and all of its residues are removed or stripped, typically using acetone, propriety resist strippers or oxygen plasma (see §4.14).

4.1.2 Thin Film Deposition

In bulk micromachining, photoresist is often unable to withstand the etching required. Consequently, a thin film of a material resistant to the etchant is often needed to act as an etch mask during the etching of the underlying material. There are a considerable number of materials which the etchants do not attack significantly as compared to the substrate material. Typically, this masking layer will subsequently be stripped away after the underlying material has been fully etched.

There are a number of different techniques, such as chemical vapour deposition, sputtering, epitaxy and evaporation among others, that facilitate the deposition or formation of very thin films (of the order of micrometers or less) of different materials on a silicon wafer (or other suitable substrates). These films can then be patterned using photolithographic techniques and suitable etching techniques. Common materials include silicon dioxide (oxide in short), silicon nitride (nitride in short), polycrystalline silicon (polysilicon or poly in short), and aluminium. In addition to these common thin film materials, a number of other materials can also be deposited as thin films, including noble metals such as gold. Noble metals, however, will contaminate microelectronic circuitry causing it to fail, so any silicon wafers with noble metals on them have to be processed using equipment specially set aside for the purpose. Noble metal films are often patterned by a method known as "lift off", rather than wet or dry etching.

In this project, silicon dioxide and silicon nitride, deposited by thermal oxidation and low-pressure chemical vapour deposition (LPCVD) respectively, are used as masking materials. Brief descriptions of their properties and corresponding deposition methods are given in the following sections.

4.1.2.1 Oxidation of Silicon

Silicon dioxide is typically grown thermally in a quartz furnace at a temperature ranging from 800°C to 1200°C, as is the case for this project. The silicon wafers are oxidised by either a supply of oxygen or water vapour, which are popularly known as dry and wet oxidation respectively. The chemical reactions are:

$$\textit{Dry oxidation}: \; Si(s) + O_2(g) \rightarrow SiO_2(s)$$

$$\textit{Wet oxidation}: \; Si(s) + 2H_2O(g) \rightarrow SiO_2(s) + 2H_2(g)$$

All gas phase oxidation processes involve gas transport of oxidant to the surface, diffusion through the existing layer and the oxidation reaction itself. Dry oxidation results in a stochiometric oxide that has a higher density and is usually pinhole free. In contrast, the water in wet oxidation causes a loosening effect on the oxide, making it prone to impurity diffusion. Consequently, the properties of an oxide produced in a dry oxygen atmosphere are superior but it requires significantly more time to oxidise the same thickness using dry oxidation as compared to wet oxidation. Hence, for small thickness (1000Å or less), dry oxidation is usually used whereas wet oxidation is preferred for thicker oxide.

The oxidation of silicon differs from a straight deposition process due to the consumption of the substrate surface while forming the film, which results in excellent adhesion. In the process, a layer of the silicon is used to react to form the oxide (Fig. 4-3). The oxidising species has to diffuse through the already existing oxide for reaction to take place at the interface. The high temperature aids the diffusion of the oxidant through the surface oxide layer to the silicon interface. For thin oxides, this is not the limiting factor; instead, the process is reaction rate limited. However, when the oxide becomes thicker, the oxidation is diffusion limited. The oxide growth can be described with a differential equation from the Deal-Grove model [92] which solution shows that a thin oxide grows linearly with time, whereas a thick oxide grows only with the square root of the oxidising time. Consequently, oxide is thermally grown up to a thickness of about 2 µm only.

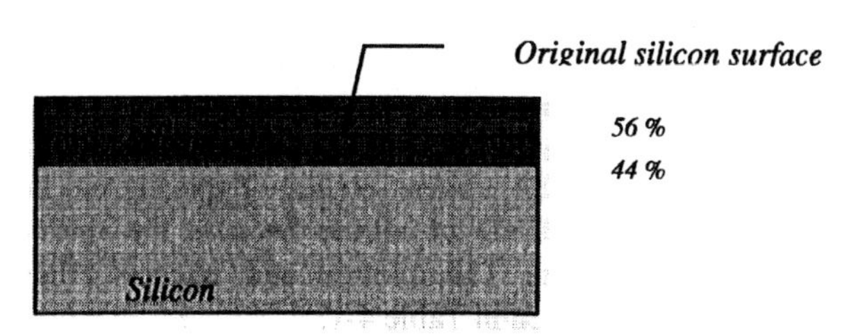

Figure 4-3. Silicon/silicon dioxide interface

Due to molecular mismatch (one silicon atom in the oxide layer takes up nearly twice as much space as in single-crystalline silicon) and thermal expansion differences, the resulting SiO2 film is under compressive stress. This has to be kept in mind for further processing. If the oxide is stripped on one side of the wafer, it may result in bending of the wafer. Wafer curvature can be fatal for processes such as anodic bonding. The relevant etch rates of silicon oxide in the various etchants are given in Table 4-1.

4.1.2.2 Low Pressure Chemical Vapour Deposition of Silicon Nitride

In the chemical vapour deposition process (CVD), the constituents of a vapour phase, often diluted with an inert carrier gas, undergo reactions at a hot surface to deposit a solid film, typically in a reaction chamber with temperature and pressure control [93-94]. The reactants are absorbed on the heated substrate surface and the atoms undergo migration and film-forming reactions. Gaseous by-products are in turn desorbed and removed from the chamber. The reactions forming the solid material may occur on or near the heated surface (heterogeneous reactions) or in the gas phase (homogeneous reactions). The latter lead to gas phase cluster deposition and result in poor adhesion, low density and high defect films, hence heterogeneous reactions are usually preferred. The most favourable end product of these physical and chemical interactions on the substrate surface is a stochiometrically correct film.

The CVD method is very versatile and works at low or atmospheric pressure (LPCVD and APCVD respectively) and at relatively low temperatures. In addition to thermal energy, other energy sources (especially for low temperature applications) such as radio frequency (RF), photo radiation, or laser radiation can be used to enhance the process; these commonly known as plasma-enhanced CVD (PECVD), photo-assisted CVD and laser-assisted CVD (LCVD) respectively.

For the purpose of nitride deposition, LPCVD and PECVD are by far the more popular choices [95]. In this project, LPCVD is used because it yields a stochiometric nitride layer and offers superior uniformity and purity. The nitride thus deposited is also an excellent barrier against oxygen to prevent the underlying silicon from being oxidised. Therefore, it serves well as a mask for local oxidation (LOCOS). In addition, LPCVD nitride is almost not etched in aqueous potassium hydroxide (KOH), thus making it an ideal mask material for deep KOH silicon etches. Typically, a layer of oxide is deposited prior to LPCVD of nitride because the nitride layer is under significant tensile stress. The relevant etch rates of silicon nitride in the various etchants are given in Table 4-1.

4.1.3 Wet Etching

Wet etching is a blanket name encompassing the removal of material by immersing the wafer in a liquid bath of chemical etchant which can be generalized into two broad categories, namely, isotropic and anisotropic etchants. Isotropic etchants etch the material at the same rate in all directions whereas anisotropic etchants etch the silicon wafer at different rates in different directions. At a particular etchant composition and temperature, the etch rate is strongly dependent on crystal direction and dopant concentration.

This dependency makes it possible to produce a large variety of silicon structures in a controllable and reproducible manner. The anisotropic character of etching silicon allows us, starting with a mask opening aligned along crystallographic orientations, to control precisely the shape and dimensions of microstructures. Due to this greater control of the shapes produced, anisotropic etchants are collectively the more popular choice for wet etching of silicon.

Single crystal silicon has a diamond cubic crystal structure with several low index planes that can be chosen for wafer orientation. Commercially available silicon wafers typically have surfaces orientated in either the (100) or (110) plane, which are commonly known as (100)-type and (110)-type wafers respectively. In this project, (100)-type wafers are used and all subsequent descriptions are in reference to (100)-type wafers, unless otherwise stated.

The three most commonly planes used for MEMS devices are shown in Fig. 4-4. The atomic densities of these planes are quite different, with the highest atomic density for the {111} planes, medium density for the {100} planes and the lowest density for the {110} planes. These differences in atomic density are the common explanation for the anisotropic nature of chemical wet etching of silicon as shown in Fig. 4-5(b), that is, the higher the atomic density the slower the etching. The {111} plane is etched much slower than the {100} plane and hence acts more or less as an etch-stopping plane. The area to be etched is defined by a mask that typically consists of silicon oxide or silicon nitride.

Anisotropic wet etching of silicon is based on the differences in etching speeds of different crystal planes of single crystalline silicon. The two most used anisotropic etchants are EDP (ethylenediamine-pyrocatechol-water) and KOH (potassium hydroxide-water). The anisotropy of silicon etching in certain alkaline solutions stems from the crystal structure of silicon, and the shapes that are realizable are restricted to those that are bounded at least in part by slowly etching planes [96-97]. In this project, silicon wafers with (100) surface and aqueous solution of potassium hydroxide (KOH, 55%, 60°C) are used. In contrast to typical KOH concentration (30 to 40%), such a high KOH molarity is used in this project is because flat {100} surfaces and a smaller under-etching of convex corners can be obtained. In addition to that, three other etchants, namely, hydrofluoric acid (HF, 49%), buffered hydrofluoric acid (BHF, HF:NH_4F - 6:1, ambient) and phosphoric acid (H_3PO_4, 85%, 155°C), are also used, but for the etching of borosilicate glass, silicon dioxide and silicon nitride respectively. These are all isotropic etching.

When a masked silicon wafer is immersed into the etchant, the silicon is not only etched vertically, but also laterally. The final geometry and wall profile of the etched structure is very much dependent on the mask geometry and its orientation to the crystalline planes. Generally, there are two types of geometry, namely, concave and convex mask geometry (Fig. 4-6).

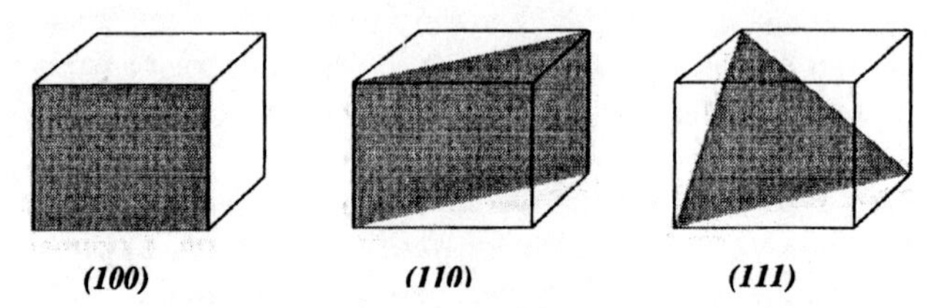

Figure 4-4. Low crystallographic index planes of silicon

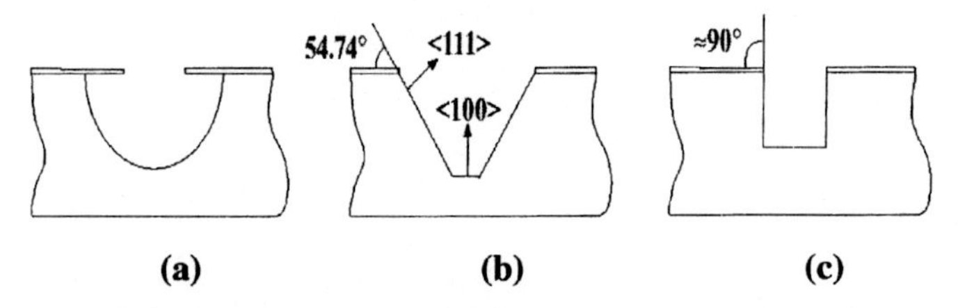

Figure 4-5. Cross-sectional views of (a) isotropic etch, (b) anisotropic etch dependent on crystal planes and (c) directionally dependent anisotropic etch

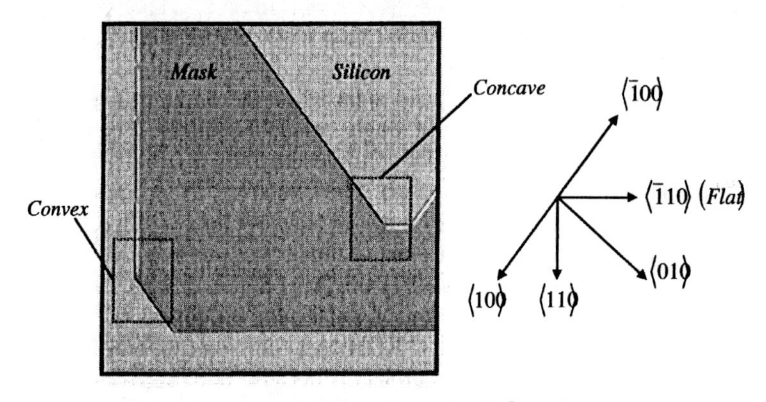

Figure 4-6. Concave (<180°) and convex (>180°) mask geometry

4.1.3.1 Concave Mask Geometry

The forms resulting from etching with a concave mask geometry largely determined by the slow-etching planes (i.e. {111} planes). These forms can be controlled by appropriately positioning the mask opening with respect to the <110> and <100> crystallographic directions for forms bounded by the slow etching {111} and {100} planes respectively (Fig. 4-7).

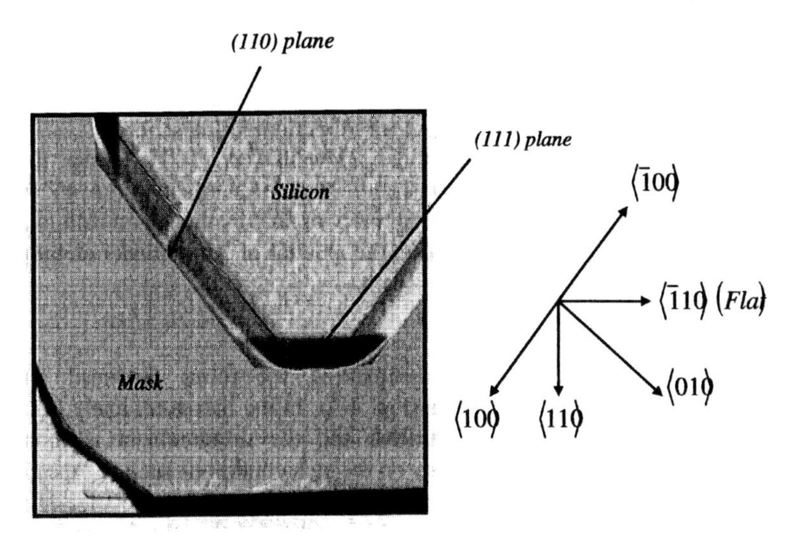

Figure 4-7. Planes occurring at concave corners during KOH etching

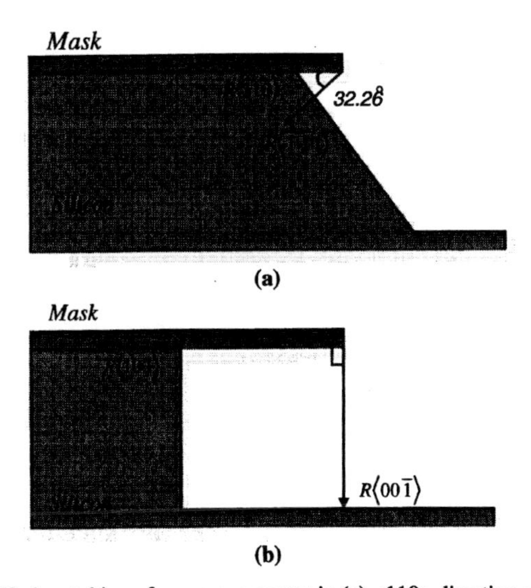

Figure 4-8. Under-etching of a concave corner in (a) <110> direction and (b) <100> direction

If one wants to exploit the emergence of the {111} faces alone, one has to define a mask opening on the wafer with edges along the <110> direction (i.e. parallel and perpendicular to the primary flat). Etching then results in an etch pit of pyramidal form. After prolonged etching, the form will always be bounded by {111} planes no matter what the mask opening looks like. Since the etch rate of the {111} planes is about 400 times smaller than the etch rate in the $\langle 00\bar{1} \rangle$ direction, these planes act as an etch stop (Fig. 4-8(a)). The lateral etch rate in the corresponding <110> direction is almost zero.

On the other hand, to obtain vertical sidewalls (i.e. {100} planes), the mask opening must be aligned with respect to the <100> direction (i.e. 45° to the primary flat). Due to the cubical symmetry of silicon, the lateral etch rate is the same as the vertical etch rate (i.e. the amount of lateral under-etching is the same as the etch depth) (Fig. 4-8(b)).

4.1.3.2 Convex Mask Geometry

In the case of convex mask geometries, the forms produced are dominated by the fast etching planes (Fig. 4-9). In the literature, many fast-etching directions have been identified, and the disagreement is most probably due to the strong dependency on the etch conditions such as type of etchant, temperature, concentration of etchant, doping of the substrate and amount of stirring. Nevertheless, it is widely accepted that the fast-etching planes for convex structures are the {411} planes [96].

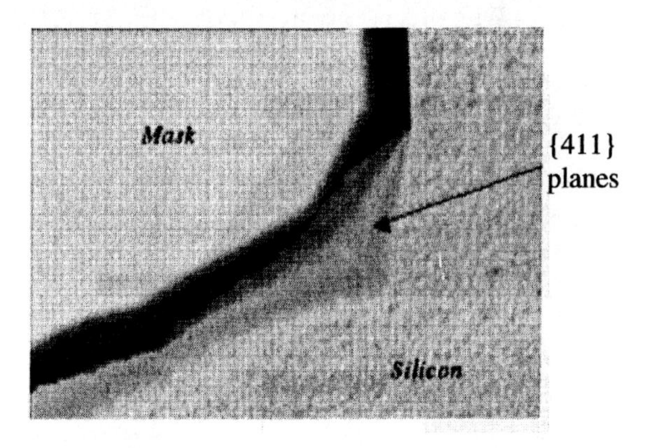

Figure 4-9. Planes occurring at convex corners during KOH etching

The etch ratio between the {411) and (100} planes, $R\langle 411 \rangle / R\langle 100 \rangle$, is found to be dependent on the KOH concentration but temperature independent in the range from 60°C to 100°C [96]. Generally, $R\langle 411 \rangle / R\langle 100 \rangle$ decreases with KOH concentration. As can be seen from Fig. 4-10, the border where the (411) and (100) planes intersect moves in the $\langle 1\bar{\tfrac{1}{4}}0 \rangle$ direction. The angle between this direction and the etch direction of the (411) plane is 13.6°. Therefore, this border moves with a speed of:

$$R_{\langle \bar{1}\frac{\bar{1}}{4}0\rangle} = \frac{1}{\cos(13.6^\circ)} \cdot R_{(411)} \tag{4-1}$$

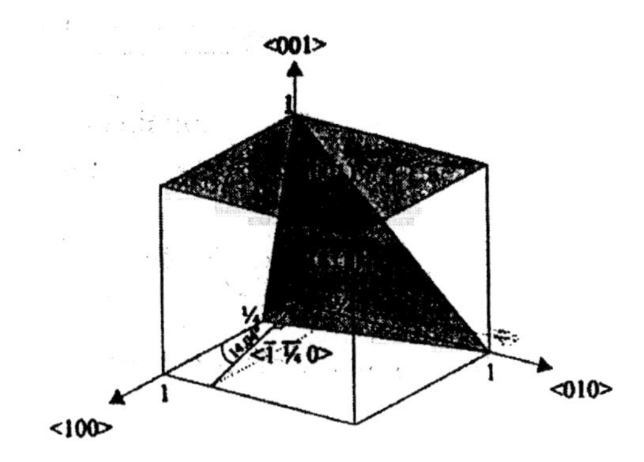

Figure 4-10. Orientation of the {411} plane

In addition, in the etching of a convex corner, the largest under-etching occurs in the direction 45° to the primary flat. Two planes from the {411} family meet each other forming a corner that moves for example in the <100> direction (Fig. 4-11). This corner moves with the speed:

$$R_{\langle \bar{1}00\rangle} = \frac{1}{\cos(14.04^\circ)} \cdot R_{\left(\bar{1}\frac{\bar{1}}{4}1\right)} \tag{4-2}$$

Corresponding, the under-etching in the $\langle \bar{1}\bar{1}0\rangle$ direction is given by:

$$R_{\langle \bar{1}\bar{1}0\rangle} = \frac{1}{\sin(30.96^\circ)} \cdot R_{\left(\bar{1}\frac{\bar{1}}{4}1\right)} \tag{4-3}$$

For the standard 55%60°C KOH solution used in this project, an etch ratio of $R\langle 411\rangle / R\langle 100\rangle \sim 1.3$ can be obtained from the results given in Fig. 4-12. This result in corner speeds in the $\langle \bar{1}00\rangle$ and $\langle \bar{1}\bar{1}0\rangle$ directions of:

$$R_{\langle \bar{1}00\rangle} = 1.38 \cdot R_{(100)} \tag{4-4}$$

and

$$R_{\langle\bar{1}10\rangle} = 2.60 \cdot R_{(100)} \tag{4-5}$$

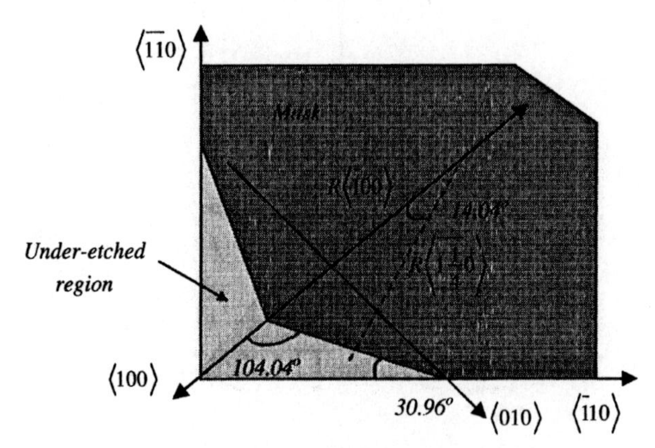

Figure 4-11. Under-etching of a convex corner

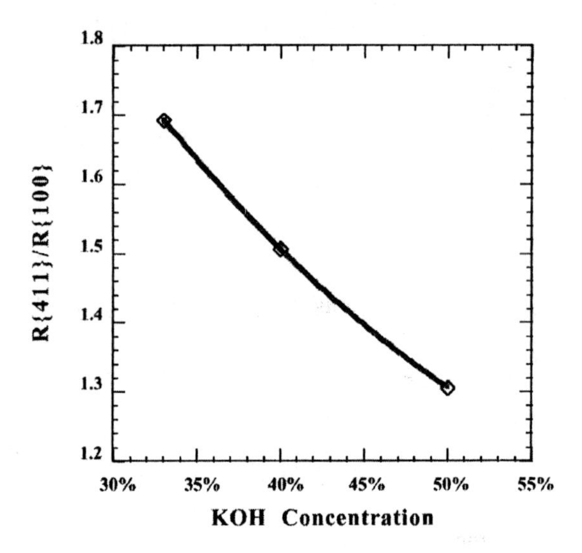

Figure 4-12. Etch ratio between {411} and {100} planes [98]

The amount of under-etching of the convex corners can hence be calculated for different etch depths. This is important because it enables the author to design a valve-seat that will not be excessively under-etched for the required etch depth.

4.1.4 Dry Etching

Dry etching encompasses a family of methods by which a solid-state surface is etched in the gas phase, physically by ion bombardment, chemically by a chemical reaction with a reactive species at the surface, or combined physical and chemical mechanisms. Dry etching techniques generally involve the use of plasma (a hot ionised gas) driven chemical reactions and/or energetic ion beams to remove material and they often yield finer patterns than wet etching. They also offer greater safety, as large quantities of corrosive acids are not required.

Dry etching of silicon is based on ions in plasma removing the silicon and it is essentially mechanical in nature since the silicon is bombarded by high-energy ions, which hence gives it the name "ion milling" or "sputtering". Normally, this process is combined with the utilization of reactive gases in the plasma that also react chemically with the silicon (i.e. reactive ion etch or RIE). Generally, chemical etching is isotropic (anisotropy is a result of the crystalline nature of silicon) and mechanical sputtering is anisotropic. This anisotropy can be enhanced by the choice of gases and other process parameters [99]. The anisotropy of the RIE etch results in vertical walls perpendicular to the wafer surface as illustrated in Fig. 4-5c. A plasma consists totally or partly of ionised gas. The gas is ionised when a sufficiently high voltage is applied over the gas. The plasma generation is initiated by thermally emitted electrons from the negative electrode. Electrons can also come from the ionised atoms generated when the electrical voltage is applied. These free electrons are accelerated by the applied voltage. They can be ionised by collisions between the electrons and the gas molecules or atoms. The etching is performed in a chamber at a pressure of 0.5 to 25 Pa. Most reactive ion etches are based on chlorine or fluorine processes. Common gases are CF_4, SF_6 and Cl_2 [99]. Mask materials include resist (polymer), silicon oxide and metal. In some cases, a buried silicon oxide is used as a stopping layer. The selectivity is not very high for resist or oxide but can be improved by appropriate selection of process gases. Normal etch rates for silicon etching are about 100 to 450 Å/min [100]. These selectivities and etch rates have limited the use of RIE to small etch depths. Nevertheless, in recent years, deep reactive ion etching processes (DRIE) have been developed. These have both higher etch rates of up to several micrometers per minute for silicon and a much better selectivity with respect to the mask material. Commercially available equipment has been developed which makes it possible to etch to quite large depths, straight through 500 μm silicon wafers. These developments of the DRIE process made it possible to design structures with high aspect ratios. The main drawback is the high costs of both the equipment and process, in addition to the fact that batch processing is sill not possible at this point of time. In this project, CF_4-O_2 (9:1) and O_2 are used for reactive ion etching of the silicon nitride mask and ashing the photoresist respectively. The various relevant etch rates of materials used in this project are tabulated in Table 4-1.

Table 4-1. Relevant etch rates

Etchant Conditions	***Material***		
	Silicon	Si_3N_4	SiO_2
6:1 BHF Ambient	-	9 Å/min	900 Å/min
KOH (55%) 60°C	9.89 μm/h	-	54 nm/h
H_3PO_4 (85%) 155°C	-	10 nm/min	< 10 Å/min
CF_4-O_2 (9:1) 250W 250 mTorr 13.5 MHz	Unknown	600 Å/min	100 Å/min

4.2 Process Flow

The proposed micropump is of a layered structure with two *Pyrex*™7740 glass wafers (4") and a silicon wafer (4") anodically bonded together. Two passive valves and a pump chamber are patterned in the silicon wafer. The top glass wafer is chemically etched (using 49% hydrofluoric acid (HF)) to the desired profile while inlet and outlet holes are laser-drilled in the bottom glass wafer. Commercially available piezoceramic discs are glued to the top glass wafer to form the micropump actuator unit.

For defining the structures in the silicon substrate, KOH is the primary wet anisotropic etchant. Buffered hydrofluoric acid (BHF) is used for etching the SiO_2 layer while RIE (CF_4+O_2) is used to plasma etch the nitride masking layer. In the following sections, the process flow for fabricating the proposed micropump is discussed in detail. Details of the process parameters are given in Appendix D.

4.2.1 Silicon Process

Step 1 Standard cleaning

The wafers are first cleaned using the standard cleaning process (Appendix D) as preparation for subsequent thermal oxidation step. Prior to all oxidation processes, the wafer must be cleaned.

Step 2 1.2 μm wet oxidation (420 minutes)

A 1.2 μm thick layer of oxide is first thermally grown at 1100°C. The wafer is subjected to 30 minutes of dry oxidation, followed by 6 hours of wet oxidation and finally another 30 minutes of dry oxidation. Even though the quality of the oxide formed by dry oxidation is superior, the oxidation rate is much slower. Hence, wet oxidation has to be used. This layer of oxide will eventually form the valve-seats.

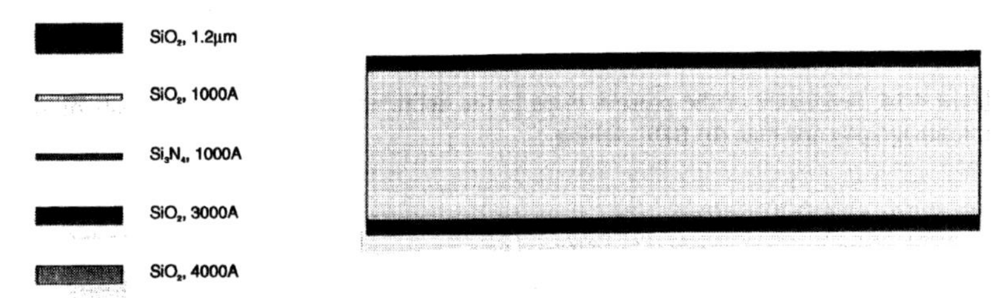

Figure 4-13. Schematic of wet oxidation of wafer

The grown oxide is measured with an ellipsometer and the thickness was found to range from 1.08 to 1.16 μm.

Step 3 & 4 Definition of inlet valve-seats

Photolithography is then carried out to define the inlet valve-seats on the backside. As this is the first mask, the mask is aligned to the wafer flat and alignment marks are brought onto the wafer. The alignment in this step is critical because all subsequent patterns will be affected. The wafer is then etched in BHF for 15 minutes to form the valve-seats. The etch rate of the oxide in BHF is 900 Å/min and the 1.2 μm thick oxide should be removed in 13.3 minutes. The excess time is to ensure that the oxide is sufficiently etched. In order to determine if the oxide has been fully removed, a small amount of water is sprinkled onto the wafer to test the nature of the surface. If the surface is hydrophobic (i.e. water collegiate into droplets on the surface), then the oxide has been fully removed since silicon dioxide is hydrophilic. In addition, the water on the hydrophobic silicon areas can be blown away with nitrogen easily.

Step 5 & 6 Definition of outlet valve-seats

Similarly, photolithography is carried out to define the outlet valve-seats on the backside. The wafer is then etched in BHF for 15 minutes (1.2 μm thick oxide) to form the valve-seats.

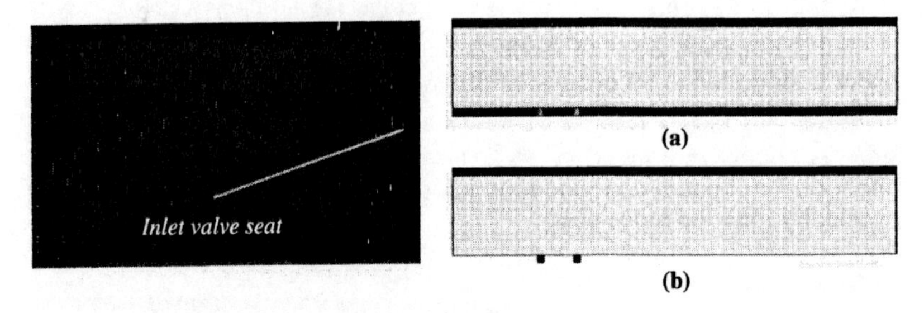

Figure 4-14. Schematic of the process steps in the definition of the inlet valve-seats by (a) photolithography and then (b) BHF etching

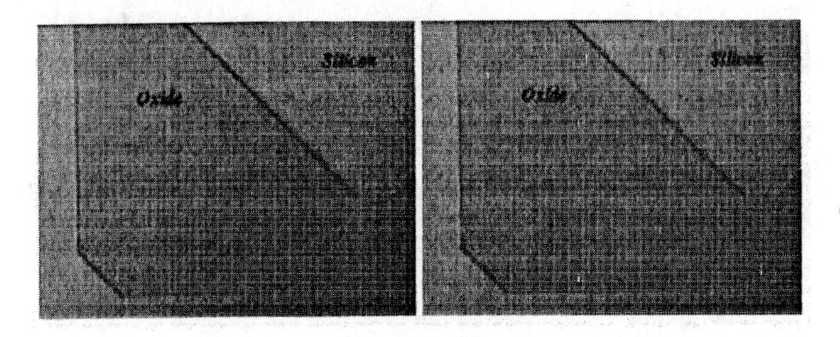

Figure 4-15. Inlet valve-seat

Step 7 1000Å dry oxidation (40 minutes)

The wafer is then cleaned and dry oxidised for 1000Å of oxide. This layer of oxide will form the masking layer for the subsequent etching step.

Step 8 LPCVD deposition of 1000Å nitride (21 minutes)

A layer of LPCVD nitride with a thickness of 1000Å is then deposited. As Si_3N_4 is highly resistant to KOH, this nitride layer will be used as the masking layer for the subsequent KOH etching step.

Step 9 & 10 Photolithography (front-side) and RIE (2 minutes)

Photolithography is then carried out to pattern the pump chamber, outlet channel and outlet valve chamber. To transfer the pattern to the nitride layer, RIE (CF_4+O_2 - 9:1 plasma) is carried out to etch the nitride for about 2 minutes. The etch rate of Si_3N_4 in CF_4+O_2 plasma is about 600 Å/min. Excess dry etching is carried out to ensure that the nitride is fully removed.

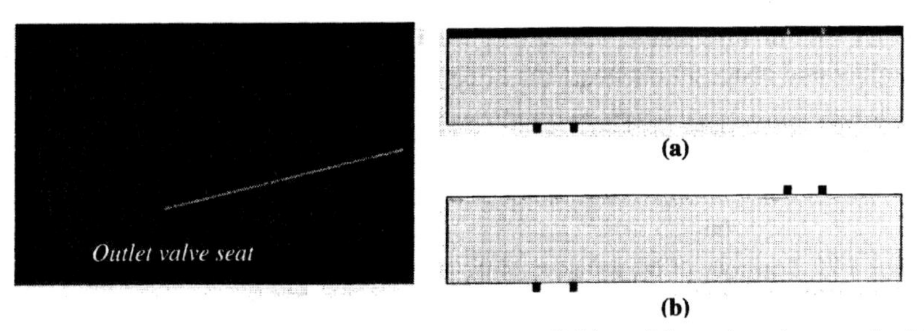

Figure 4-16. Schematic of the process steps in the definition of the outlet valve-seats by (a) photolithography and (b) BHF etching

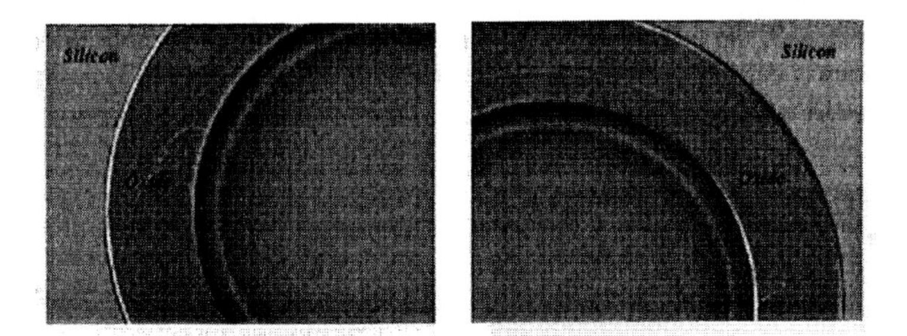

Figure 4-17. Outlet valve-seat

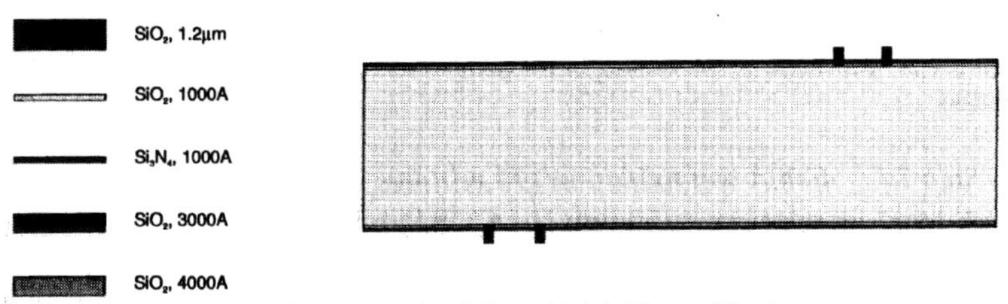

Figure 4-18. Schematic of the oxide/nitride masking layer

Step 11 BHF etching (1½ minutes)

The wafer is etched in BHF for 1½ minute to remove the 1000Å thick oxide. The mask is now defined for the KOH etch.

Step 12 & 13 KOH etching (1 hour 7 minutes) and BHF etching (15 minutes)

The opened silicon areas are etched with KOH to form the pump chamber, outlet channel and outlet valve chamber. In the KOH solution

(55% 60°C) used by the author, an etch rate of 9.89 μm/h was measured. This is slightly different from that found in the literature [97] which reported an etch rate of 11.2 μm/h. The author wants a depth of 10 μm for the pump chamber and hence an etching time of roughly 1 hour 7 minutes is needed.

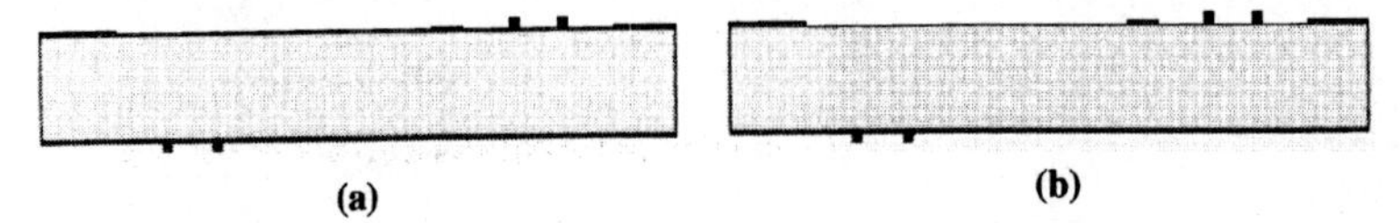

Figure 4-19. Schematic of the definition of the oxide/nitride masking layer by (a) RIE and then (b) BHF oxide etching

Following the KOH etch, the wafer is then etched in BHF for 15 minutes to remove overhanging oxide at the shallow edges and outlet valve-seat areas (Fig. 4-22), left from the preceding KOH etch. If this overhanging oxide is not removed, the subsequent new layer of nitride may not be able to cover these areas entirely. The overhanging silicon nitride does not cause problems because it will be removed in *step 15*.

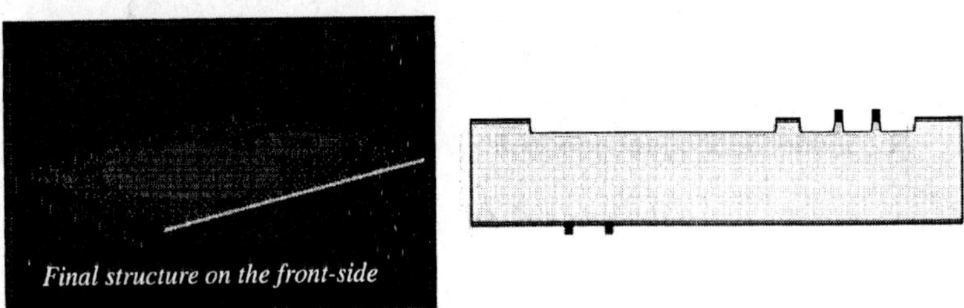

Figure 4-20. Schematic of the etching of the pump chamber, outlet channel and outlet valve chamber

Step 14 4000Å wet oxidation (98 minutes)

A local or selective oxidation (LOCOS1) of 4000Å is then carried out. Only areas not covered by silicon nitride will be oxidised. This step is necessary to have a hydrophilic surface in the shallows at the end of the fabrication process.

Step 15 & 16 RIE (2 minutes) and BHF etching (1½ minute)

All the nitride (1000Å) is removed by RIE for 2 minutes in order to define a new mask subsequently. In addition, the oxide (1000Å) that was used as a mask previously is etched away in BHF for 1½ minute. It has to be taken into consideration in the design that the oxide thickness in the shallows and the oxide thickness of the valve-seats will also be reduced in these steps.

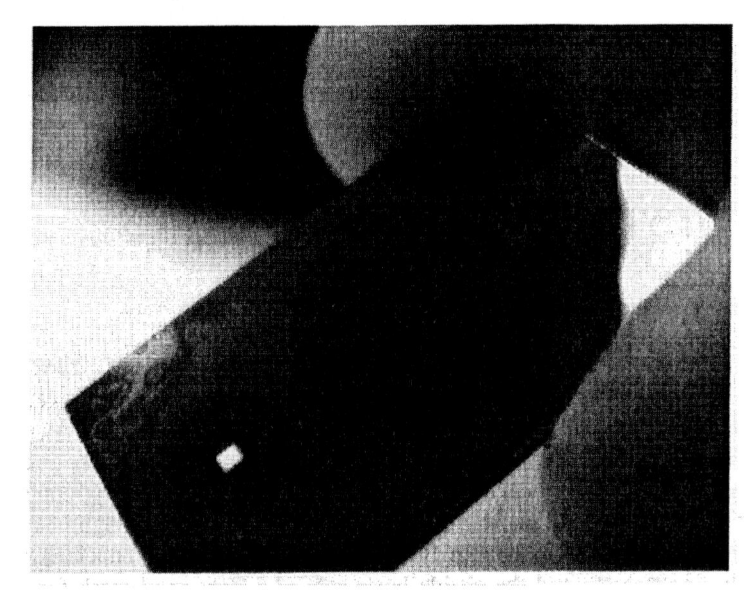

Figure 4-21. Photograph of pump chamber, outlet channel and outlet valve chamber

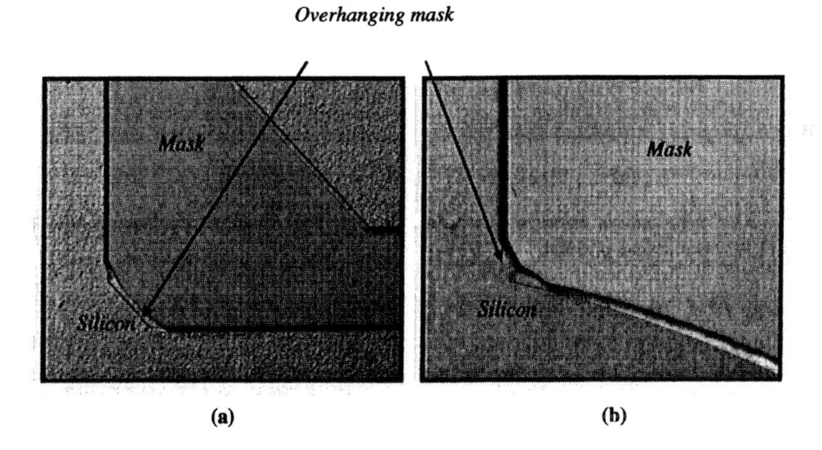

Figure 4-22. Undercutting of the oxide/nitride mask at (a) valve-seat area and (b) outlet channel

Figure 4-23. Schematic of the (a) oxidation of shallows followed by (b) RIE and BHF etching

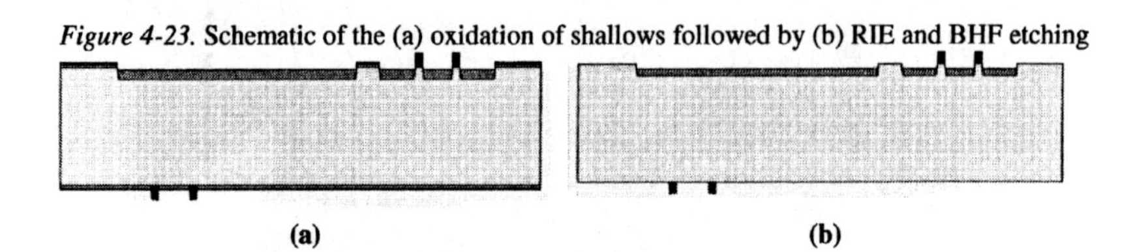

to remove preceding oxide/nitride mask layer

Step 17 & 18 1000Å dry oxidation (40 minutes) and LPCVD deposition of 1000Å nitride (21 minutes)

A 1000Å thermal silicon oxide is grown and a 1000Å LPCVD silicon nitride is deposited. These layers will serve as a mask for the deep silicon etch in step 24 and the local oxidation (LOCOS2) in step 25. The etch rate of LPCVD silicon nitride in BHF is about 9 Å/min and almost negligible in KOH. Therefore, the oxide can be opened in BHF without attacking the nitride significantly and the nitride layer acts as a very good mask for the prolonged KOH etching. Consequently, only one oxide/nitride mask is needed for both the local oxidation and deep silicon etch.

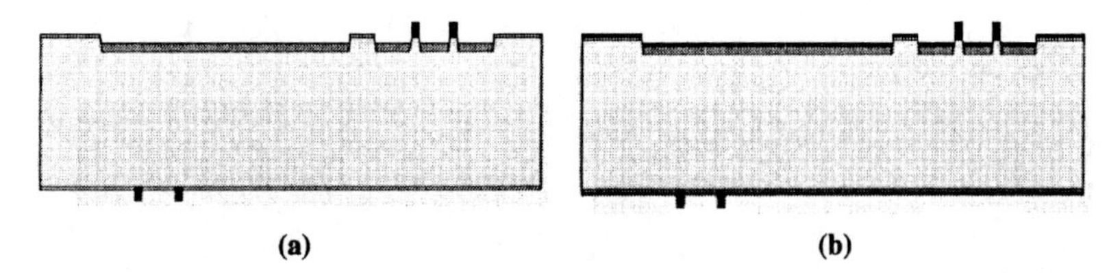

Figure 4-24. Schematic of the second oxide/nitride mask layer by (a) dry oxidation followed by (b) LPCVD nitride deposition

Step 19 & 20 Photolithography (front-side) and RIE (2 minutes)

Photolithography and RIE are then carried out to transfer the pattern for the inlet and outlet cavities onto the nitride layer on the front-side. Due to the 10 μm depth, the resist is spin coated at a slower spinning rate in order to achieve a thicker coat of resist. This is critical in ensuring that all the shallow edges will be covered.

Step 21 & 22 Photolithography (backside) and RIE (2 minutes)

Photolithography and RIE are carried out to define the nitride mask pattern for the inlet and outlet valves on the backside.

Step 23 BHF etching (3½ minutes)

The opened oxide areas, which are around 3000Å thick, are then etched away in BHF for 3½ minutes. The mask is now defined for the deep silicon etch.

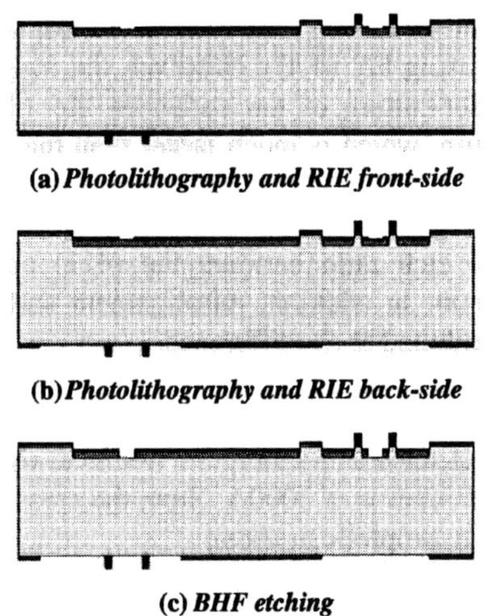

(a) Photolithography and RIE front-side
(b) Photolithography and RIE back-side
(c) BHF etching

Figure 4-25. Schematic of the process steps for the definition of the mask for deep silicon etch

Step 24 KOH etching (25 hours)

The opened silicon areas are then etched away in KOH for 25 hours. This etch will form the inlet and outlet valves and cavities. The targeted etch depth is 250 μm. This will leave the valve membrane to have a thickness of 40 μm. Measurements with the surface profiler yield values ranging from 247 μm to 255 μm.

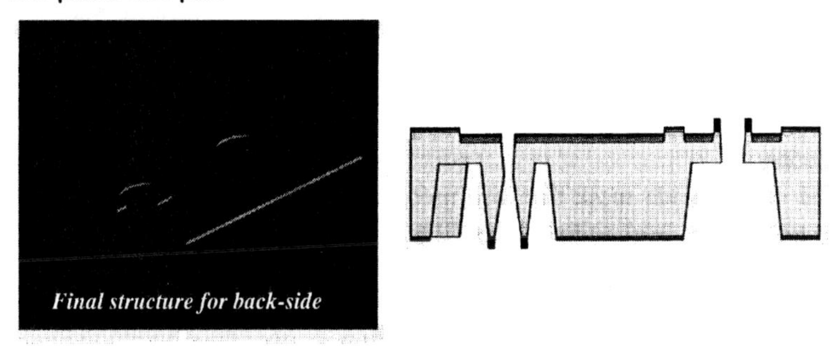

Figure 4-26. Schematic of the etching of the inlet and outlet valves and cavities

For a concave mask geometry (such as the inner area of the inlet valve), the maximum under-etching is in the direction 45° to the flat (i.e. <100>

direction). Theoretically, the etch in this direction is the same as the depth (i.e. 250 μm). The measured value is 257 μm.

For a convex mask geometry (such as the outer area of the inlet valve), there will be severe undercutting of the mask. The maximum undercutting occurs in the direction 45° to the flat. The theoretical undercutting in this direction is 345 μm (using Eq. (4-4)). However, the measured value is 492 μm. In addition, under-cutting of the corner in the $\langle 110 \rangle$ direction was measured to be 747 μm, which is much larger than the calculated value of 650 μm. The discrepancies can be attributed to the fact that the etch rate for the solution which the author used is different from that quoted in the literature. Hence, the etch ratio between the (411) and (100) planes is expected to be different. In addition, misalignment and inaccuracy in the measurements also attributed to the difference.

Step 25 3000Å wet oxidation (85 minutes)

The wafer is then cleaned and undergoes wet local oxidation (LOCOS2) to obtain an oxide of thickness of 3000Å. Now, the areas that underwent the deep etch will also be hydrophilic in nature.

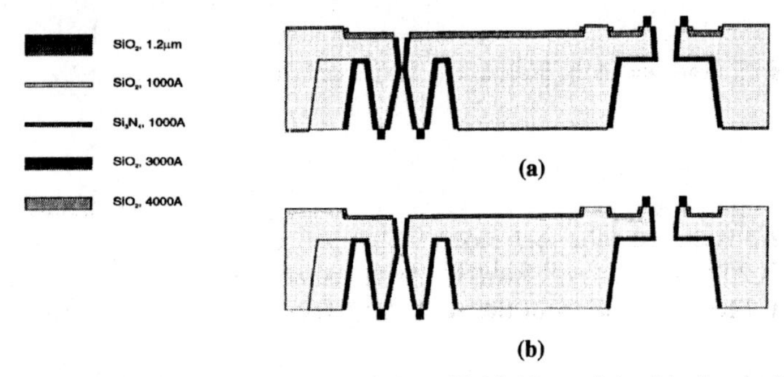

Figure 4-27. Schematic of the removal of the oxide/nitride mask by (a) phosphoric acid etching and (b) BHF etching

Step 26 & 27 Phosphoric acid etching (10 minutes) and BHF etching (1 minute)

The oxide/nitride mask that was used for the deep etch is removed by using phosphoric acid followed by BHF. To ensure that all the nitride is removed, phosphoric acid is used instead of plasma. The oxide is etched with BHF for 1000Å leaving only oxide in the shallows and deep etches. This completes the process for the silicon layer of the micropump. Pictures of the completed silicon structure are given in Fig. 4-28.

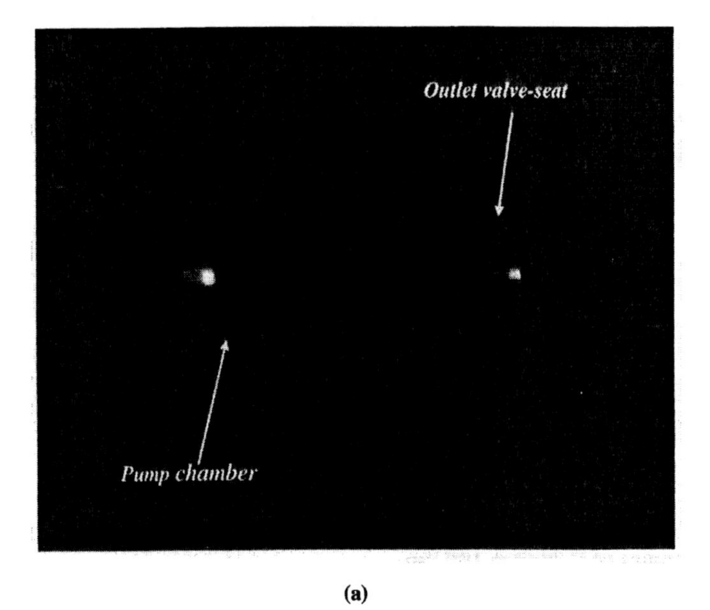

(a)

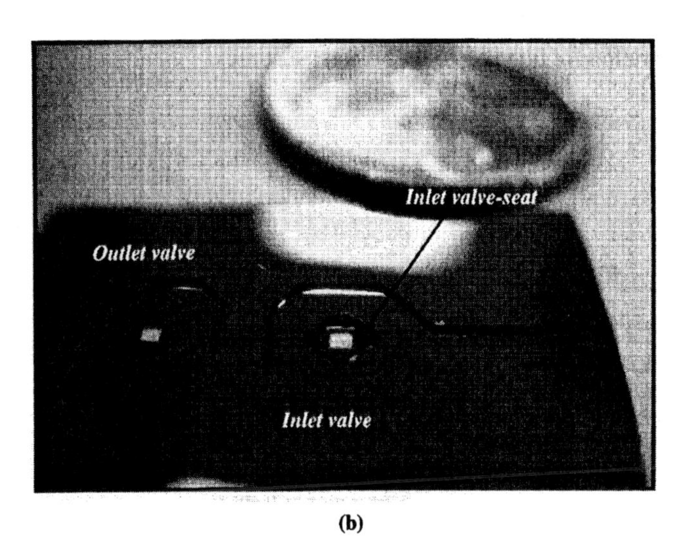

(b)

Figure 4-28. Photographs of completed silicon structure showing (a) front-side and (b) back-side

4.2.2 Glass Process For Top Glass Wafer

Step 1 Standard cleaning

The wafers are first cleaned using the standard cleaning process as preparation for subsequent LPCVD step.

Step 2 LPCVD Polysilicon Deposition (5000Å)

A 5000Å thick polysilicon layer is deposited to serve as the mask material for the deep etching of the glass substrates.

Step 3 & 4 Photolithography & Plasma Etching

Standard photolithography is performed to transfer the pattern onto the polysilicon layer. The polysilicon layer is then plasma etched to open the windows for the deep etch.

Step 5 BHF etching (400 μm)

The glass wafers are then etched in BHF for a depth of 400 μm to give membranes of thickness 100 μm.

Step 6 KOH etching

The polysilicon layer is stripped using KOH and the wafers are cleaned in preparation for bonding to the silicon substrates.

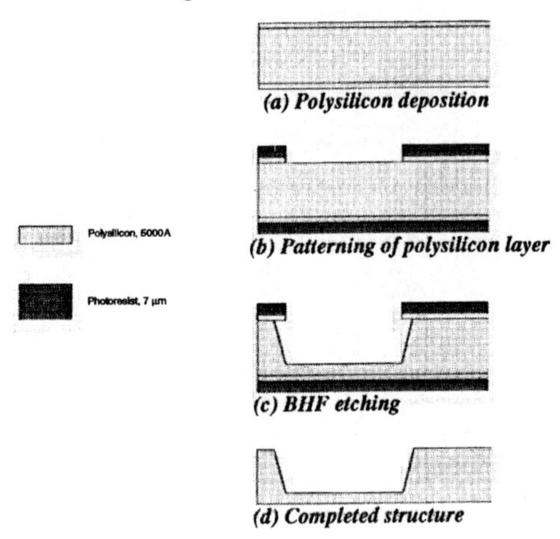

Figure 4-29. Schematic of the process steps for top glass wafer

4.2.3 Glass Process For Bottom Glass Wafer

In order to define the inlet and outlet for the micropump, holes (1 mm in diameter) are laser drilled in the bottom glass wafer. There are three methods commonly used for drilling of glass, namely, ultrasonic drilling, sand blasting and laser drilling. Ultrasonically drilled holes embody deep trenches and sharp edges but the surface of the wafer is not damaged during the drilling process. Ultrasonic drilling also does not result in microcracks, edge chipping, and formation of molten particles during the drilling process. The main disadvantage of sand blasting is the widening of the holes during the process. The diameter (e.g., of holes) at the entrance of the beam is larger than at the exit. The reason is that the photoresist used for masking is also removed. The widening depends on the wafer thickness. Nevertheless, it offers the possibility of machining several wafers at the same time (i.e. batch processing). Laser machining causes microcracking in the holes and loose particles are formed on the surface during the ablation process. These must be removed by a grinding and polishing step before anodic bonding, since for anodic bonding the surfaces of the materials must be polished, very planar and free from particles. In addition, it is also too time consuming to drill one hole at a time.

In this project, laser drilling is used because of the availability of the equipment and capability. In addition, laser drilling has very good geometric tolerances and it is a very straightforward process, similar to normal mechanical drilling. A sample of the holes drilled is shown in Fig. 4-30.

As mentioned above, laser drilling often results in loose particles, sharp edges and deep trenches (Fig. 4-30(a)). In order to reduce these defects and to smoothen the holes for better flow property, the drilled glass wafers are etched in HF as a finishing step. Figures 4-31 and 4-32 show the SEM of the walls and entrances of the holes before and after HF treatment respectively. It can be seen that the walls are generally flatter and loose particles are removed after HF treatment. However, the deep trenches are still clearly visible. The removal of the loose particles, especially those protruding out from the glass surface, is vital because the subsequent anodic bonding process requires very flat surfaces. Though microcracks cannot be removed completely in the laser drilling process, they should be minimised because the stresses that build up during the subsequent bonding and dicing processes can cause these cracks to propagate, and in the worst-case scenario crack the entire glass wafer.

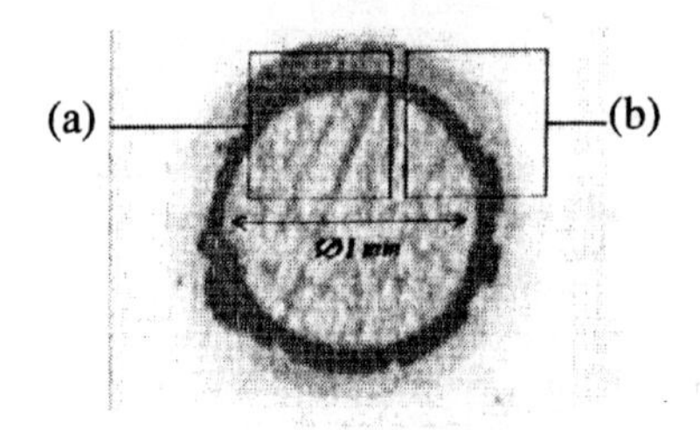

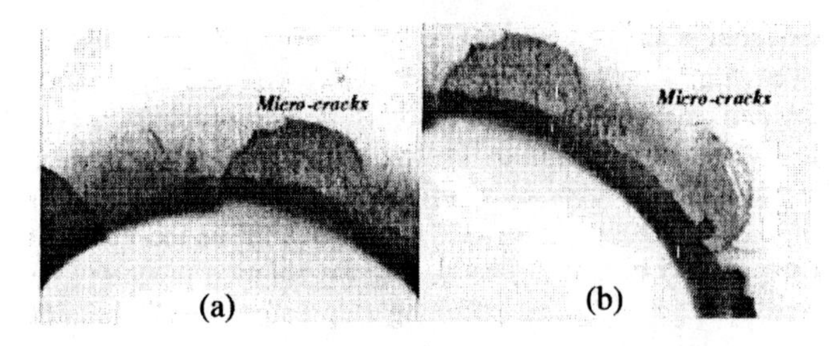

Figure 4-30. Laser drilled holes in Pyrex™ 7740 borosilicate glass wafer

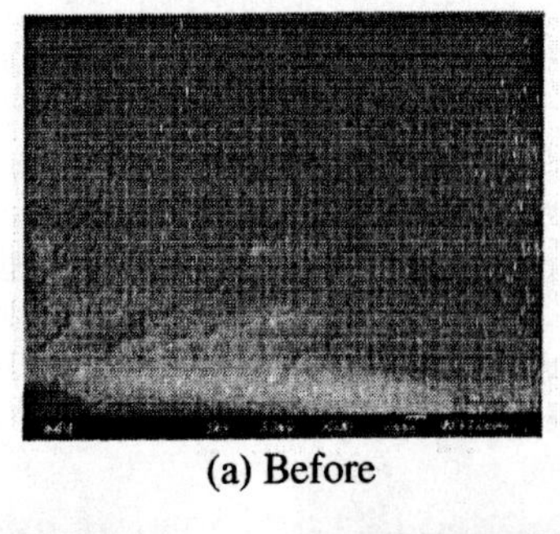

(a) Before

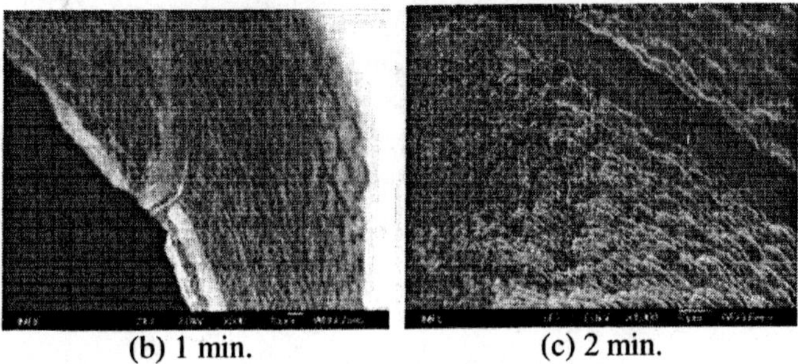

(b) 1 min. (c) 2 min.

Figure 4-31. SEM micrographs of the walls of laser drilled holes (a) before HF treatment, (b) after 1 minute of etching in HF and (c) after 2 minute of etching in HF

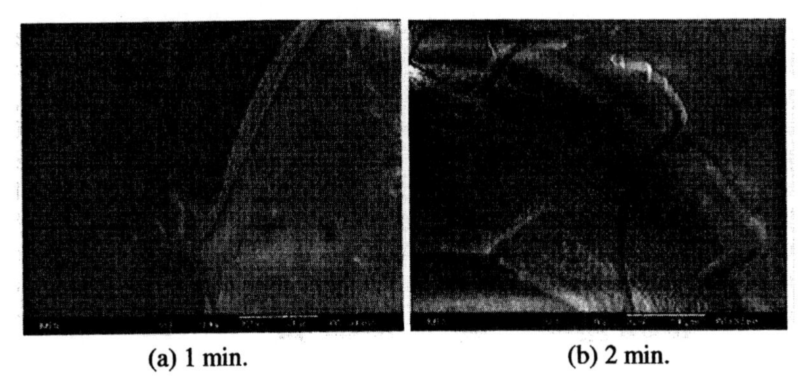

(a) 1 min. (b) 2 min.

Figure 4-32. SEM micrographs of the entrance of laser drilled holes (a) after 30 seconds of etching in HF and (b) after 2 minute of etching in HF

4.2.4 Wafer Bonding

Wafer bonding, as the name implies, refers to the mechanical fixation of two or more wafers to one another. It is a technology that is of great importance to micromachining, especially in areas which require three-dimensional structures in a technology that is pre-dominantly planar in nature otherwise. Historically, some of the earliest uses of wafer-to-wafer bonding were for packaging of pressure sensors. These wafer-to-wafer bonds were performed at low temperatures (less than 450 C) and involved either field-assisted silicon-to-glass bonding (anodic bonding) or a eutectic bond between silicon wafers using a gold thin film. In the literature, many different terms are used for the various bonding processes, namely (as taken from [101]):

Direct bonding: This term refers to the bonding of wafers without an intermediate layer. A large number of different materials can be bonded this way.

Silicon fusion bonding (SFB): This term refers to the bonding of two silicon wafers with or without a thin layer of thermal or native oxide.

Direct silicon bonding: This term refers to the bonding of bare silicon wafers (with a native oxide layer).

Indirect bonding: This term refers to the bonding of wafers with an intermediate layer. The use of conventional glue belongs to this type of bonding.

Field-assisted bonding: This term refers to bonding in the presence of an electric field. Anodic bonding of a silicon wafer to a Pyrex™ wafer belongs to this category.

In this project, the layered structure of the micropump is realised by the anodic bonding of the silicon wafers and Pyrex™ glass wafers. Anodic bonding offers the advantages of low process temperature (about 450°C), low residual stress, and less stringent requirements on the surface quality of the wafers as compared to SFB. Pyrex™ 7740 borosilicate glass is by far the most popular choice for anodic bonding with silicon because the thermal expansion coefficients of the two materials are matching in the range of temperatures suitable for anodic bonding; this minimizes the generation of thermal stress. Besides, the positive sodium ions present in Pyrex™ 7740 is necessary to build up a space charge region which is central to the whole anodic bonding process.

Anodic bonding has become a key process in microsensor and microactuator technology and is widely used for the hermetic sealing of micromachined devices since its introduction in 1969 [102]. Other terms commonly used for this technique are electrostatic bonding or field assisted bonding. Anodic bonding is based on joining an electron conducting material (e.g. silicon) and a material with ion conductivity (e.g. alkali-containing glass). The bonding mechanism is assisted by heating at 180-500 °C (near the annealing point but well below the melting point of the glass) and the application of an external electric field typically in the range 200-1000 V. The glass used is normally a sodium glass, such as Pyrex™ 7740 from Corning, TEMPAX™ from Schott or SD-2™ from HOYA, the main criterion being that it must be slightly conducting at the chosen bonding temperature. A standard pre-treatment (cleaning) of the wafers to be bonded may be needed (e.g. 5 minutes in H_2O_2:H_2SO_4, 1:2.5 by volume). When the external electric field is applied at the elevated temperature, the positive sodium ions in the glass migrate towards the negative pole and create a space charge (depletion) region adjacent to the glass-silicon interface. The elevated temperature causes the positive sodium ions to become quite mobile to facilitate the migration process. The voltage drop over this depletion layer creates a large electric field that pulls the wafers into intimate contact. The elevated temperature also enables covalent bonds to form between silicon and oxygen atoms at the surfaces of the two materials, resulting in an oxide layer. A set-up for anodic bonding is shown in Fig. 4-33. Anodic bonding is less sensitive to small particles or roughness of the surface (less than one micron) as compared to SFB.

The bonding process can be monitored by observing the current. The temperature and voltage are kept constant during the bonding process. A soon as the voltage is switched on, a current peak occurs, indicating the drift of the sodium ions to the cathode and the build-up of the space charge region. The current will subsequently decreases and stabilizes. The bond will normally be completed when the current has reached about 10-30% of its initial peak value. Normal bonding time is about 4-10 minutes but up to half an hour may be needed. The voltage should be applied for a time-period long enough to allow the current to settle at a steady state minimized level.

In this project, a "chip-level" anodic bonder, according to the schematic as illustrated in Fig. 4-34, is set-up to carry out device level anodic bonding (Fig. 4-34).

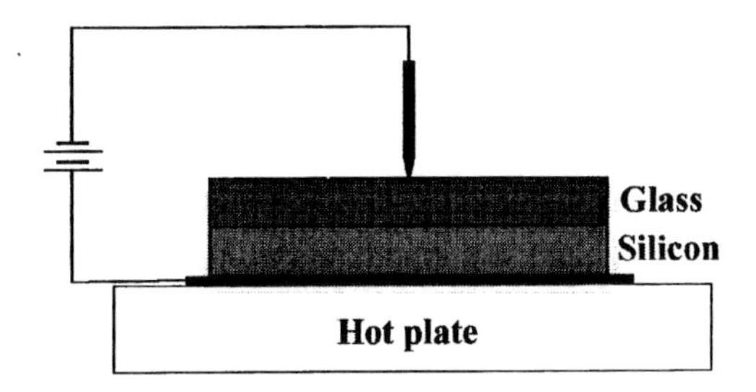

Figure 4-33. Schematic set-up for anodic bonding of glass to silicon

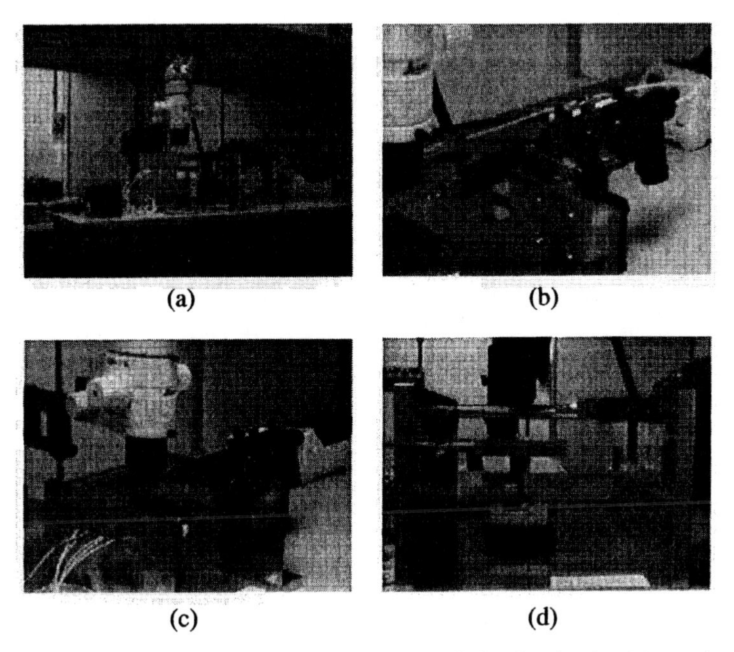

Figure 4-34. Photographs of in-house "chip-level" anodic bonder showing (a) overview of entire bonder, (b) electrodes, (c) microscope and (d) hot-plate and translation table

For the proposed micropump, three substrates have to be bonded by means of a two-electrode bonding process (Fig. 4-35). The bottom glass substrate (negative polarity) is first bonded to the silicon substrate (positive polarity). This is followed by the bonding of the top glass substrate (maintained at negative polarity) to the bonded structure (positive polarity). Due to pole reversal of the bonded bottom glass substrate during the second bonding step, yellowish brown sodium silicate will be formed at the interface with the silicon substrate. The formation of this sodium silicate is not known to adversely affect the quality of the bond though [103]. An alternative to this method is to use a three-electrode bonding process, which maintains the silicon substrate at positive polarity and all the glass substrates at negative polarity throughout the whole bonding process. Though the three-electrode process does not result in the formation of the unwanted sodium silicate at the interface, the two-electrode bonding method is much more easier to carry out. In addition, it can be difficult for the electrode to access the central silicon substrate during the second bonding step. The second glass substrate to be bonded has to be smaller than the silicon substrate for the positive electrode to make contact with the latter. Photographs of the bonded structure are given in Fig. 4-36. In the figure, due to the uneven surface resulting from the laser drilling, certain areas cannot be bonded. The bonded areas can easily identified, as they are dark grey in colour.

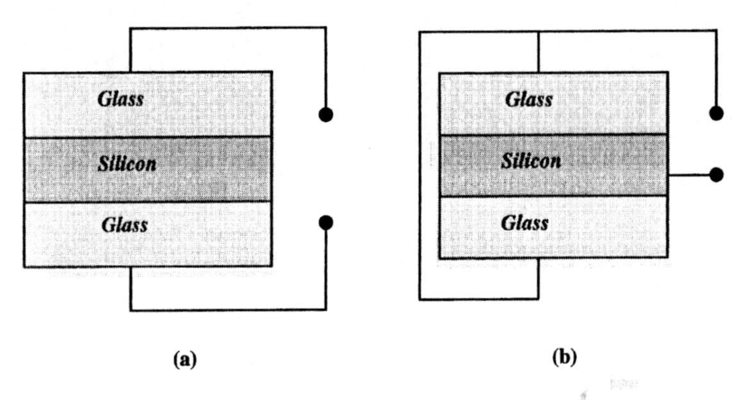

Figure 4-35. Schematic of (a) two-electrode with pole reversal bonding process and (b) three-electrode bonding process

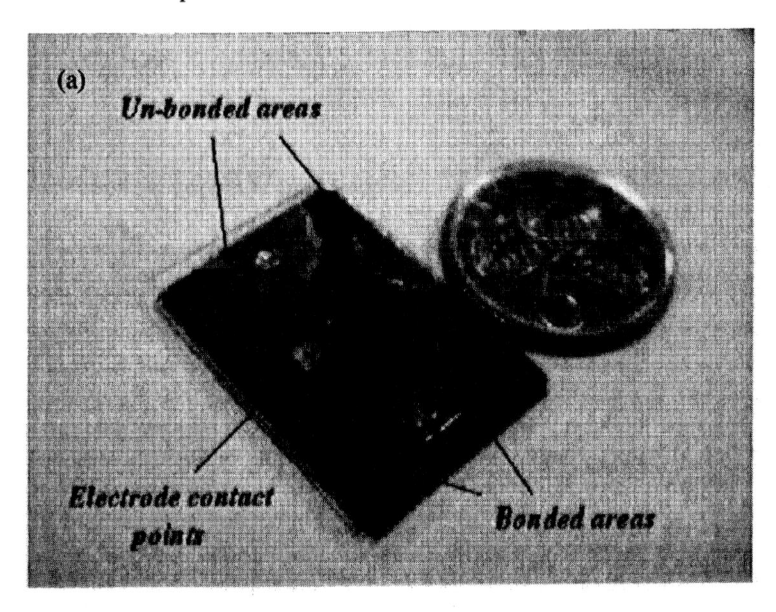

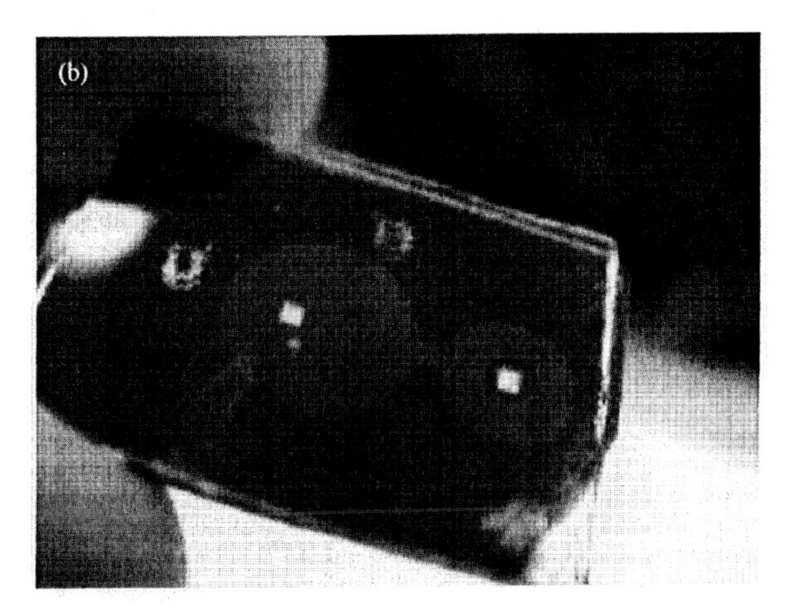

Figure 4-36. Photographs of bonded structures showing (a) poor bond and (b) good bond

Chapter 5

VERIFICATION AND TESTING

5.1 Experimental Set-up

An experimental set-up has been specially designed and built to evaluate the performance and yield of the prototype micropumps. The schematic for the set-up is given in Fig. 5-1 and photographs of the actual set-up is given in Figs. 5-2 to 5-4. The set-up can also be used to test other common microfluidics devices such as microvalves, microflow-sensor and microchannels.

The main performance measures of concern to the author are the flow rate and pressure head developed. To measure the flow rate, two methods are used, namely, by a flow meter (*MicroFlow* by *Bronkhorst™*) or by the gravimetric method using an electronic balance (*AT261DR Analytical Balance* from *Mettler Toledo™* with a resolution of 0.01 mg). The former method is meant for measuring low flows of 1 μl/min or less whereas the latter method is for larger flow rates. In addition, using the flow meter enables one to measure instantaneous flow rate as opposed to average flow rate using the gravimetric method.

To measure the pressure head developed across the micropump, two gauge pressure transducers *(HEISE DXD Digital Pressure Transducer* from *Dresser™*) are used. The pressure transducers are capable of measuring up to a pressure of 15 psi (about 1 Bar).

The measuring equipment is linked up to a PC for the acquisition and processing of data by means of the PC. This not only help to minimise human contact (which may result in human error) and also over the capability of prolonged period of data acquisition without the need for constant monitoring. This is especially useful when using the gravimetric method for determining flow rate.

In micropump testing, a performance measure of interest is the dependency of the flow rate on backpressure. In order to create a backpressure (or over-pressure) at the outlet of the pump, a pressure regulator-pressure chamber configuration has been designed and implemented into the set-up. This helps to create a stable, reproducible and controllable backpressure. The pressure controller (*DPI515 pressure controller* from *Druck*™) is able to regulate the pressure with a resolution of 0.01 mBar and provides a backpressure up to 300 mBar.

Another feature of this set-up is the priming procedure. In contrast to the traditional priming methods of flushing the system with a liquid with low surface tension (such as ethanol or isoproponal) or applying an over-pressure at the inlet and rinsing the system for a prolong period of time until trapped gases are sufficiently removed, the author uses a procedure based on carbon dioxide purge to prime the system instead [104]. As compared to the conventional method, this carbon dioxide purge method offers a faster, more reproducible and less tedious means of priming the system. Instead of washing air bubbles out of the system, the underlying principle behind this method is to take advantage of the higher solubility of carbon dioxide in the priming liquid (i.e. water). In fact, carbon dioxide is about thirty times more soluble in water as compared to air (Fig. 5-5). Therefore, any enclosed carbon dioxide bubbles can be easily removed by dissolution in the rinsing water. The procedure is to first remove air from the entire system (by means of a vacuum pump) and then filling it with carbon dioxide; the subsequent rinsing with water will then dissolve all the carbon dioxide entrapped in the system.

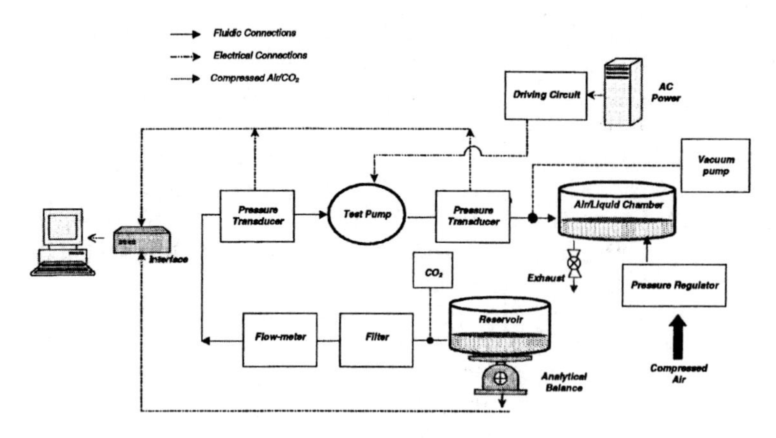

Figure 5-1. Schematic of experimental set-up

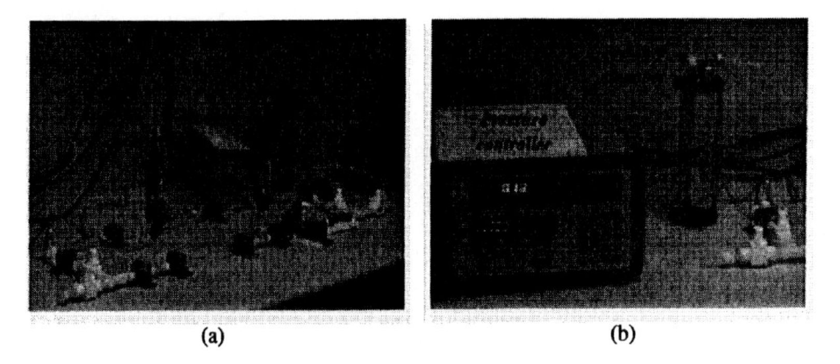

Figure 5-2. Photographs of experimental set-up showing (a) overview of set-up and (b) pressure control set-up

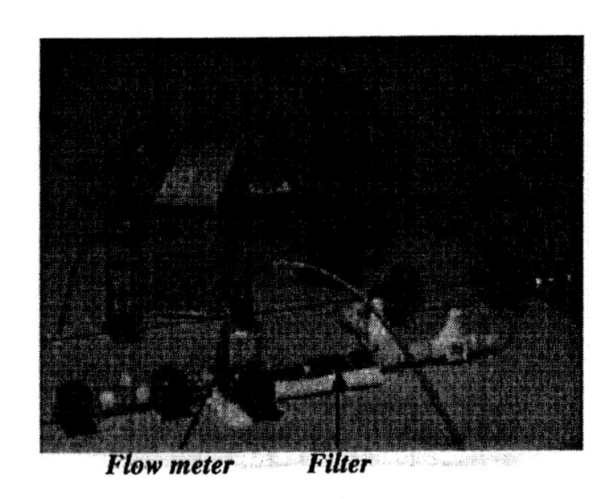

Readout for flow meter

Figure 5-3. Photographs of experimental set-up showing input to set-up

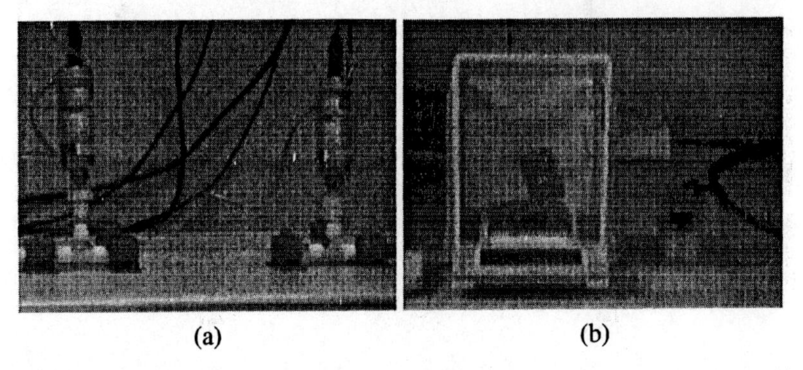

(a) (b)

Figure 5-4. Photographs of experimental set-up showing measurement instruments (a) pressure transducers and (b) set-up for gravimetric measurement

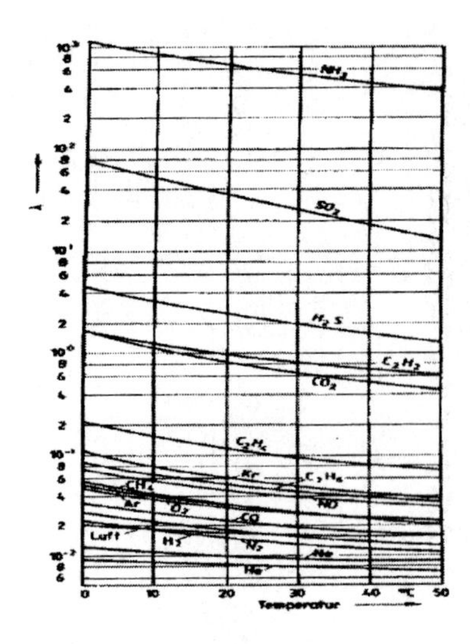

Figure 5-5. Solubility coefficient λ ($m^3/(kg.hPa)$) for different gases in water, as a function of temperature (taken from [104])

5.2 Piezoelectric Actuator Unit Model Verification

5.2.1 Analytical Model Verification

For verification purposes, experiments are conducted on several piezoelectric actuators which are taken from commercial piezoelectric buzzers. The maximum deflections of two types of actuator units under various input voltage are measured using a GIROD-TAST™ dial indicator.

The indicator has a resolution of 2 μm and a measuring range of 4 mm. In order to obtain more measurement data, two actuator units of different geometries are tested. Their dimensions are as follow:

1. Actuator a. Brass membrane with diameter of 15 mm and thickness of 0.35 mm. Piezoceramic disc of diameter 10 mm and thickness 0.2 mm.
2. Actuator b. Brass membrane with diameter of 20 mm and thickness of 0.35 mm. Piezoceramic disc of diameter 15 mm and thickness of 0.2 mm.

Figure 5-6. Experimental set-up for measuring the deflection of the actuator unit

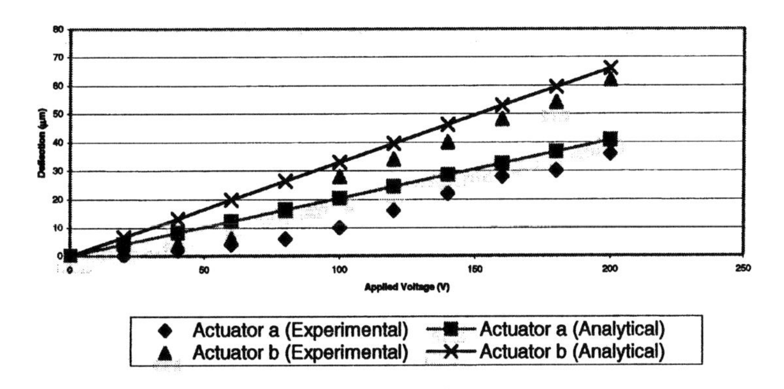

Figure 5-7. Variation of the maximum deflection of the actuator unit with input voltage

Table 5-1. Comparison of experimental and analytical results for the maximum deflection of a piezoelectric bimorph

Applied Voltage (Volt)	*Actuator a (μm)*		*Actuator b (μm)*	
	Experimental	*Analytical Model*	*Experimental*	*Analytical Model*
0	0	0	0	0
20	0	4.078	2	6.601
40	2	8.155	4	13.203
60	4	12.233	6	19.805
80	6	16.311	16	26.406
100	10	20.388	28	33.008
120	16	24.466	34	39.609
140	22	28.544	40	46.211
160	28	32.622	48	52.812
180	30	36.699	54	59.414
200	36	40.777	62	66.015

From Eqs. (3-29) and (3-30), the deflection is linearly proportional to the applied voltage. However, the experimental results show that the deflection does not varies linearly with the input voltage at low voltages. In addition, it is observed that the discrepancy between the theoretical and experimental results is significant at low voltages. The reasons are:

1. The resolution of dial meter is only 2 μm and it is not precise enough to measure the small deflections of the actuator membrane when driven at low voltages;
2. This is a contact measurement technique and the detector head, being in contact with the membrane surface, exerts a small force on the latter. The membrane must overcome this force before being about to deform. Consequently, the measured deflections at low voltages are inaccurate.

It can also be observed from Eqs. (3-29) and (3-20) that the stroke of the actuator unit can be increased by:

1. Increasing the actuation voltage. The breakdown electrical field strength (1200V/mm) places a limitation on this.
2. Increasing the diameter of the membrane. However, this will increase the overall dimension of the pump.
3. Selecting a piezoelectric material with a high piezoelectric constant.
4. Decreasing the thickness of the piezoceramic disc in order to increase the electrical field strength, which is inversely proportional to the thickness. However, this method will reduce the actuating force and breakdown field strength.

5.2.2 FEM Model Verification

In order to verify the FEM model, the deflections of a piezoceramic actuator at different input voltages were measured using a *Polytec™ Scanning Vibrometer* (Model: PSV 300). The piezoceramic bimorph consists of a circular piezoceramic disc (PXE5, 10 mm in diameter and 200 mm in thickness) glued onto a square silicon membrane (15 x 15 x 0.3 mm^3). The piezoceramic bimorph, test set-up and a sample of the results are shown in Fig. 5-8.

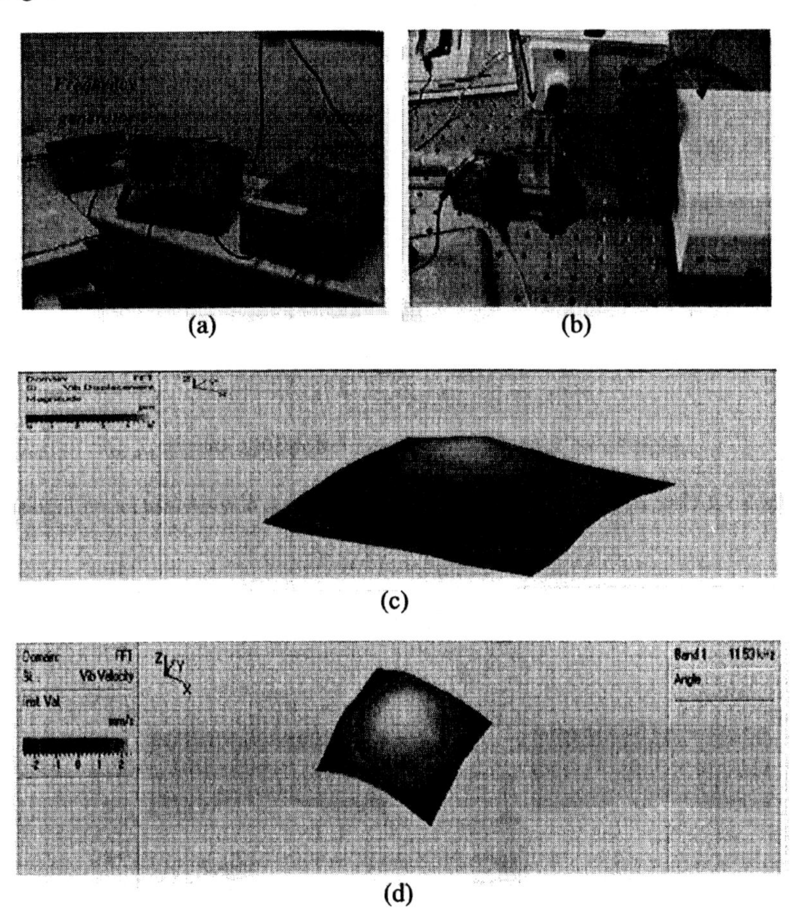

Figure 5-8. Photographs of (a) piezoceramic actuator driving circuit, (b) experimental set-up, (c) deflection of piezoelectric actuator at 100V input and (d) fundamental mode of piezoelectric actuator unit

A FEM model with fully clamped boundary condition for the piezoceramic bimorph is built and simulated (Fig. 5-9). The maximum deflections for various driving voltages ranging from 0 to 180 V are obtained (Fig. 5-10). A comparison of the experimental and simulation results is given in Fig. 5-11.

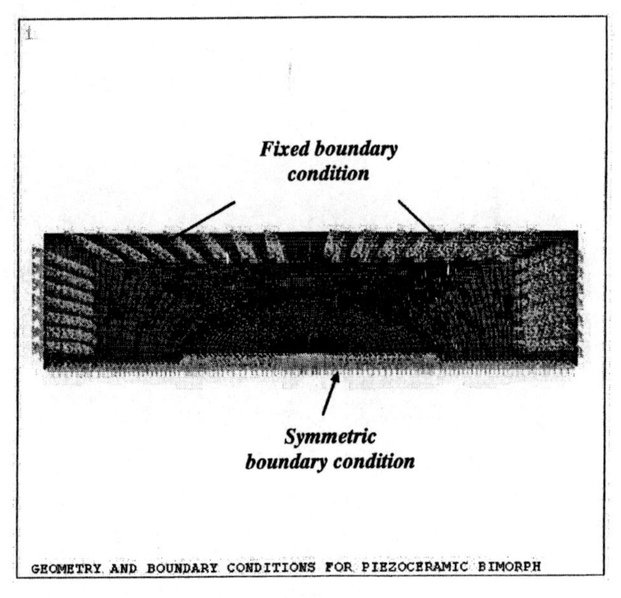

Figure 5-9. Geometry and boundary conditions of piezoceramic bimorph used for verification

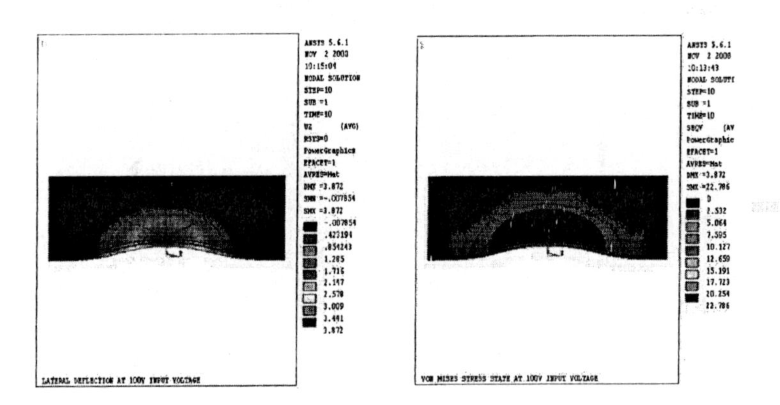

Figure 5-10. (a) Deflection and (b) Von Mises stress distribution in piezoceramic actuator for 100 V input voltage

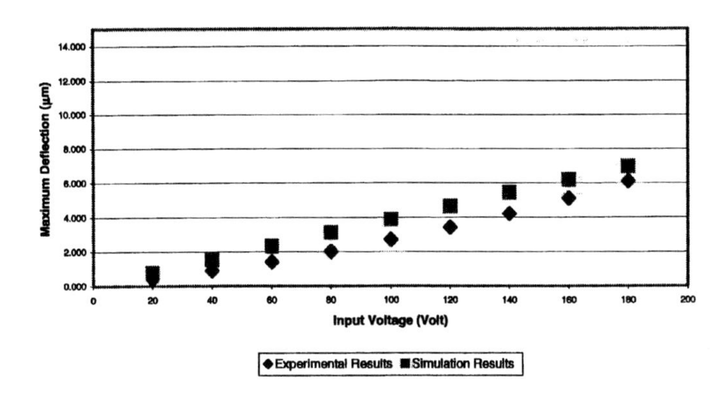

Figure 5-11. Comparison of experimental and simulation results for the piezoceramic bimorph

From Fig. 5-11, it can be seen that there is reasonable agreement between the experimental and simulated results. However, the difference is significant at lower voltages (93% difference at 20V but 14% at 180V) because of the limitation in the vibrometer, which is unable to read sub-micron displacements accurately. The discrepancies can also be attributed to the non-linear effects which are not taken into consideration in the constitutive equations used in the FE formulation, imperfect clamping of the test bimorph (i.e. not a perfect fully fixed boundary condition) and the offset of the piezoceramic disc from the centre of the membrane.

5.3 Preliminary Functional Tests

Preliminary tests on the micropumps were not able to yield any pumping action with neither water nor air at zero backpressure. At an actuation voltage of up to 250V, only a movement of the piezoelectric disc could be observed. Due to a lack of a suitable mask for the HF etching of the membrane in the glass substrate given the available equipment, no processing on the top glass substrate was possible. As a result, the glass membrane was too thick for the piezoelectric actuator to generate sufficient pressure to open the valves. The thicker glass membrane also meant that the actuator was much stiffer hence resulting in much smaller stroke volumes than the simulated results given in Chapter 4. As a consequent, the current pump was not self-priming. In order to verify this hypothesis, FEM simulations were carried out to determine the compression ratio when the glass membrane is 500 μm thick instead of the targeted 100 μm (Fig. 5-12).

For the current micropump, under a 250 V input voltage, the compression ratio is given by (Eq. 3-44):

$$\varepsilon = \frac{0.00044 \times 250}{1.145} \approx 0.096$$

In addition:

$$\frac{|\Delta P_{crit}|}{P_a} = \frac{18500}{100000} = 0.185$$

Clearly, in this case: $(\varepsilon = 0.096) < \left(\frac{|\Delta P_{crit}|}{P_a} = 0.185 \right)$

Therefore, one can conclude that the current micropump is not self-priming. In addition, as the compression ratio is significantly smaller than $|\Delta P_{crit}| / P_a$, hence the pump is very sensitive to the presence of bubbles in the pump. The entrapment of air bubbles in the pump despite flushing the pump with ethanol was another contributing factor to the prototype being unable to pump.

The pump was found to be able to pump water when an under-pressure was applied at the outlet by means of a syringe. A flow rate of 60 μl/min was measured at an actuation voltage of 260V (peak-to-peak) and a driving frequency of 100 Hz. This test showed that the proposed design is feasible as the valves are able to open and close accordingly for pumping to take place. However, improvements have to be made to the actuator before pumping can be achieved under normal conditions.

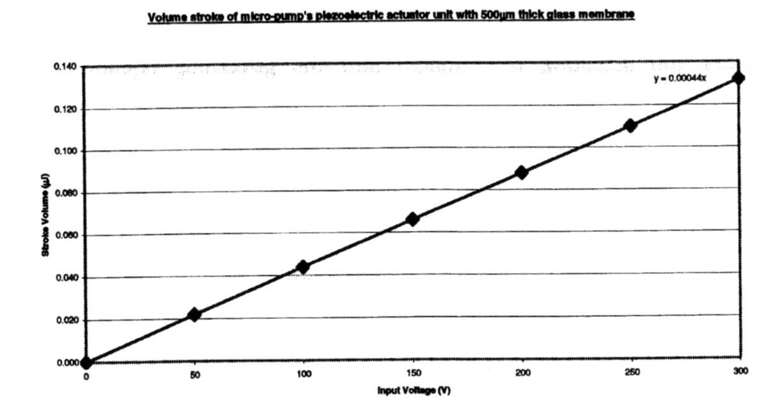

Figure 5-12. Stroke volume of fabricated micropump

5.4 Conclusions And Recommendations

5.4.1 Conclusions

In this project, the design rules for the design of a reciprocating positive-displacement micropump are developed; design rules or criteria for general micropump operation, switching of the valves, self-priming capability and bubble-tolerance are developed. As to date, research in micropumps and microfluidics, in general, is still a relatively new field. Consequently, the theoretical understanding behind the development of several microfluidics devices is not thoroughly established. Through this project, it is aimed that a proper set of design rules, or at least a framework for design rules development, can be established. As micropumps design is application driven, the proposed micropump is designed according to specifications for application in an insulin delivery system. Nevertheless, the concepts and methodologies developed can be implemented into the design of other types of micropumps, such as valves-less pumps or reciprocating pumps based on electrostatic actuation.

In addition to design rules, a simulation model of the micropump has also been developed. Based on a semi-numerical-analytical model, the main performance measures can be determined prior to the actual realisation of the pumps. This is especially important in shortening the development cycle and minimising the need for lengthy development of the actual prototype. The *coupled-field FEM* (i.e. electro-mechanical and structure-fluid interactions) method was used to carry out functional element scale modelling. A governing differential equation is developed analytically for system level modelling after integration of the various elemental models (i.e. valves, channels and actuating membranes) into the governing equation. This equation is then solved by numerical means to obtain the characteristics of the micropump.

Process development was also carried out in this project to realise an actual micropump prototype. The proposed prototypes, made from silicon and borosilicate glass wafers, are fabricated using typical micromachining (such as photolithography, thin film deposition, wet and dry etching among others), wafer bonding and laser ablation technologies. The proposed process can be readily adapted for the mass manufacturing of micropumps in silicon. Preliminary functional tests on the prototypes did not yield any pumping action with neither water nor air at zero backpressure. The main contributing factor was the actuating membrane which was too thick for the piezoelectric actuator to generate sufficient pressure for pumping as a suitable masking material for the processing of the glass substrate to the required thickness was not available. Another contributing factor was the presence of air bubbles in the pump. Nevertheless, pumping was observed when an under-pressure was created at the outlet. In this case, a flow rate of 60 μl/min was measured at an actuation voltage of 260V (peak-to-peak) and a driving frequency of 100 Hz. This demonstrated the feasibility of the proposed design and fabrication process.

This project presented in detail the development of a reciprocating micropump. Design rules and a simulation model were developed based on which a prototype micropump was realised by means of bulk micromachining and wafer bonding technologies.

5.4.2 Recommendations

The most important area for further work is to determine a suitable mask material for the glass etching process. As to date, polysilicon and chrome/gold masks are the most common choices but they are limited due to their low selectivities in BHF or HF. Further research can be conducted to explore other materials such as aluminium. Once a suitable mask material is found, only then can a prototype of the required dimensions be realised following which, detail experimentations and verification of the models proposed in this project can be carried out.

Another major area for further work is to determine optimised geometric parameters for the micropump after a detail verification and refinement of the proposed micropump model. The equations and models developed in the current work can be extended to obtain optimised solutions. ANSYS™ offers a module catered for optimisation problems and the ANSYS™ scripts written in the current project can be readily adapted to obtain optimised solutions. In addition, the development of a full micropump model utilizing structural-piezoelectric-fluid coupling seems an interesting area for future work, bearing in mind that such a model has not been realised by anyone at this point of time to the author's knowledge.

In addition, other alternatives to the fabrication process can also be explored. Provided that the resources are available, DRIE can be used to pattern the valves to obtain vertical walls and minimal undercutting of the valve seats. This will enable a greater control of the final geometry of the valves. In addition, controlled etch stop methods such as boron implantation or electrochemical etching can be explored to have better control of the etch depth. The current etch stop by timing of the KOH etching cannot be used if one wants to obtain membranes of thickness 20 μm or less. Glass processing is also an area needing much research. Currently, gold/chromium and polysilicon masks are the only materials which are known to withstand the HF etching, so other masking materials can be explored. In addition, the isotropic etching nature of HF leads to poor control of the lateral under-etching, resulting in poor control of the final geometry. Recently, there are reports in the literature of using certain etchant compositions (such as addition of HCl to HF) to minimize under-etching and the use of DRIE to obtain vertical walls in glass. These are areas worth further investigation.

Another area of further research is the utilization of other materials for microfluidics devices. As mentioned in an earlier chapter, the author believes that the use of polymers (such as polycarbonate and polyimide) will be the future focus of researchers in this area. With the advent of polymeric fabrication technologies catered especially for microsystems (such as LIGA, micro-stereolithography and micro-thermoplastic replication), truly 3D polymeric structures with tolerances well suited to microsystems will be able to be realised.

REFERENCES

[1] P. Gravesen, J. Brandebjerg, and O. Søndergard Jensen. "Microfluidics - a review," Journal of Micromech. & Microeng., Vol. **3**, 168 (1993).
[2] "Market analysis for micro-systems 1996-2002" by NEXUS.
[3] L.J. Thomas Jr. et al, United States Patent US3963380 "Micro pump powered by piezoelectric disk benders" (1975).
[4] N.F. Moody et al, United States Patent US04152098 "Micropump" (1979).
[5] P.R. Goudy Jr. et al, United States Patent US04379681 "Fluid Pump With Dual Diaphragm Check Valves" (1983).
[6] M. J. Kell, United States Patent US04411603 "Diaphragm Type Blood Pump For Medical Use" (1983).
[7] D.L. Millerd, United States Patent US04734092 "Ambulatory Drug Delivery Device" (1988).
[8] Shinichi Kamisuki et al, United States Patent US05259737 "Micropump With Valve Structure" (1993).
[9] Harald Van Lintel, United States Patent US05759014 "Micropump" (1998).
[10] J.G. Smits, European Patent EP134614A1 "Piezo-electrical micropump" (1984).
[11] H. Miyazaki et al, European Patent EP0435653A1 "Micropump" (1990).
[12] Shinichi Kamisuki et al., United States Patent US05259737 "Micropump With Valve Structure" (1993).
[13] Harald T. G. van Lintel et al, United States Patent US05271724 "Valve Equipped With A Position Detector And A Micropump Incorporating Said Valve" (1993).
[14] Nicolaas Frans de Rooij et al, United States Patent US05462839 "Process For The Manufacture Of A Micromachined Device To Contain Or Convey A Fluid" (1995).
[15] Harald Van Lintel, United States Patent US05759014 "Micropump" (1998).
[16] Paul J. Mulhauser et al, United States Patent US05919167 "Disposable Micropump" (1999).
[17] S. Shoji and M. Esashi. "Micro-flow devices and systems," J. Micromech. & Microeng, Vol. **4**, 157 (1994).
[18] H.T.G. Van Lintel, F.C.M. Van De Pol and S. Bouwstra. "A piezoelectric micropump based on micro-machining of silicon," Sensors and Actuators A, Vol. **15**, 153 (1989).
[19] M. Esashi, S. Shoji, A. Nakano. "Normally close microvalve and micropump fabricated on a silicon wafer," Proc. MEMS '89, 29 (1989).
[20] R. Zengerle, J. Ulrich, S. Kluge, M. Richter and A. Richter. "A bi-directional silicon micropump," Sensors and Actuators A, Vol. **50**, 81 (1995).
[21] R. Linnermann, P. Woias, C.D. Senfft and J.A.D. Ditterich. "A self-priming and bubble-tolerant piezoelectric micropump for liquids and gases," Proc. MEMS'98, 532 (1998).
[22] D. Maillefer, H. van Lintel, G. Rey-Mermet and R. Hirschi. "A high performance silicon micro-pump for implantatable drug delivery system," Proc. MEMS'99, 541 (1999).
[23] K.P. Kamper, J. Dopper, W. Ehrfeld and S. Oberbeck. "A self-filling low-cost membrane micro-pump," Proc. MEMS'98, 432 (1998).
[24] F.C.M. Van De Pol, H.T.G. Van Lintel, M. Elwenspoek and J.H.J. Fluitman. "A thermopneumatic micropump based on micro-engineering techniques" Sensors and Actuators A, Vol. **21**, 198 (1990).

[25] M.J. Zdelblick and J.B. Angell. "A microminiature electric-to-fluidic valve," Proc. 4th Int. Conf. Solid-State Sensors and Actuators (Transducer '87), Tokyo, Japan, 827 (1987).
[26] J.W. Judy, T. Tamagawa and D.L. Polla. "Surface-machined micromechanical membrane pump," Proc. MEMS'91, 182 (1991).
[27] R. Zengerle, A. Richter and H. Sandmaier. "A micro membrane pump with electrostatic actuation," Proc. MEMS '92, 19 (1992).
[28] R. Rapp, W.K. Schomburg, D. Mass, J. Schulz and W. Stark. "LIGA micropump for gases and liquids," Sensors & Actuators A, Vol. **21**, 57 (1994).
[29] B. Bustgens, W. Bacher, W. Menz and W.K. Schomburg. "Micropump manufactured by thermoplastic molding," Proc. MEMS'94, 18 (1994).
[30] W.L. Benard, H. Kahn, A.H. Heuer, and M.A. Huff. "A titanium-nickel shape-memory alloy actuated micro-pump," Proc. Transducer'97, 361 (1997).
[31] W.L. Benard, H. Kahn, A.H. Heuer, and M.A. Huff. "Thin-Film Shape-Memory Alloy Actuated Micropumps," J. MEMS, Vol. **7**, 245 (1998).
[32] J.G. Smits. "Piezoelectric micropump with microvalves," Proc. 8th University/Government/Industry Microelectronics Symposium, 92 (1989).
[33] J. G. Smits. "Piezoelectric micropump with three valves working peristaltically," Sensors & Actuators A, Vol. **21**, 203 (1990).
[34] J.A. Folta, N.F. Raley, E.W. Hee. "Design, fabrication and testing of a miniature peristaltic membrane pump," 5th Technical Digest Solid-State Sensor and Actuator Workshop, 186 (1992).
[35] H. Mizoguchi, M. Ando, T. Mizuno, T. Takagi and N. Nakajima. "Design and fabrication of light driven micropump," Proc. MEMS '92, 31 (1992).
[36] F.C.M. Van De Pol. "A pump based on micro-engineering techniques," PHD dissertation, University of Twente, Enschede, The Netherlands, 1989.
[37] E. Stemme and G. Stemme. "A Valve-less Diffuser/Nozzle based Fluid Pump," Sensors and Actuators A, Vol. **39**, 159 (1993).
[38] T. Gerlach, M. Schuenemann, and H. Wurmus. "A new micropump principle of the reciprocating type using pyramidic micro flow channels as passive valves," J. Micromech. & Microeng., Vol. **5**, 199 (1995).
[39] T. Gerlach and H. Wurmus. "Working principle and performance of the dynamic micropump," Sensor and Actuators A, Vol. **50**, 135 (1995).
[40] A. Olsson, G. Stemme and E. Stemme. "Simulation studies of diffuser and nozzle elements for valve-less micropumps," Proc. Solid State Sensors and Actuators (Transducer '97), Vol. **2**, 1039 (1997).
[41] R.L. Bardell, N.R. Sharma, F.K. Forster, M.A. Afromowitz and J.P. Robert. "Designing high-performance micro-pumps based on no-moving-parts valves," MEMS ASME 1997, DSC-Vol. **62**/HTD-Vol. **354**, 47 (1997).
[42] A. Olsson, O. Larsson, J. Holm, L. Lundbladh, O. Ohman and G. Stemme. "Valve-less diffuser micropumps fabricated using thermoplastic replication," Proc. MEMS ' 97, 305 (1997).
[43] F.K. Forster, L. Bardell, M.A. Afromowitz, N.R. Sharma, and A. Blanchard. "Design, fabrication and testing of fixed-valve micro-pumps," Proc. ASME Fluids Engineering Division ASME 1995, IMECE, FED-Vol. **234**, 39 (1995).
[44] N. Tesla, United States Patent Office US1329559 "Valvular Conduit" (1920).
[45] W. Van Der Wijngaart, H. Andersson, P. Enoksson, K. Noren and G. Stemme. "The first self-priming and bi-directional valve-less diffuser micropump for both liquid and gas," Proc. MEMS'2000, 674 (2000).
[46] O.M. Stuetser. "Ion drag pumps," J. Appl. Phys., Vol. **31**, 136 (1960).

[47] O.M. Stuetser, United States Patent US3398685. "Ion drag pumps" (1968).
[48] A. Richter and H. Sandmaier. "An electrohydrodynamic micropump," Proc. MEMS '90, 99 (1990).
[49] G. Fuhr, R. Hagedorn, T. Muller, W. Benecke and B. Wagner. "Pumping of water solutions in microfabricated electrohydrodynamic systems," Proc. MEMS'92, 25 (1992).
[50] J.R. Melcher and G.L. Taylor. "Traveling wave bulk electroconvection induced across a temperature gradient," Phys. Fluids, Vol. **10**, 1178 (1967).
[51] S.L. Zeng, C.H. Chen, J.C. Mikkelsen Jr and J.G. Santiago. "Fabrication and characterization of electrokinetic micro pumps," Proc. Thermal and Thermomechanical Phenomena in Electronic Systems 2000 (ITHERM 2000), Vol. 2, 31 (2000).
[52] A. Manz, C.S. Effenhauser, N. Burggraf, D.J. Harrison, K. Seiler and K. Fluri. "Electroosmotic pumping and electrophoretic separations for miniaturized chemical analysis systems," J. Micromech. & Microeng., Vol. **4**, 257 (1994).
[53] T. Kikuchi, T. Ujiie, T. Ichiki and Y. Horiike. "Fabrication of quartz micro-capillary electrophoresis chips for health care devices," Proc. Microprocesses and Nanotechnology Conf. 1999, 178 (1999).
[54] R.M. Moroney, R.M. White and R.T. Howe. "Ultrasonic induced microtransport," Proc. MEMS'91, 277 (1991).
[55] Miyazaki, T. Kawai and M. Ararag. "A piezo-electric pump driven by a flexural progressive wave," Proc. MEMS'91, 283 (1991).
[56] J. Jang and S.S. Lee. "Theoretical and experimental study of MHD (magnetohydrodynamic) micropump," Sensors and Actuators A, Vol. 80, 84 (2000).
[57] J.W. Waanders. "Introduction piezoelectric ceramics," Philips Components (2000).
[58] IEEE. "IEEE Standard on Piezoelectricity," ANSI/IEEE Std. 176-1987, The Institute of Electrical and Electronics Engineers, Inc., New York (1988).
[59] M. Richter, R. Linnemann and P. Woias. "Robust design of gas and liquid micropumps," Sensors and Actuators A, Vol. **68**, 480 (1998).
[60] D. Accoto, O.T. Nedelcu, M.C. Carrozza and P. Dario. "Theoretical analysis and experimental testing of a miniature piezoelectric pump," Proc. International Symp. Micromechatronics and Human Sci., 261 (1998).
[61] O.T. Nedelcu and V. Moagar-Poladian. "Modelling of the piezoelectric micropump for improving the working parameters," Proc. MSM'99, 1999.
[62] Christopher J Morris and Fred K Forster. "Optimization of a circular piezoelectric bimorph for a micropump driver," J. Micromech. Microeng., Vol. **10**, 459 (2000).
[63] S.P. Timoshenko. "Theory of pates and shells," 2nd ed. (McGraw-Hill, 1959).
[64] W.C. Young. "Roark's formulas for stress & strain," 6th ed. (McGraw-Hill, 1989).
[65] H.J. Bathe. "Finite element procedures in engineering analysis" (Prentice-Hall, New Jersey, 1982).
[66] E.P. EerNisse. "Variational method for electroelastic vibration analysis," IEEE Trans. Son. Ultrason. SU-14, 153 (1967).
[67] H.F. Tiersten. "Linear piezoelectric plate vibrations: Elements of the linear theory of piezoelectricity and the vibrations of piezoelectric plates" (Plenum Press, New York, 1969).
[68] Y. Kagawa and G.M.L. Gladwell. "Finite element analysis of flexure-type vibrators with electrostrictive transducers," IEEE Trans. Son. Ultrason. SU-17, 41 (1970).
[69] J.T. Oden and E. Kelley. "Finite element formulation of general electrothermoelasticity problems," Int. J. Num. Meth. Eng. Vol. **3**, 161 (1971).
[70] H. Allik and T.J.R. Hughes. "Finite element method for piezoelectric vibration," Int. J. Num. Meth. Eng. Vol. **2**, 151 (1970).

[71] G.H. Schmidt. "Application of the finite element method to the extensional vibrations of piezoelectric plates," Conference on Mathematics of Finite Elements and Applications, edited by J.R. Whiteman, 351-361 (Academic Press, New York, 1972).
[72] H. Allik, K.M. Webman, and J.T. Hunt. "Vibrational response of sonar transducers using piezoelectric finite elements," J. Acoust. Soc. Am. Vol. **56**, 1782 (1974).
[73] ANSYS. "ANSYS revision 5.0, technical description of capabilities," Swanson Analysis Systems, Inc., USA (1992).
[74] N. Rodamaker and C.R. Newell. "Finite element analysis of quartz angular rate sensor," Proc. ANSYS Conf., Swanson Analysis System, Inc., 3.35-3.38 (1989).
[75] C.J. Morris and F.K. Forster. "Optimization of a circular piezoelectric bimorph for a micropump driver," J. Micromech. Microeng., Vol. **10**, 459 (2000).
[76] Abaqus. "Abaqus/Standard Verification Manual Version 5.4," Hibbit, Karlsson & Sorensen, USA (1994).
[77] ANSYS. "ANSYS release 5.6, Theory reference," Swanson Analysis Systems, Inc., USA, 11th ed. (1999).
[78] J. Ulrich and R. Zengerle. "Static and dynamic flow simulation of a KOH-etched microvalve using the finite-element method," Sensors & Actuators A, Vol. **53**, 379 (1996).
[79] M. Koch, A.G.R. Evans and A. Brunnschweiler. "Coupled FEM simulation for the characterization of the fluid flow within a micromachined cantilever valve," J. Micromech. & Microeng., Vol. **6**, 112 (1996).
[80] O. Francais, I. Dufour and E. Sarraute. "Analytical static modelling and optimization of electrostatic micropumps," J. Micromech. & Microeng., Vol. **7**, 183 (1997).
[81] A. Cozma and R. Puers. "Electrostatic actuation as a self-testing method for silicon pressure sensors," Sensors & Actuators A, Vol. **60**, 32 (1997).
[82] Q.L. Gong , Z.Y. Zhou, Y.H Yang and X.H Wang. "Design, optimization and simulation on microelectromagnetic pump," Sensors & Actuators A, Vol. **83**, 200 (2000).
[83] R. Zengerle and M. Richter. "Simulation of microfluid systems," J. Micromech. Microeng., Vol. **4**, 192 (1994).
[84] F.C. Van der Pol, P. C. Breeveld and J.H.J. Fluitman. "Bond graph modelling of an electro-thermo-pneumatic micropump," Technical Digest of MME '90, 19 (1990).
[85] P. Voigt, G. Schrag and G. Wachutka. "Electrofluid full-system modelling of a flap valve micropump based on Kirchhoffian network theory," Sensors & Actuators A, Vol. **66**, 9 (1998).
[86] T. Bourouina and J.P. Grandchamp. "Modelling micropumps with electrical equivalent networks," J. Micromech. & Microeng., Vol. **6**, 398 (1996).
[87] J.F. Douglas, J.M. Gasiorek and J.A. Swaffield. "Fluid Mechanics," 2nd ed. (Pitman, Great Britain, 1985).
[88] J. Ulrich, H. Fuller and R. Zengerle. "Static and dynamic flow simulation of a KOH-etched micro valve," Proc. Transducers '95- Eurosensors IX, 17 (1995).
[89] M. Koch, N. Harris, R. Maas, A. G REvans, N.M. White and A. Brunnschweiler. "A novel micropump design with thick-film piezoelectric actuation," Meas. Sci. Technol., Vol. **8**, 49 (1997).
[90] H. Seidel. "The mechanism of anisotropic silicon etching and its relevance for micromachining," Proc. Transducers '87, 120 (1987).
[91] M. Madou. "Fundamentals of Microfabrication," 1st ed. (CRC Press, New York, 1997).
[92] W.R. Runyan and K.E. Bean. "Semiconductor integrated circuit processing technology," Addison Wesley, U.S. (1990).

[93] A. Sherman. "Chemical Vapour Deposition for Microelectronics" (Noyes, New York, 1997).

[94] T. Kamins. "Polycrystalline Silicon for Integrated Circuit Applications" (Kluwer, Deventer, 1988).

[95] A. Stoffel, A. Kovacs, W. Kronast and B. Muller. "LPCVD against PECVD for micromechanical applications," J. Micromech. & Microeng., Vol. **6**, 1 (1996).

[96] H.L. Offereins, G.K. Mayer, H. Sandmaier and K. Kuhl. "Fabrication of non-underetched convex corners in anisotropic etching of (100)-silicon in aqueous KOH with respect to novel micromechanic elements," J. Electrochemical Society, Vol. **137**, 3947 (1990).

[97] H. Seidel, L. Csepregi, A. Heuberger and H. Baumgartel. "Anisotropic etching of crystalline silicon in alkaline solutions I. Orientation dependence and behaviour of passivation layers," J. Electrochemical Society, Vol. **137**, 3613 (1990).

[98] Peter Enoksson. "Novel resonant micromachined silicon devices for fluid applications," Ph. D. dissertation, Royal Institute of Technology (1997).

[99] S. M. Sze. "VLSI Technology" (McGraw-Hill, New York, 1988).

[100] W. H. Ko and J. T. Sumito, "Semiconductor Integrated Circuit Technology and Micromachining, Ch. 5," Fundamentals and General Aspects, Vol. **1**, Sensors A Comprehensive Survey, T. Granke and W. H. Ko, Eds. (VCH, Weinheim, 1989).

[101] M. Elwenspoek and H.V. Jansen. "Silicon micromachining," 1st ed. (Cambridge University Press, New York, 1998).

[102] G. Wallis and D.I. Pomerantz. "Field-assisted glass-metal sealing," J. Applied Phys., Vol. **40**, 3946 (1969).

[103] M. Harz. "Anodic bonding for the third dimension," J. Micromech. & Microeng. Vol. **2**, 161 (1992).

[104] R. Zengerle, M. Leitner, S. Kluge and A. Ritcher. "Carbon dioxide priming of micro liquid systems," Proc. MEMS '95, 340 (1995).

APPENDIX A

(Formulation of the analytical model for the micro-pump piezoelectric actuator unit.)

The governing partial differential equation for the bending of plates (in polar coordinates) that gives the deflection surface of a laterally loaded plate is given by [63]:

$$\nabla^4 w = \left(\frac{\partial^2}{\partial r^2} + \frac{1}{r}\frac{\partial}{\partial r} + \frac{1}{r^2}\frac{\partial^2}{\partial \theta^2}\right)\left(\frac{\partial^2 w}{\partial r^2} + \frac{1}{r}\frac{\partial w}{\partial r} + \frac{1}{r^2}\frac{\partial^2 w}{\partial \theta^2}\right) = \frac{q}{D} \tag{A-1}$$

where w is the lateral displacement; r is the radial distance in polar coordinate; θ is the angular coordinate; q is the intensity of the lateral load over the plate and D is the flexural rigidity given by $D = Eh^3/12(1-\upsilon^2)$ with E being the Young's modulus, h the thickness and υ the Poisson's ratio of the material.

When the load is symmetrically distributed with respect to the centre of the plate, the deflection is independent of θ, and Equation (A-1) reduces to:

$$\left(\frac{d^2}{dr^2} + \frac{1}{r}\frac{d}{dr} +\right)\left(\frac{d^2 w}{dr^2} + \frac{1}{r}\frac{dw}{dr} +\right) = \frac{q}{D}$$
$$\frac{1}{r}\frac{d}{dr}\left\{r\frac{d}{dr}\left[\frac{1}{r}\frac{d}{dr}\left(r\frac{\partial w}{\partial r}\right)\right]\right\} = \frac{q}{D} \tag{A-2}$$

Upon integration, the general equation for deflection of a circular plate is obtained:

$$w = \frac{qr^4}{64D} + \frac{C_1 r^2}{4} + C_2 \log_e\left(\frac{r}{R}\right) + C_3 \tag{A-3}$$

where C_1, C_2 and C_3 are the constants of integration which can be determined by substitution of the appropriate boundary conditions and R is the radius of the plate.

The derivation of the governing partial differential equation (i.e. Equation A-1) is based on the following assumptions:

1. The plate is thin, that is, the thickness is no more than ¼ of the least transverse dimension.
2. The deflection is small being that the maximum deflection is no more than ½ of the thickness.
3. The plate is flat, uniform in thickness and made of a homogeneous isotropic material.
4. Plate is nowhere stressed beyond the elastic limit.
5. There is no deformation in the middle plane of the plate. This plane remains neutral during bending.
6. Points of the plate lying initially on a normal-to-the-middle plane of the plate remain on the normal-to-the-middle surface of the plate after bending
7. The normal stresses in the direction transverse to the plate can be disregarded.

Using these assumptions, all stress components can be expressed by the deflection w of the plate, which is function of the two coordinates in the plane of the plate. This function satisfies a linear partial differential equation, which, together with the boundary conditions, completely defines w.

Now, for a circular plate without any central hole, C_2 effectively reduces to a zero term. In addition, in the absence of any lateral load, the first term on the RHS of Equation (A-3) disappears. Hence, Equation (A-3) reduces to:

$$w = \frac{C_1 r^2}{4} + C_3 \qquad \text{(A-4)}$$

As above-mentioned, the stresses and moments can be expressed in terms of w and they are given by:

$$M_r = -D\left[\frac{d^2 w}{dr^2} + \frac{\upsilon}{r}\frac{dw}{dr}\right] \qquad \text{(A-5)}$$

$$M_t = -D\left[\frac{1}{r}\frac{dw}{dr} + \upsilon\frac{d^2w}{dr^2}\right] \tag{A-6}$$

where M_r is the radial bending moment per unit length and M_t is the tangential moment per unit length.

For the case of a plate under pure bending, the following boundary conditions are applicable:

$$M_r = M_o \tag{A-7a}$$

$$w\big|_{r=b} = 0 \tag{A-7b}$$

From Equation (A-5) and (A-7a):

$$M_r = -D\left[\frac{C_1}{2} + \frac{\upsilon}{r}\frac{C_1 r}{2}\right] = M_o$$

Therefore:

$$C_1 = -\frac{2M_o}{D(1+\upsilon)}$$

Using boundary condition (A-7b) and substituting in C_1:

$$C_2 = \frac{M_o b^2}{2D(1+\upsilon)}$$

Hence, the deflection of a circular plate under pure bending is given by:

$$w = \frac{M_o}{2D(1+\upsilon)}\left(b^2 - r^2\right) \tag{A-8}$$

In order to adapt Equation (A-8) into the case of a bilaminar circular plate under pure bending, an equivalent plate flexural stiffness D_e and Poisson's ratio υ_e can be used provided the plate deforms into a spherical surface and the edge restraints do not prevent motion parallel to the surface of the plate [64]. This restriction effectively assures that bending moments are constant in magnitude at all locations in the plate and in all directions. These equivalent material properties are given by:

$$\text{Equivalent } D_e = \frac{c_p^E t_p^3}{12\left(1-\upsilon_p^2\right)} K_1 \tag{A-9}$$

$$\text{Equivalent } \upsilon_e = \upsilon_p \frac{K_2}{K_1} \tag{A-10}$$

where

$$K_1 = 1 + \frac{c_m t_m^3 \left(1-\upsilon_p^2\right)}{c_p^E t_p^3 \left(1-\upsilon_m^2\right)} + \frac{3\left(1-\upsilon_p^2\right)\left(1+\frac{t_m}{t_p}\right)^2 \left(1+\frac{c_p^E t_p}{c_m t_m}\right)}{\left(1+\frac{c_p^E t_p}{c_m t_m}\right)^2 - \left(\upsilon_p + \upsilon_m \frac{c_p^E t_p}{c_m t_m}\right)^2} \tag{A-11}$$

$$K_2 = 1 + \frac{\upsilon_m c_m t_m^3 \left(1-\upsilon_p^2\right)}{\upsilon_p c_p^E t_p^3 \left(1-\upsilon_m^2\right)} + \frac{3\left(1-\upsilon_p^2\right)\left(1+\frac{t_m}{t_p}\right)^2 \left(1+\frac{\upsilon_m c_p^E t_p}{\upsilon_p c_m t_m}\right)}{\left(1+\frac{c_p^E t_p}{c_m t_m}\right)^2 - \left(\upsilon_p + \upsilon_m \frac{c_p^E t_p}{c_m t_m}\right)^2} \tag{A-12}$$

where υ_p is the Poisson's ratio of the piezoelectric disc; υ_m is the Poisson's ratio of the piezoelectric disc and c_m is the Young's modulus of the membrane.

In the case which the Poisson's ratios of the two materials are equal:

$$K_1 = K_2$$

and

$$\upsilon_e = \upsilon_p = \upsilon_m$$

Upon substitution of D_e and υ_e into Equation (A-8):

$$\begin{aligned} w &= \frac{M_o}{2D_e(1+\upsilon_e)}\left(b^2 - r^2\right) \\ &= \frac{c_p^E d_{31} t_m}{4D_e(1+\upsilon_e)}\left(b^2 - r^2\right)\cdot U \end{aligned} \tag{A-13}$$

Now, the inner bilaminar plate will induce shearing forces, Q_o, along the inner diameter of the outer annular plate. The general deflection equation for the outer portion, which is the case of an annular plate fully clamped at the outer edges and under a uniform annular line load at the inner edges, is given by [64]:

$$w_{II} = w_b + \theta_b r F_1 - Q_o \frac{r^3}{D_m} F_2 \tag{A-14}$$

$$w_b = \frac{Q_o a^3}{D_m}\left(\frac{C_3 L_2}{C_4} - L_1\right) \tag{A-15}$$

$$\theta_b = \frac{Q_o a^2}{D_m C_4} L_2 \tag{A-16}$$

where w_{II} is the lateral displacement of the outer portion; r is the radial distance in polar coordinate; θ_b is the slope at the inner edge; Q_o is the unit load along the edges of the plate; D_m is the flexural rigidity of the membrane; C_3, C_4, L_1 and L_2 are constants given by:

$$C_3 = \frac{1+\upsilon}{2}\frac{b}{a}\ln\left(\frac{a}{b}\right) + \frac{1-\upsilon}{4}\left(\frac{a}{b} - \frac{b}{a}\right)$$

$$C_4 = \frac{1}{2}\left[(1+\upsilon)\frac{b}{a} + (1-\upsilon)\frac{a}{b}\right]$$

$$L_1 = \frac{b}{4a}\left\{\left[\left(\frac{b}{a}\right)^2 + 1\right]\ln\left(\frac{a}{b}\right) + \left(\frac{b}{a}\right)^2 - 1\right\}$$

$$L_2 = \frac{b}{4a}\left[\left(\frac{b}{a}\right)^2 - 1 + 2\ln\left(\frac{a}{b}\right)\right]$$

and F_1 and F_2 are functions given by:

$$F_1 = \frac{1+\upsilon}{2}\frac{b}{r}\ln\left(\frac{r}{b}\right) + \frac{1-\upsilon}{4}\left(\frac{r}{b} - \frac{b}{r}\right)$$

and

$$F_2 = \frac{b}{4r}\left\{\left[\left(\frac{b}{r}\right)^2 + 1\right]\ln\left(\frac{r}{b}\right) + \left(\frac{b}{r}\right)^2 - 1\right\}\langle r-b\rangle^0$$

where the singularity function brackets $\langle\ \rangle$ indicate that the expression contained within the brackets must be equated to zero unless $r > b$ after which they are treated as any other brackets.

Before the application of Equation (A-14), the magnitude of the shearing forces has to be determined. This can be found from the condition of continuity along the circle $r = b$, from which it follows that both portions of the plate have, at that circle, the same slope, that is:

$$\left.\frac{dw_I}{dr}\right|_{r=b} = \left.\frac{dw_{II}}{dr}\right|_{r=b}$$

which implies from Equations (A-14) and (A-16) that:

$$\frac{M_o b}{D_e(1+\upsilon_e)} = \frac{Q_o a^2}{D_m C_4} L_2$$

Hence:

$$Q_o = \frac{C_4 M_o b}{L_2 a^2 (1+\upsilon_e)} \frac{D_m}{D_e} \tag{A-17}$$

The substitution of Equation (A-17) will completely define the governing equation for the outer annular plate. In order to completely define the equation governing the deflection of the inner bilaminar plate, w_b (Equation (A-15)) must be superimposed into Equation (A-13), which yields:

$$\begin{aligned} w_I &= \frac{M_o}{2D_e(1+\upsilon_e)}\left(b^2 - r^2\right) + \frac{Q_o a^3}{D_m}\left(\frac{C_3 L_2}{C_4} - L_1\right) \\ &= L_3 \frac{t_m}{4}\left(b^2 - r^2\right) \cdot U + \frac{t_m ab}{2} \frac{C_4 L_3}{L_2}\left(\frac{C_3 L_2}{C_4} - L_1\right) \cdot U \end{aligned} \tag{A-18}$$

where

$$L_3 = \frac{c_p^E d_{31}}{D_e(1+\upsilon_e)}$$

The equation governing the deflection of the outer annular portion is given by:

$$\begin{aligned} w_{II} &= w_b + \theta_b r F_1 - Q_o \frac{r^3}{D_m} F_2 = \frac{t_m ab}{2} \frac{C_4 L_3}{L_2}\left(\frac{C_3 L_2}{C_4} - L_1\right) \cdot U \\ &+ L_3 \frac{t_m b}{2} \cdot F_1 \cdot U \cdot r - \frac{t_m b}{2a^2} \frac{C_4 L_3}{L_2} \cdot F_2 \cdot U \cdot r^3 \end{aligned} \tag{A-19}$$

APPENDIX B

Material properties of PXE5™

Philips Components — Product specification

Piezoelectric Ceramics — Survey of grades

2 SURVEY OF GRADES

The following grades, consisting of modified lead zirconate titanates, are distinguished according to their electrical and mechanical properties and field of application.

Table 1 gives typical values measured on discs Ø16 × 1 mm at 21 ±1 °C, 24 hours after poling.

The properties of components manufactured with these grades depend upon the dimensions of the product, the method of manufacture and on the applied voltage level. Therefore, a meaningful interpretation of the properties of the material is best made in consultation with the supplier.

Table 1 Typical values

PROPERTY AND SYMBOL	PXE 5	PXE 52	PXE 59	PXE 21	PXE 41	PXE 42	PXE 43	UNIT
Thermal data								
Curie temperature	285	165	360	270	315	325	300	°C
Specific heat	420	420	420	420	420	420	420	J/kg K
Thermal conductivity	1.2	1.2	1.2	1.2	1.2	1.2	1.2	W/m K
Mechanical data								
Density ρ_m	7.8	7.8	7.9	7.8	7.9	7.8	7.8	10^3 kg/m^3
Compliance:								
s_{33}^E	18	20	18	19	15	15	13	10^{-12}/Pa
s_{11}^E	15	16	16	15	12	13	11	10^{-12}/Pa
s_{55}^E	39	–	45	–	37	–	–	10^{-12}/Pa
Poisson's ratio σ	0.3	0.3	0.35	0.3	0.3	0.3	0.3	
Mechanical quality factor for radial mode Q_m^E	75	65	80	75	1200	750	1000	
Frequency constants:								
N_P^E	1975	1925	1970	2000	2175	2200	2350	Hz m or m/s
$N_3^D = 1/2\upsilon_3^D$	1850	1800	2060	1900	2000	2015	2050	Hz m or m/s
$N_1^E = 1/2\upsilon_1^E$	1450	1400	1400	–	1620	–	–	Hz m or m/s
$N_5^E = 1/2\upsilon_5^E$	930	–	900	–	950	–	–	Hz m or m/s
Compressive strength	>600	>600	>600	>600	>600	>600	>600	10^6 Pa
Tensile strength	80	80	100	80	80	80	80	10^6 Pa

Philips Components — Product specification

Piezoelectric Ceramics — Material grade specifications

PROPERTY AND SYMBOL	PXE 5	PXE 52	PXE 59	PXE 21	PXE 41	PXE 42	PXE 43	UNIT
Electrical data								
Relative permittivity ($\varepsilon_0 = 8.85 \times 10^{-12}$ F/m):								
$\varepsilon^T_{33}/\varepsilon_0$	2100	3900	1850	2000	1225	1325	1000	
$\varepsilon^T_{11}/\varepsilon_0$	1800	3300	1650	–	1400	–	–	
Resistivity ρ	5	1	5	5	1	1	1	10^{10} Ωm
Time constant $\rho\varepsilon^T_{33}$ (25 °C)	>300	>500	>100	>25	>7	–	–	minute
Dielectric loss factor tan δ	20	16	17	15	2.5	3.5	2	10^{-3}
Electro-mechanical data								
Coupling factor:								
k_p	0.68	0.70	0.66	0.64	0.64	0.61	0.53	
k_{33}	0.75	0.80	0.71	0.74	0.74	0.70	0.66	
k_{31}	0.38	0.39	0.37	0.37	0.38	0.34	0.30	
k_{15}	0.66	–	0.68	–	0.70	–	–	
Piezoelectric charge constants:								
d_{33}	500	700	460	450	325	315	230	10^{-12} C/N or m/V
d_{31}	–215	–280	–195	–200	–150	–130	–100	10^{-12} C/N or m/V
d_{15}	515	–	550	–	480	–	–	10^{-12} C/N or m/V
Piezoelectric voltage constants:								
g_{33}	24	20	28	25	30	27	27	10^{-3} Vm/N or m^2/C
g_{31}	–10	–10	–13	–12	–12	–11	–11	10^{-3} Vm/N or m^2/C
g_{15}	33	–	37	–	39	–	–	10^{-3} Vm/N or m^2/C
Time stability (%)								
Coupling factor k_p	–0.5	–0.6	–0.1	–1.5	–1.5	–2.5	–2.0	relative change per time decade (days)
Permittivity ε^T_{33}	–1.0	–1.0	–2.0	–2.0	1.0	–6.0	–4.5	
Frequency constant N^E_p	0.5	0.3	+0.1	0.5	0.5	1.5	1.0	
Quality factor Q^E_m	–	–3.0	+0.1	–	10.0	–	–	
Dielectric loss factor tan δ	–	–	–0.1	–	–10	–	–	

Philips Components — Product specification

Piezoelectric Ceramics — PXE 5

3 PXE 5 SPECIFICATIONS

SYMBOL	TYPICAL VALUE	UNIT
d_{33}	500	10^{-12} m/V
d_{31}	−215	10^{-12} m/V
g_{33}	24	10^{-3} Vm/N
g_{31}	−10	10^{-3} Vm/N
k_p	0.68	
$\frac{\varepsilon_{33}^{T}}{\varepsilon_0}$	2100	
N_P^E	1975	Hzm
S_{33}^E	18	10^{-12}/Pa
T_C	285	°C

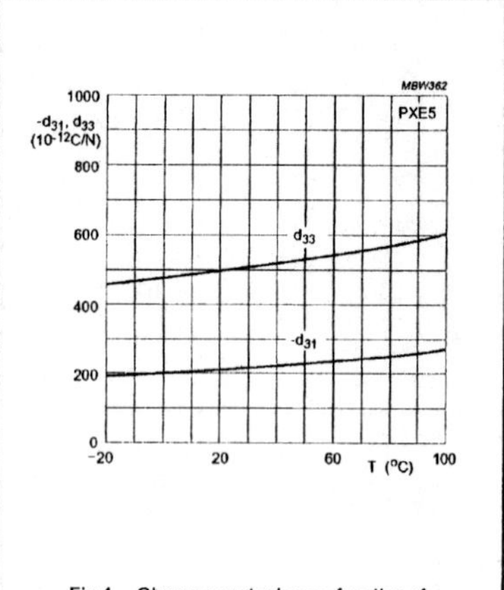

Fig.1 Charge constants as a function of temperature.

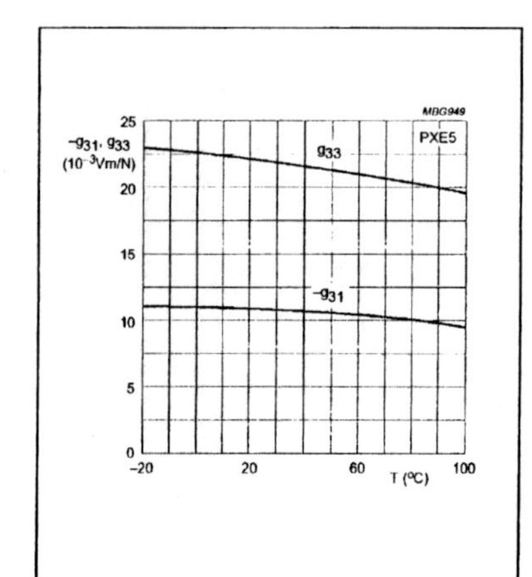

Fig.2 Voltage constants as a function of temperature.

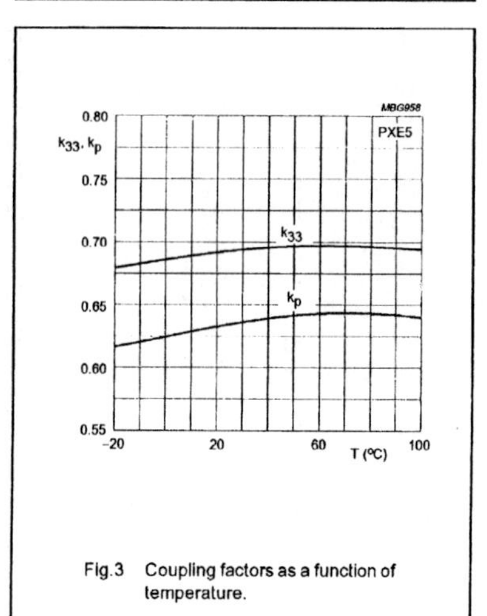

Fig.3 Coupling factors as a function of temperature.

Piezoelectric Ceramics — PXE 5

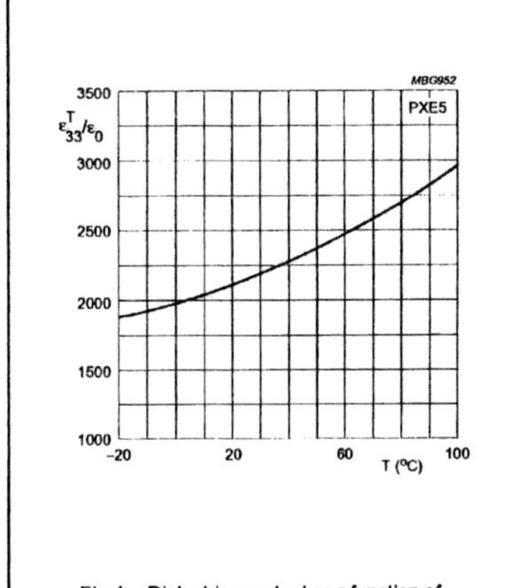

Fig.4 Dielectric constant as a function of temperature.

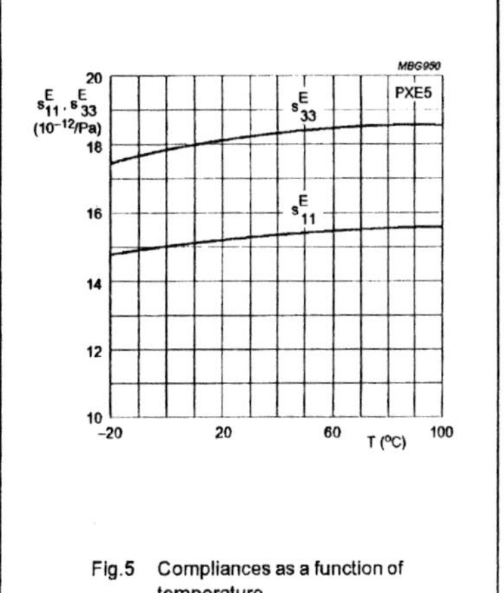

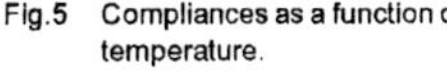
Fig.5 Compliances as a function of temperature.

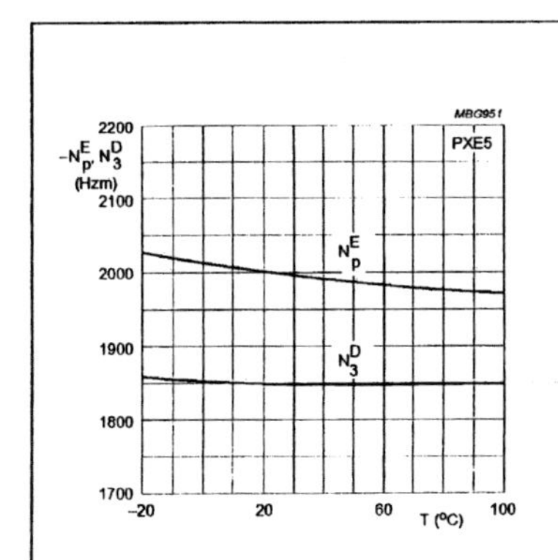

Fig.6 Frequency constants as a function of temperature.

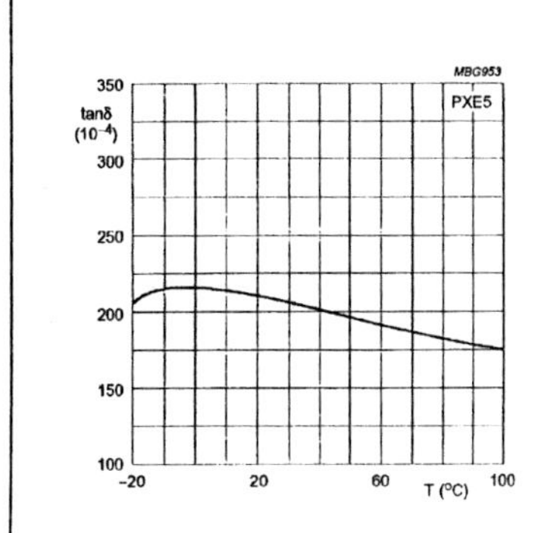

Fig.7 Dielectric loss factor as a function of temperature.

Philips Components — Product specification

Piezoelectric Ceramics — Discs

1 DISCS

1.1 Discs with nickel electrodes

Our standard range of discs with nickel electrodes is available in grade PXE 5. Other grades and sizes are available on request. The positive pole is marked.

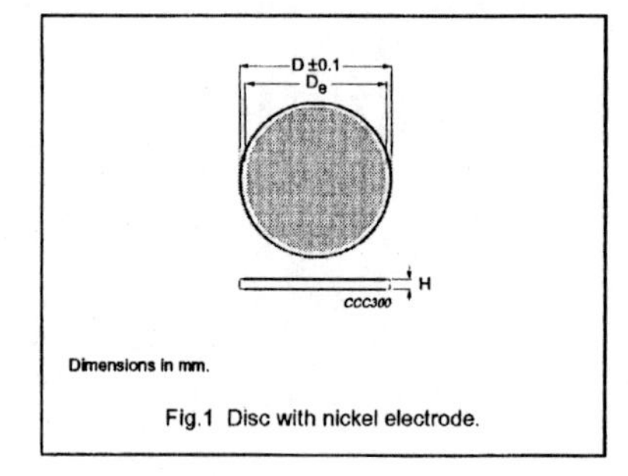

Fig.1 Disc with nickel electrode.

Disc data, nickel electrode; see Fig.1

GRADE	D (mm)	D_e (mm)	H (mm)	CAPACITANCE (pF)	TYPE NUMBER
PXE 5	5	5.0 ±0.1	0.3 ±0.03	1220 ±25%	DSC5/0.3-PX5-N
	5	5.0 ±0.1	0.5 ±0.03	750 ±25%	DSC5/0.5-PX5-N
	5	5.0 ±0.1	1.0 ±0.03	375 ±25%	DSC5/1-PX5-N
	5	5.0 ±0.1	2.0 ±0.1	185 ±25%	DSC5/2-PX5-N
	10	9.0 ±0.3	0.2 ±0.03	5900 ±25%	DSC10/0.2-PX5-N
	10	9.0 ±0.3	0.5 ±0.03	2360 ±25%	DSC10/0.5-PX5-N
	10	9.5 ±0.3	1.0 ±0.03	1390 ±25%	DSC10/1-PX5-N
	10	9.5 ±0.3	2.0 ±0.1	700 ±25%	DSC10/2-PX5-N
	10	9.5 ±0.3	3.0 ±0.1	460 ±25%	DSC10/3-PX5-N
	10	9.5 ±0.3	5.0 ±0.1	280 ±25%	DSC10/5-PX5-N
	16	15.0 ±0.3	0.2 ±0.03	17000 ±25%	DSC16/0.2-PX5-N
	16	15.5 ±0.3	0.5 ±0.03	6800 ±25%	DSC16/0.5-PX5-N
	16	15.5 ±0.3	1.0 ±0.03	3620 ±25%	DSC16/1-PX5-N
	16	15.5 ±0.3	2.0 ±0.1	1800 ±25%	DSC16/2-PX5-N
	16	15.5 ±0.3	3.0 ±0.1	1200 ±25%	DSC16/3-PX5-N
	20	19.0 ±0.3	0.2 ±0.03	27000 ±25%	DSC20/0.2-PX5-N
	20	19.5 ±0.3	0.5 ±0.03	10800 ±25%	DSC20/0.5-PX5-N
	20	19.5 ±0.3	1.0 ±0.03	5700 ±25%	DSC20/1-PX5-N
	20	19.5 ±0.3	2.0 ±0.1	2850 ±25%	DSC20/2-PX5-N
	25	24.0 ±0.3	0.2 ±0.03	42925 ±25%	DSC25/0.2-PX5-N
	25	24.5 ±0.3	0.5 ±0.03	17200 ±25%	DSC25/0.5-PX5-N
	25	24.5 ±0.3	1.0 ±0.03	9000 ±25%	DSC25/1-PX5-N
	25	24.5 ±0.3	2.0 ±0.1	4500 ±25%	DSC25/2-PX5-N

APPENDIX C

(*MATLAB*™ script for the definition of piezoelectric material parameters)

```
MATERIAL DEFINITION OF PXE5
HEXAGONAL CLASS CRYSTAL FERROELECTRIC CERAMIC
format long
%COMPLIANCE
SE11=15E-6
SE22=15E-6
SE33=18E-6
SE55=39E-6
%POISSON RATIO
V12=0.3
V13=0.3
V23=0.3
%PERMITTIVITY
PTO=8.854E-6
PT11=1800*PTO
PT22=PT11
PT33=2100*PTO
%PIEZOELECTRIC CONSTANTS
D33=500E-6
D31=-215E-6
D32=D31
D15=515E-6
D24=D15
%DENSITY
DES=7800E-18
%
%
%COMPLIANCE MATRIX
SE12=(-V12)*SE22
SE13=(-V13)*SE33
SE21=SE12
SE23=(-V23)*SE33
SE31=SE13
SE32=SE23
SE44=SE55
SE66=1/((1/SE11)/(2*(1+V13)))
SEM=[SE11 SE12 SE13 0 0 0;SE21 SE22 SE23 0 0 0;SE31 SE32 SE33
0 0 0;0 0 0 SE44 0 0;0 0 0 0 SE55 0;0 0 0 0 0 SE66]
```

```
%STIFFNESS MATRIX
CEM=inv(SEM)
%
%
%PIEZOELECTRIC D MATRIX
DM=[0 0 0 0 D15 0;0 0 0 D24 0 0;D31 D32 D33 0 0 0]
format long
%PIEZOELECTRIC E MATRIX
EM=DM*CEM
%PERMITTIVITY MATRIX
PTM=[PT11 0 0;0 PT22 0;0 0 PT33]
PSM=PTM-DM*EM'
******************************RESULTS********************
**********
SE11 =
  1.500000000000000e-005
SE22 =
  1.500000000000000e-005
SE33 =
  1.800000000000000e-005
SE55 =
  3.900000000000000e-005
V12 =
 0.30000000000000
V13 =
 0.30000000000000
V23 =
 0.30000000000000
PTO =
  8.854000000000000e-006
PT11 =
 0.01593720000000
PT22 =
 0.01593720000000
PT33 =
 0.01859340000000
D33 =
  5.000000000000000e-004
D31 =
 -2.150000000000000e-004
D32 =
 -2.150000000000000e-004
```

```
D15 =
   5.150000000000001e-004
D24 =
   5.150000000000001e-004
DES =
   7.800000000000001e-015
SE12 =
  -4.500000000000000e-006
SE13 =
  -5.400000000000000e-006
SE21 =
  -4.500000000000000e-006
SE23 =
  -5.400000000000000e-006
SE31 =
  -5.400000000000000e-006
SE32 =
  -5.400000000000000e-006
SE44 =
   3.900000000000000e-005
SE66 =
   3.900000000000000e-005
SEM =
  1.0e-004 *
  Columns 1 through 4
 0.15000000000000 -0.04500000000000 -0.05400000000000                 0
-0.04500000000000  0.15000000000000 -0.05400000000000                 0
-0.05400000000000 -0.05400000000000  0.18000000000000                 0
                0                 0                 0  0.39000000000000
                0                 0                 0                 0
                0                 0                 0                 0

  Columns 5 through 6
                0                 0
                0                 0
                0                 0
                0                 0
 0.39000000000000                 0
                0  0.39000000000000

CEM =
  1.0e+004 *
```

```
  Columns 1 through 4
9.45115490570036  4.32294977749523  4.13223140495868                 0
4.32294977749523  9.45115490570036  4.13223140495868                 0
4.13223140495868  4.13223140495868  8.03489439853076                 0
               0                 0                 0  2.56410256410256
               0                 0                 0                 0
               0                 0                 0                 0

  Columns 5 through 6
               0                 0
               0                 0
               0                 0
               0                 0
 2.56410256410256                0
               0  2.56410256410256

 DM =
  1.0e-003 *

  Columns 1 through 4
               0                 0                 0                 0
               0                 0                 0  0.51500000000000
-0.21500000000000 -0.21500000000000  0.50000000000000                 0

  Columns 5 through 6
 0.51500000000000                0
               0                 0
               0                 0
 EM =

  Columns 1 through 4
               0                 0                 0                 0
               0                 0                 0 13.20512820512821
  -8.95316804407713      -8.95316804407713      22.40587695133149
0
  Columns 5 through 6

 13.20512820512821               0
               0                 0
               0                 0
```

```
PTM =
  0.01593720000000                  0                  0
                 0   0.01593720000000                  0
                 0                  0   0.01859340000000

PSM =
  0.00913655897436                  0                  0
                 0   0.00913655897436                  0
                 0                  0   0.00354059926538
```

APPENDIX D

(Process Parameters)

Standard Cleaning

Step	*Solvent*	*Temperature*	*Time*
1	Nitric acid 100%	Ambient	10 min
2	D.I. water (cascade)	Ambient	8 min
3	BHF [7:1]	Ambient	5 sec
4	D.I. water (cascade)	Ambient	8 min
5	Fuming Nitric acid 70%	115°C	10 min
6	D.I. water (cascade)	Ambient	8 min

Note: All steps are performed on the wet bench.

Dry Oxidation

Step	*Process*	*Equipment*	*Conditions*
1	Standard Cleaning	Wet bench	
2	Dry Oxidation	Furnace (N_2 and O_2)	1100°C, time dependent on target thickness

Wet Oxidation

Step	*Process*	*Equipment*	*Conditions*
1	Standard Cleaning	Wet bench	
2	Dry Oxidation	Furnace (N_2 and O_2)	1100°C, 30min
3	Wet Oxidation	Furnace (N_2 and H_2O)	1100°C, time dependent on target thickness
4	Dry Oxidation	Furnace (N_2 and O_2)	1100°C, 30min

Photolithography

Step	*Process*	*Equipment*	*Conditions*
1	Dehydration	Oven	120°C, 30 min

2	HMDS	HMDS chamber	25 min
3	Pre-bake	Hotplate	90°C, 90 sec
4	Spin-coating	Spin coater	AZ-7720 Positive photoresist 2500 rpm, 40 sec
5	Alignment & Exposure	Mask aligner	I-line Lamp intensity: 14 mW/cm^2 Exposure time: 6 sec
6	Post-bake	Hotplate	110°C, 60 sec
7	Development	Wet bench	AZ-300 MIF Developer, 60 sec
8	Inspection	Microscope	
9	Hard-bake	Hotplate	120°C, 2 min
10	Stripping	Wet bench	Acetone/ Isoproponal/D.I/ water Ambient, 5 min

LPCVD Nitride Deposition

Step	*Process*	*Equipment*	*Conditions*
1	Standard Cleaning	Wet bench	
2	LPCVD	Furnace (NH_3 and SiH_2Cl_2)	800°C, time dependent on target thickness

Plasma Etching

Step	*Process*	*Equipment*	*Conditions*
1	Nitride plasma etching (RIE)	Plasma Etcher (CF_4 and O_2, 9:1)	250 W 150 mTorr
2	Resist Ashing	Plasma Etcher (O_2)	200 W 150 mTorr

Note: Process time dependent on material and thickness to be etched.

Wet Etching

Process	*Etchant*	*Conditions*
KOH etching	KOH, 55%	60°C
BHF etching	BHF, 7:1	Ambient

HF etching	HF, 49%	Ambient
H_3PO_4 etching	H_3PO_4, 85%	155°C with reflux

Note: All steps are performed on the wet bench.
Process time dependent on material and thickness to be etched.

<u>Laser Drilling</u>

Process	*Equipment*	*Conditions*
Laser drilling	ESI® 5250 Laser µVia Drill (Nd:YAG laser)	Laser power: 1.95 W
		Laser rap. rate: 3 kHz
		Laser beam moving velocity: 1 mm/min
		No. of repetitions: 15
		Effective laser diameter: 15 µm
		Inner diameter: 950 µm
		Revolutions: 6

<u>Anodic Bonding</u>

Process	*Equipment*	*Conditions*
Anodic bonding	"Chip-scale" Anodic bonder	Temperature: 400°C
		Voltage: xxx kV

PART II

Chapter 6

DEVELOPMENT OF INTEGRATED MICROFLUIDIC DEVICES FOR GENETIC ANALYSIS

Robin H. Liu and Piotr Grodzinski
Microfluidics Laboratory, PSRL, Motorala Labs, USA

Abstract: Biotechnology, in conjunction with semiconductor and microelectronics, would have a tremendous impact on new solutions in gene and drug discovery, point-of-care systems, pharmacogenomics, environmental and food safety applications. A combination of micro-fabrication techniques and molecular biology procedures has a potential to produce powerful, miniature analytical devices aiding further development of genetic analysis. These devices, called microfluidic lab chips, are capable of processing minute volumes of genetic sample in order to pursue on-chip reactions and detect reaction outcomes. Microfluidics for biotechnology applications requires development of inexpensive, high volume fabrication techniques and reduction of biochemical assays to the chip format. In this chapter, we will discuss design, fabrication, and testing of plastic microfluidic devices for on-chip genetic sample preparation and DNA microarray detection. Plastic microfabrication methods are being used to produce components of a complete micro-system for genetic analysis. The detailed discussion on the development of micromixers, microvalves, cell capture and micro-PCR (polymerase chain reaction) devices, and biochannel hybridisation arrays is given. We also describe a path to further individual component integration.

Key words: Microfluidics, DNA analysis, plastic fabrication, biochips.

F.E.H. Tay (ed.), Microfluidics and BioMEMS Applications, 143–183.

1. INTRODUCTION

1.1 Genetic analysis

Genes are the hereditary units that regulate the functions of living organisms [1]. The entire genetic content of any organism, known as its genome, is encoded in strands of deoxyribonucleic acid (DNA). Cells carry out their normal biological functions through the genetic instructions encoded in their DNA in a process known as gene expression [2], which results in the production of proteins. Differences in living organisms result from variability in their genomes, which can be caused by differences in the sequences of genes or in the levels of gene expression. For example, diseases may be caused by a mutation of a gene that alters a protein or the gene's level of protein expression.

Genetic analysis, the study of genes and their functions, has created opportunities to fundamentally alter the fields of human biology and medicine through the discovery of new therapeutic targets, the development of novel drugs that interact with those targets and an improved ability to diagnose and manage disease [3]. Interest in understanding the relationships between genes and disease has generated a worldwide effort to identify and sequence genes of many organisms, including three billion nucleotide pairs and 100,000 genes within the human genome [4]. Due to the breakthrough of the Human Genome project, genomics is expected to fundamentally change our healthcare and improve the quality of human life. Genetic analysis will enable researchers to not only further understand the function and role of specific genes in disease, but also to tailor therapeutics based on individual genetic makeup. To deliver on this promise, a fast and inexpensive method for acquiring genetic information from clinical samples is required.

In recent years, new technologies for genetic analysis have emerged, such as automated electrophoresis [5], microarrays [6], and mass spectrometry [7]. Electrophoresis has been used in a variety of genetic analysis applications, including methods for determining gene sequences, analysing gene expression levels, identifying variations in gene sequences and relating variability in gene sequence or expression to disease states or drug response. Electrophoresis uses electric field to separate purified genetic samples (typically charged molecules such as DNA) based on their different migration rates. Microarrays are chips in which different strands of DNA have been arranged on the chip surface. The microarray chips can be used to determine whether a sample contains DNA strands with sequences, which match those capture probes on the array. Microarray chips are used in applications including differential gene expression, polymorphism screening and mutation detection. Mass spectrometry is utilized primarily in certain genetic analysis applications involving the separation of short DNA strands on the basis of size.

All techniques for analysing nucleic acids, including electrophoresis, mass spectrometry and microarrays, require a significant number of steps for preparing a sample prior to analysis. Conventional bench-top analytical systems, which utilize various manual and time-consuming laboratory processes, including centrifugation, thermocycling, filtration, measuring, mixing and dispensing, are prone to human errors that compromise the data and result in increased costs. Currently used biochemical diagnostic procedures require highly trained personnel with extensive expertise, expensive equipment in laboratory settings, and time. The development of rapid evaluation techniques permitting for early detection of infection and general health monitoring will be a tremendous asset to contemporary medical applications. The recent developments in miniaturized total chemical analysis systems (μ-TAS) and microfluidics technology have the potential to change the ways analysis in medicine, environmental sciences, and agriculture is being done [8]. The miniaturization of chemical reactions to the chip level would lead to significantly increasing analysis throughput, the reagents volume reduction, decrease of the analysis time, enhancing accuracy, reducing complexity and lowering operating costs.

1.2 Microfluidics

The research on microfluidics involves the development of miniaturized devices, miniaturized systems, and new applications related to the handling of fluids (includes liquids and gases). Over the last ten years, there has been a dramatically increasing interest in the development and realization of microfluidic systems using Micro-Electro-Mechanical System (MEMS) technique, although study on microfluidics has begun about 20 years ago when a micro gas chromatographic air analyser was made in Stanford [9] and an ink jet printing nozzle array was made in IBM [10]. The current rapid growth of the field is partly due to the rapid developments in MEMS-related fluid devices, ranging from sensors, such as pH and temperature, to fluid actuators, such as pumps, mixers, and valves [11], and also the need for complex (bio)chemical analysis system and drug delivery [12]. New generations of complete micromachined systems are now making possible advanced bioanalytical instruments with new levels of performance and capability.

1.2.1 Fluidic Theory

Microfluidic devices reduce fluid mechanics to micro-level. Fluids can be classified into two: Newtonian fluid, (e.g., air, most aqueous solutions), of which the shear stress is linearly related to the velocity, and non-Newtonian fluid (e.g., blood, molten polymers). Most fluids encountered in microfluidic applications may be treated as Newtonian for purposes of first approximation. The pressure-driven momentum governing equation for Newtonian fluid is Navier-Stokes equation [13]:

$$\rho\frac{Dv}{Dt} = -\nabla p + \mu\nabla^2 v + \rho b \tag{1}$$

where ρ is fluid density, p is applied pressure, v is velocity vector, μ is fluid viscosity and b is body force unit. The left-hand side of the equation represents "inertial forces" and the right hand side represents the forces on the fluid due to applied pressure, viscosity, and body force (such as gravity force). One of the most important scaling factors to characterize fluid properties is Reynolds number:

$$\mathrm{Re} = \frac{inertial \quad force}{viscous \quad force} = \frac{\rho\frac{Dv}{Dt}}{\mu\nabla^2 v} \approx \frac{\rho v H}{\mu} \tag{2}$$

where H is the characteristic dimension size of a micro-device, v is inlet fluid velocity. The Reynolds number gives the ratio between inertial forces and viscous forces in a flow. The definition used in a microchannel is Re = $(Q/A)Dh/\nu$, where Q is the volumetric flow rate through the channel, A is the cross-sectional area and Dh is the hydraulic diameter of the channel (4A/wetted perimeter of the channel), and ν is the kinematic viscosity of the fluid.

Microfluidic systems often operate in low Reynolds number domain (100 > Re > 0.001) due to small dimension feature of the devices. With small Reynolds number, fluid often flows as laminar and turbulence is not likely to occur. Viscous force dominates and inertial force is negligible.

There are many advantages when a device goes from macroscale to microscale, including the promise of performing some analytical functions more rapidly and accurately, and handling smaller samples. However, scaling theory shows that miniaturization of a device is not a simple matter of reducing its size. Some effects scale well and provide improved performance while others are not as efficient as they are in macroscopic level. For example, the surface-to-volume ratio of devices increases at small scales, rendering the surface force more favourable than volume (pressure) forces. High surface-to-volume ratio also gives rise to fast chemical reactions because the reagents can practically act on the entire small sample at once. At the same time, however, surface adsorption of biomolecules will be significant, easily resulting in clogging and fouling of biomolecular fluid in microsystems.

1.2.2 Lab Chips

Microfluidic chips (also called "lab chips"), which contain interconnected fluidic microchannel networks, reaction chambers and valves, can carry out conventional biochemical measurement with increased speed and reliability at reduced cost. They hold the potential to the on-chip realization of micro total analysis system (μ–TAS). Liquid phase separation techniques were among the first to be miniaturized onto microchip. Capillary electrophoresis, the key to many biological assays, was the first analytical separation technique demonstrated on microchip [14]. Using this technique, enzyme assays [15], restriction enzyme mapping of DNA [16] and DNA sequencing [5] have all been accomplished in integrated microfluidic systems. Parallel processing (96 sample at a time) has been built into the device platform for capillary array DNA sequencing applications [17]. Thermal cell lysing have also been accomplished on chip [18]. PCR amplification based on single well thermal cycling [19] and continuous flow through defined temperature zones [20] have been demonstrated in silicon devices. The successful miniaturization of the individual analytical tools paved the way to subsequent realization of integrated micro-devices with multiple functional units. PCR amplification reactors fabricated on silicon were coupled with glass capillary electrophoresis chips and a rapid assay for genomic Salmonella DNA was performed in less than 45 minutes [21]. The process of cell lysing, multiplex PCR and electrophoretic analysis was integrated on a single glass chip [18].

A true lab chip needs to perform all analytical functions including sample preparation, mixing steps, chemical reactions, separation, and detection in an integrated microfluidic circuit. Of all the above functional units, only sample preparation has not been realized on the chip level. Most of the currently demonstrated microfluidic device components pursue single functionality and use purified DNA as an input sample. Existing microfluidic technologies work very well for highly predictable and homogeneous samples common in genetic testing and drug discovery processes. One of the biggest challenges for current lab chips, however, is to perform analysis in samples as complex and heterogeneous as whole blood or contaminated environmental fluids.

Sample preparation represents the most time-consuming and labor-intensive procedure in the whole DNA analysis. The back-end micro-device developments (such as DNA microarray biochips) have shifted the bottleneck, impeding further progress in rapid analysis devices to front-end sample preparation where the 'real' samples of bodily fluids are used. Eventually, sample preparation will need to be brought to the chip level in order to achieve a high level of automation.

In the following sections, we will discuss the approach to the development of an integrated sample preparation plastic device and focus on the description of its relevant components, such as micromixers, microvalves, cell capture and micro-PCR (polymerase chain reaction) devices. A novel detection device, biochannel arrays, resulted from a combination of DNA microarray and microfluidics technologies, will also be presented.

2. PLASTIC MICROFABRICATION

Fabrications of microfluidic devices have been reported using glass [18], silicon-glass [21] and silicon-plastic [22] techniques. An alternative approach in microfluidic systems is to use plastics as device substrates due to their low cost, versatility of material properties, and ease of batch fabrication through moulding [23], embossing [24], and casting processes [25]. Injection moulding relies on the use of mould inserts, which are being filled with melted plastic material [23]. Hot embossing also employs a mould insert that is pressed into the plastic material with the plastic material held above its glass transition temperature [24]. Casting relies on liquid plastic material, which is poured onto the master mould and allowed to polymerize. In some applications, it is desirable to integrate functional components (e.g. heaters, coolers, and detectors) into the device structure. We have developed a casting technique allowing for simultaneous incorporation of functional elements into the device structures during the fabrication process [25]. Unlike other plastic replication techniques, casting process can be carried out under near atmospheric pressure at relatively low temperature ranges and such integration scheme can be achieved. However, due to the time required for polymerisation, hot embossing and injection moulding techniques are capable of higher throughput, high precision, and low cost. Most micro-replication processes, such as hot embossing and injection moulding, consist of stamper fabrication, micromoulding, and plastic bonding (Fig. 1).

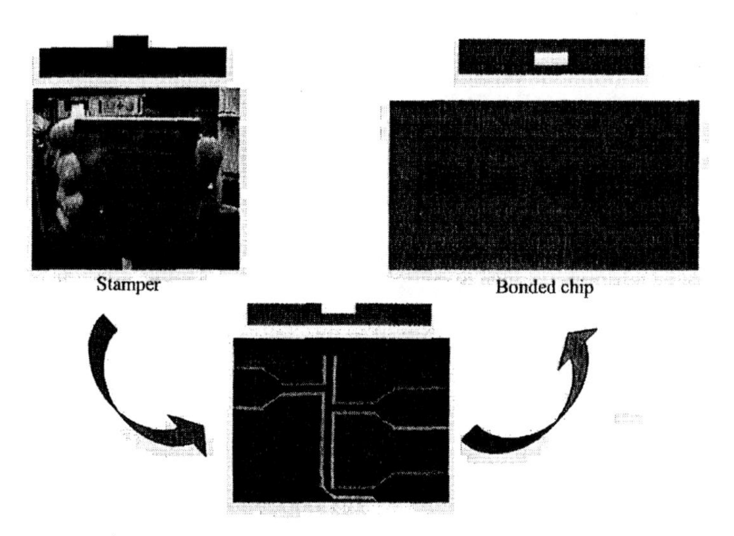

Figure 1. Process of plastic microfabrication. Microchannel patterns are replicated from Ni or Si stamper to a moulded polycarbonate (PC) layer, and finally bonded to form an enclosed microstructure

2.1 Stamper Fabrication

A variety of methods have been developed to form stampers. Photolithography of SU-8 photoresist or etch of silicon or glass to produce masters followed by a nickel electroplating remains the most common. Figure 2 shows SU-8 patterns and nickel electroplated in a SU-8 master mould. For rapid prototyping, KOH-etched Si has also been used directly as a stamper when number of produced devices in a given design is not high (<25). The positive angle of the KOH-etched Si channel sidewall (Fig. 3) provides ease of releasing the plastic device from the stamper during the de-moulding process. However, for larger volume fabrication, nickel stampers (Fig. 2(b)) through overplating process of SU-8 structures are preferred.

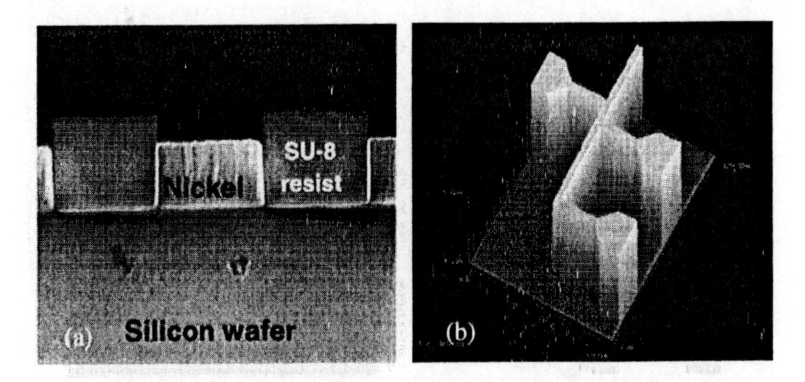

Figure 2. Formation of a Ni stamper. Left: Ni was electroplated in a SU-8 master structure followed by removing the SU-8 to achieve a complete stamper; Right: 3D image of the Ni stamper structure

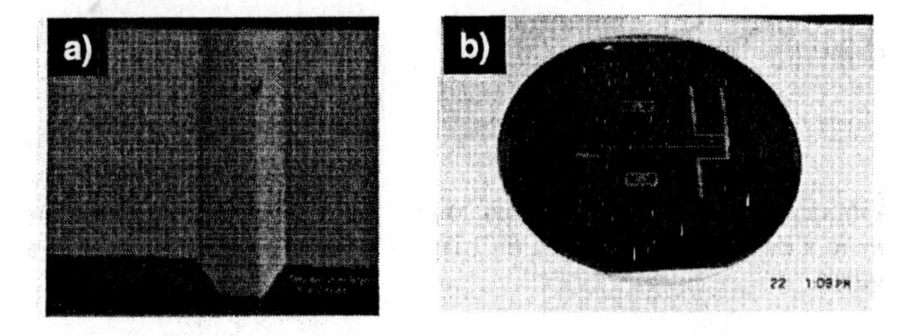

Figure 3. a) wet etched silicon stamper, b) Ni-plated stamper

2.2 Micromoulding

Following stamper fabrication, compression moulding or hot embossing can be used to produce micro features in plastic chips using the stamper. For compression moulding, plastic pellets are placed on the stamper and subsequently melted by heating up above their melting point. A ~ 5 ton compression force is then applied on the molten plastic against the stamper in a press machine. The compression force should be increased from 0 to 5 ton gradually so that the air bubbles trapped inside the molten polymer have time to migrate out. We developed a rapid compression moulding process with a cycle time as short as 3 min, which includes mould set-up, material melting, compression, and de-moulding.

For hot embossing, the stamper and the polymer substrate (a blank polymer layer) are heated in vacuum to a temperature just above the glass transition temperature *Tg* of the polymer material. The master stamper is then pressed onto the stamper with a highly controlled force, typically on the order of several kN for 60-90 seconds. While still maintaining the embossing force, the polymer is then cooled to below *Tg* to keep the thermally induced stress to a minimum. After reaching the desired low temperature, the de-embossing force is applied and the polymer substrate is separated from the master stamper. The complete cycle time from loading to unloading of a substrate can be as little as 10-15 minutes.

2.3 Plastic bonding

The bonding of the moulded plastic parts is the final fabrication step leading to the formation of a fully functional plastic micro device. The process requires a hermetic seal over large substrate areas with the ability to preserve integrity of small void structures at the same time. Conventional thermal bonding and film lamination processes can be used. During the thermal bonding process, the moulded plastic structure and a sealing plastic substrate (typically with the same material as the moulded structure) are sandwiched and then placed inside a metal bonding cartridge with vacuum and cooling capability. Once the assembly is loaded into the bonding cartridge, a vacuum of $4x10^{-2}$ Torr is applied to the bonding chamber. The loaded bonding cartridge is then placed in a hydraulic press machine, and thermal bonding is carried out at 5 °C above *Tg* of the polymer material for 10 minutes at a pressure of 4 metric tons. After bonding, the cartridge is cooled to 80 °C for 10 minutes before the bonded structure is taken out of the cartridge.

Due to the need of maintaining the integrity of biomolecules, low temperature assembly methods have also been developed, such as using an adhesive tape to enclose moulded plastic structures [26]. For many shallow structures (e.g., a microchannel with 500 μm width and 40 μm depth), the thin tape can easily sag into the channel, and in some cases adhere to the bottom of the channel (Fig. 4(a)). To address this problem, a hard plastic or glass substrate is used to support a double-sided adhesive tape from the back channel (Fig. 4(b)).

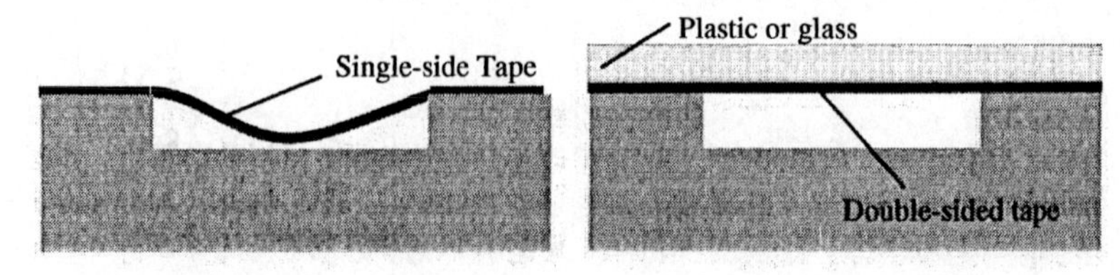

Figure 4. Schematics of a channel (a) sealed with a single-side adhesive tape; (b) bonded to a hard plastic or glass layer with a double-side tape

2.4 Fabrication of Structures with Complex Geometry

The plastic replication technique can be used to construct structures with complex geometries, such as 3D micro-dome structures inside a microchannel for DNA microarray application. To realize 3D micro-dome structures, the Si stamper is fabricated using an isotropic wet etching process. A mixture of hydrofluoric acid (HF), nitric acid (HNO_3), and acetic acid (CH_3COOH) in a ratio of 1:3:8, also referred as "HNA", is used as the etchant. One important issue of this isotropic etching process associated with the Si etching rate is the dissolution of the reaction products into the solution. If the reaction products can be transported quickly into the solution and the fresh etchant solution can be replenished and move into the etching area rapidly, the Si etching rate is high. Otherwise, the etching rate can be very slow. We utilize this mechanism to achieve different etch rates at different locations. The areas between channels (which are ridge structures as shown in Fig. 5) are larger than the areas of dome arrays (which are pit structures here). The solution can easily move in and out of the channel areas as compared to the smaller pit areas. As a result, the Si in the areas between channels is etched twice as fast as Si in the pit areas. The resulting Si stamper has 40 μm high ridges and 20 μm deep pits (Fig. 6). Following compression moulding process using the Si stamper, the resulting channel structure with micro-dome arrays is shown in Fig. 7. The channel is 40 μm deep, while the domes are 20 μm high. The 3D micro-dome structure increases the surface-to-volume ratio of the DNA capture probes if DNA oligo is dispensed on the micro-domes. This can be useful in DNA microarray biochips.

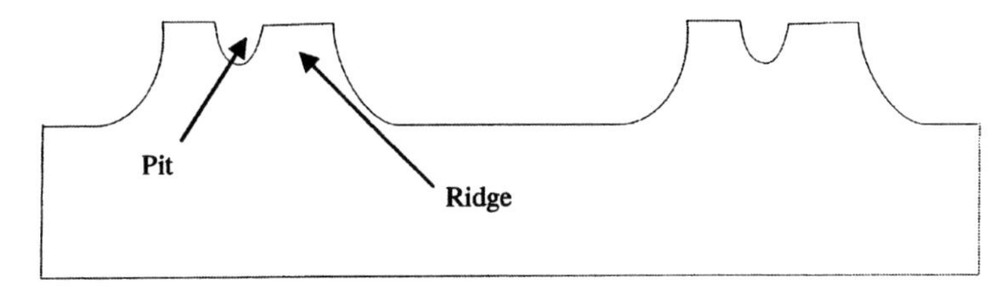

Figure 5. Schematic of the isotropic etched Si structure. Note that the ridge structures and the pits will be transformed into microchannels and micro domes in the plastic chip after the plastic compression moulding

Figure 6. SEM of the isotropic etched Si stamper (note that only the pit structures are shown here)

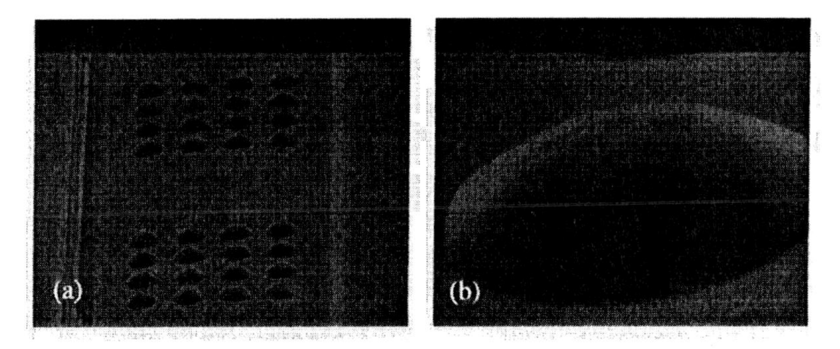

Figure 7. (a) SEM of the micro-dome structures inside a plastic channel. (b) Close view of a micro-dome structure

2.5 Metalization on Plastic

In many cases, metal thin films need to be incorporated with plastic structures to introduce extra functionalities (e.g., heating and temperature sensing). The electrical components are difficult to implement on plastic using standard silicon technology that relies on traditional photoresist processes. Traditional photoresist processes involve wet chemical processes that may degrade or attack chemical integrity of plastic materials. The use of a shadow mask (Fig. 8) to define the metal patterns allows for the elimination of the photoresist process. The shadow mask is made using standard silicon processes in silicon wafer. The placement of the silicon wafer with appropriate features over plastic substrate and subsequent exposure to the sputtered metal results in the formation of metal lines on plastic substrate. This technique gives better resolution and thickness control than other techniques, such as screen printing [27].

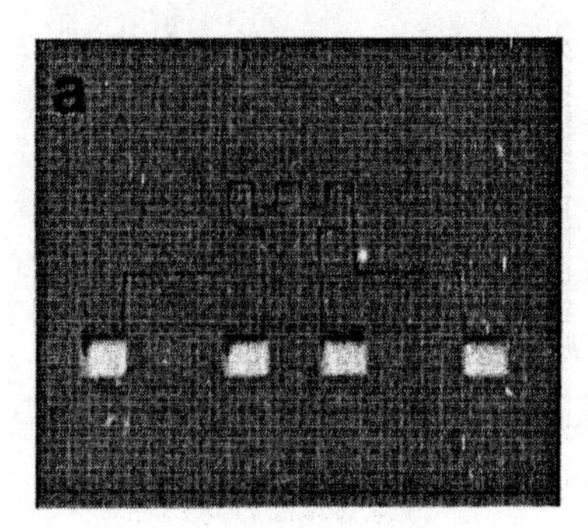

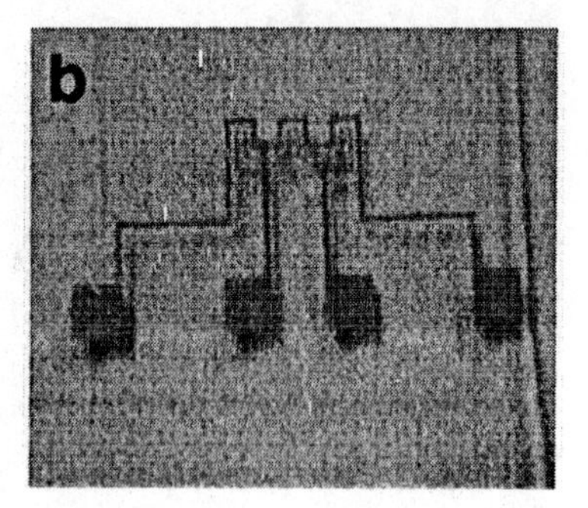

Figure 8. a) Shadow mask etched in silicon, b) metal features deposited on polycarbonate using shadow mask process

3. INTEGRATED SAMPLE PREPARATION DEVICES

A complete genetic analysis system should be capable of analysing a sample from bodily fluid (blood, urine, saliva), extracting DNA from concentrated cells, purifying and amplifying DNA, and finally detecting DNA fragments of interest. In the discussion below, we will present examples of fabricated device components and show the path to their integration.

3.1 Micromixer

Many biological processes such as antibody antigen binding and DNA hybridisation require rapid and uniform mixing. Since turbulence is not practically attainable in micro-scale channel flows or mini systems with small Reynolds numbers, mixing in microfluidic systems is typically dominated by diffusion. Unfortunately, a pure diffusion-based mixing process can be very inefficient and often takes a long time. Mixing is particularly inefficient in solutions containing macromolecules that have diffusion coefficients one or two orders of magnitude lower than that of most liquids [28]. Thus, some mechanisms must be used to enhance on-chip micromixing. Effective mixing at this scale requires that fluids be manipulated to increase the interfacial surface area between initially distinct fluid regions so that diffusion can complete the mixing process in a shorter time.

The literature contains a number of devices designed to enhance mixing on the microscale. These devices fall into one of two categories: (1) active mixers [29], [30], [31], that exert some form of active control over the flow field through such means as moving parts or varying pressure gradients, or (2) passive mixers [32], [33], [34], that utilize no energy input except the mechanism (pressure head or pump) used to drive the fluid flow at a constant rate. One example of a passive micromixer is a multi-stage multi-layer laminarization scheme developed by Branebjerg *et al.* [32]. The mixer was designed to have two flow streams guided on top of each other and kept separated by a plate until they are forced to laminate (divide and mix). The fast mixing was a result of the increased contact area and decreased diffusion length when the two liquids were stacked. One disadvantage of using such a mixer in biomolecular analysis systems is the possibility of causing potential damage to biomolecules and denaturing biomolecules (e.g., large globular proteins). Such a micro-streaming scheme will force streams through smaller passages and orifices, and possibly create very high instantaneous rates-of-strain (e.g., shear and elongation) on biological macromolecules. High rate-of-strain fields can damage proteins and particles such as cells [35]. Moreover, increased surface-to-volume ratio of this mixer can lead to clogging and fouling caused by biomolecular adsorption onto the device surface. Examples of active micromixers include those of Moroney *et al.* [31] and Evans *et al.* [30]. The former used ultrasonic travelling waves generated by a piezoelectric film and applied them to the liquid in a mixing chamber. The latter used bubble pumps by boiling at micromachined polysilicon and aluminium trace heaters, to agitate liquid and create chaotic advection. Both mixers involve use of external energy (other than hydraulic).

The complexity of most active mixing schemes makes them less attractive because they are more difficult to build, operate, and clean.

An ideal micromixer should be easy to make and integrate with other microfluidic components, should demonstrate good mixing enhancement with respect to mixing rates and quality (e.g., uniformity), and should have low rate-of-strain. In the following sections, we will discuss two micromixing approaches that meet the above criteria.

3.1.1 In-line Chaotic Mixer

A passive and highly efficient in-line "L-shaped" 3D serpentine micromixer based upon the principle of chaotic advection was recently developed [36]. Chaotic advection is a laminar stirring phenomena, in which complicated channel geometry can generate complex flow field as a result of the combination of secondary flow and separation flow instability [37]. When considering the quality and efficiency of mixing, of fundamental importance is the motion of particles within the flow domain. If a particle under consideration is so light and passive that it moves exactly with the fluid around it, the equations describing its motion are simply the advection equations: $dx/dt = V(x,t)$, where x is the particle position and V is the (given) fluid velocity field. Even for relatively simple flow fields, e.g., low Reynolds number steady flows, it is expected that solutions of the above equation will be quite complicated when the velocity field is three-dimensional. In fact, the motions of such particles can be chaotic [37]; this phenomenon is referred to as *chaotic advection.* Chaotic advection has been verified in several flows ranging from two-dimensional unsteady flow (e.g., blinking vortex flow [38]) to steady, three-dimensional flow (e.g., flows in an alternating helical coil, [39]). A chaotic advection flow field design can insure that most neighbouring paths will, in a short time, separate exponentially from one another and yield chaotic, rapid mixing.

In order to achieve chaotic advection, the geometry of a mixing channel must be 'complicated enough'. We designed a 3D "L-shaped" serpentine microchannel (Fig. 9). The channel guides the fluid streams along in-plane L-shaped geometric curves, rotates the fluid by 90°, and then guides the fluid along another in-plane L-shaped geometric curve. The planes of two successive curved segments are aligned perpendicular to each other, in order to achieve maximum stirring efficiency [40].

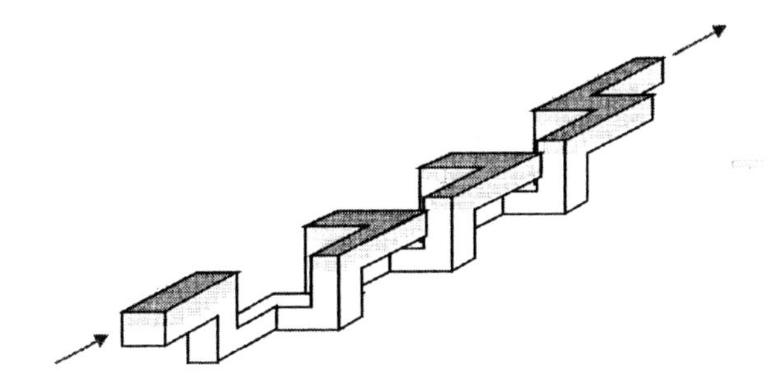

Figure 9. Schematic of a 3D L-shaped serpentine micromixer channel

The device was made by double-side KOH-etching Si or plastic/PDMS microfabrication. As shown in Fig. 10, the Si channel cross-sections are all 300 μm wide at the surface of the wafer and 150 μm deep. After the etching process was complete, the wafer was sealed by a glass cover slip using a 10 μm layer of silicone adhesive. Another mixer with similar geometry is made by PDMS microfabrication (micromoulding individual layers followed by stacking and bonding process) as shown in Fig. 11.

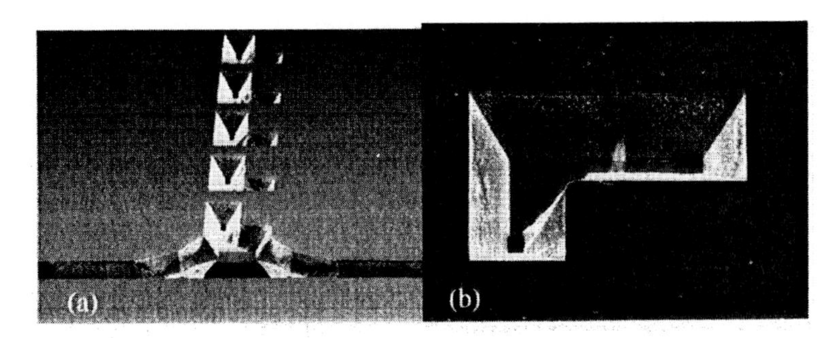

Figure 10. SEMs of a 3D "L-shaped" serpentine KOH-etched microchannel. (a) Overview of the channel including the T-junction for inlet streams. (b) A "L-shaped" curved segment (the channel is 300 μm wide on the top surface and 150 μm deep, and the orifice is 50 x 50 μm)

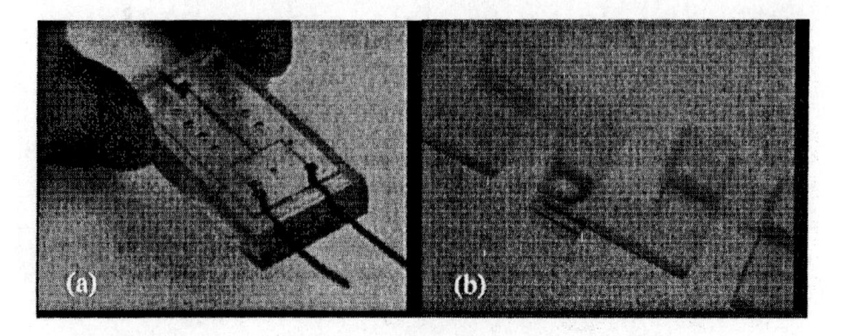

Figure 11. Photographs of a 3D "L-shaped" serpentine PDMS microchannel. (a) Overview of the channel with inlet dyed streams. (b) Close view of "L-shaped" curved segments

Flow experiments using phenolphthalein/ethyl alcohol solution in one stream and sodium hydroxide/ethyl alcohol solution in the other stream were performed in the mixer. After these two colourless streams came in contact, a chemical reaction took place resulting a red dye. Flow cross section images were obtained by focusing microscope on the vertical segments of the channel (Fig. 12). As the flow passes through the mixer cycles, the interface between the two streams becomes more and more complicated (stretched and elongated), and the streams intertwine further increasing the interface giving diffusion a larger area to thus increase the overall mixing rate.

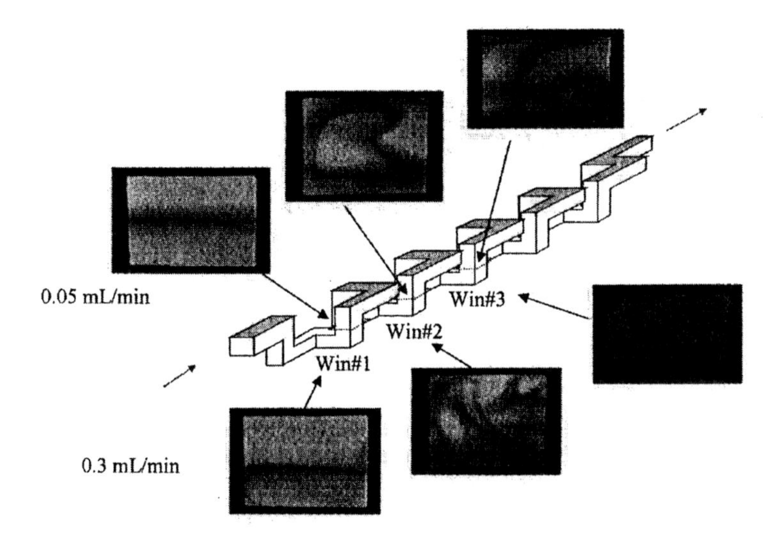

Figure 12. Photographs of reacted phenolphthalein in the mixer at different channel locations (windows) at different flowrates

Normalized average intensity values from each window (Fig. 12) in the mixer are shown in Fig. 13 for four different Re. The intensity at window #1 indicates the amount of mixing that has taken place before the streams enter the first mixing segment. The longer the streams stay in this section (i.e. the lower the Re) the more mixed they become. The mixing rate of the L-shaped mixer, which is given by the slope of the intensity curve, increases with increasing Reynolds number. An exponentially increased mixing rate was observed at Re = 36 in the L-shaped channel, suggesting that chaotic advection occurs in the mixer at Re > 36. As shown in Fig. 14, after the streams enter the L-shaped mixer, mixing at Re = 70 is essentially complete in approximately 5 ms. The use of this in-line chaotic mixer to enhance cell capture by mixing the sample solution and magnetic capture beads is discussed in Section 3.3.

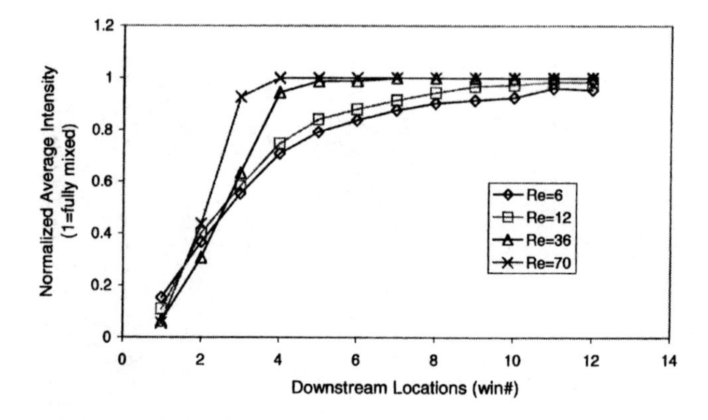

Figure 13. Plot of percentage mixing versus downstream location in the 3D "L-shaped" serpentine mixer. See Fig. 12 for a visual explanation of the downstream locations

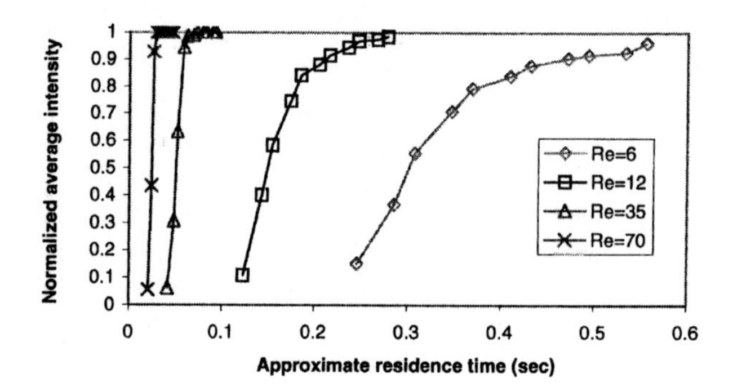

Figure 14. Normalized average intensity in each window of the 3D "L-shaped" serpentine channel as a function of residence time

When Re is larger than one, fluid inertia is still important. The chaotic mixer utilizes fluid inertia to set up the secondary flow. It is well known that in a curved pipe segment fluid inertia leads to the formation of two longitudinal vortices of opposite sign [39]. The advection of a passive scalar in steady laminar flow through a sequence of channel mixing segments leads to enhanced transverse and longitudinal stirring. Thus it is reasonable to anticipate the occurrence of enhanced mixing when a fluid is pushed at low Re through a microchannel with a contorted geometry. Conventional ideas of secondary flow instability can be applied as unsteady mixing forces to enhance mixing, despite steady state flow rates through the system. The smaller scale motions of the flow, produced via instabilities, are the primary stirrers. As Re is enhanced (the flow rate is increased), more and more such instabilities are likely to occur, and one expects mixing efficiency to go up with increased flow rate despite decreased flow residence time in the channel.

3.1.2 Acoustic Microstreaming Mixer

In-line mixers as discussed in Section 3.1.1 are useful for fluid-to-fluid mixing. However, in many applications such as hybridisation, mixing enhancement is required in a chamber for mixing between the bio-species in the solution and the chamber surface. For example, in DNA microarray hybridisation, of fundamental importance is mixing and binding between the target DNA in the sample solution and the DNA capture probes immobilized on the chip surface. To enhance such a mixing, we have developed another mixing technique based on bubble-induced acoustic microstreaming principle [41]. A piezoelectric (PZT) disk is attached to the hybridisation/detection chamber, which is designed in such a way that a desired number of air bubbles with desirable size are trapped in the solution. The air bubble in a liquid medium acts as an actuator (i.e., the bubble surface behaves like a vibrating membrane) when the bubble undergoes expansion and contraction within a sound field. The behaviour of bubbles in sound fields is determined largely by their resonance characteristics. For frequencies in the range considered here (~ kHz) the radius of a bubble resonant at frequency f (Hz) is given by the equation:

$$2\pi af = \sqrt{3\gamma P_o / \rho} \tag{3}$$

where a is the bubble radius (cm), γ is the ratio of specific heats for the gas, Po is the hydrostatic pressure (dynes/cm2) and ρ is the density of the liquid (g/cc). Using this equation with the parameter values f = 5000 Hz, γ = 1.4, Po = 106 dynes/cm2 and ρ = 1.0 g/cc one finds the radius a for resonance to be 0.65 mm. When the bubble undergoes volume change within a sound field, the frictional forces between the boundary and the liquid medium result in a bulk liquid circulation flow around the air bubble, called cavitation microstreaming or acoustic microstreaming [41]. This liquid circulation flow can be used to effectively enhance mixing. The most effective mixing enhancement is provided by particular excitation frequencies generated by a desired number of air bubbles having a size selected in accordance with the resonant frequency of the PZT transducer.

Precisely controlling the bubble size is critical for achieving repeatable and consistent mixing enhancement effect. We used microfabrication technique to machine a number of small air pockets in a polycarbonate chamber. Since polycarbonate is a hydrophobic material, air bubbles can be easily trapped in the pockets when the chamber is filled with liquid solutions. The dimension of the pockets defines the size of bubbles. Fluidic dye experiments were performed to investigate bubble-induced acoustic mixing in such a chamber. As shown in Fig. 15, a PZT disk was glued on the top surface of a polypropylene layer, which was pre-drilled with a number of air pockets (0.5 mm diameter and 0.5 mm depth) on the opposite side. These pockets were inside a reaction chamber (100 µL), facing the detection probes on a pre-treated glass slide. A double-side adhesive tape was used to bond the polypropylene layer with the glass slide and serves as a spacing gasket to define the shape and dimension of the chamber. The PZT was driven at 5 kHz and peak-to-peak 5 V. Fluidic experiment (Fig. 16) shows that rapid mixing can be achieved across the whole chamber within 1 min and 45 sec, while the mixing based on pure diffusion (i.e., without acoustic mixing) takes about two hours for the same chamber. During the experiment, we visualized that air bubbles resting on a solid surface and set into vibration by the sound field generated steady circulatory flows, resulting convection flows and thus rapid mixing. To further evaluate acoustic mixing, DNA hybridisation experiments were performed (see Section 3.5.2 for details).

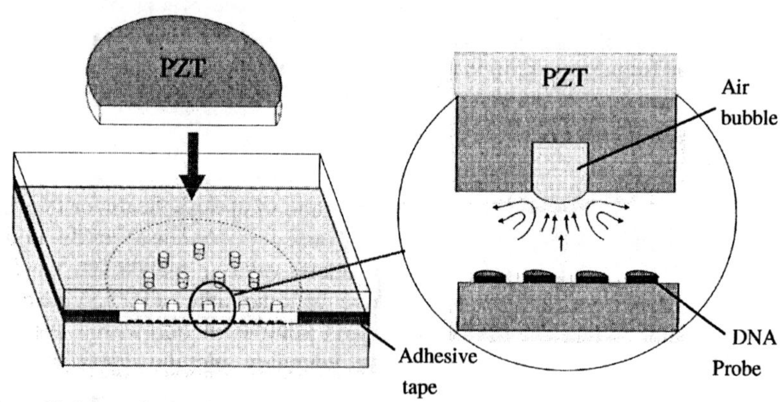

Figure 15. Schematic showing a number of air pockets in the top layer of the biochip chamber with DNA probes. (a) Overview; (b) side view

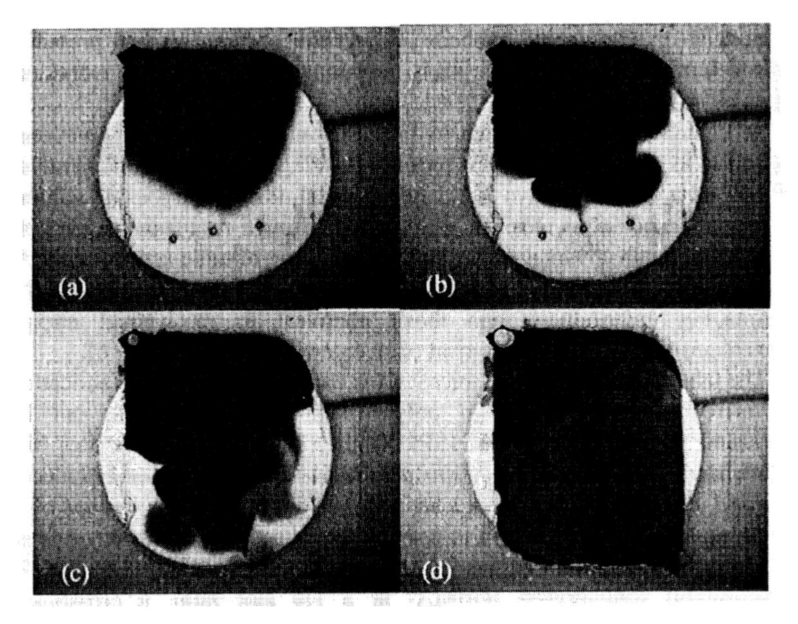

Figure 16. Snapshots showing multi-bubble induced (9 top bubbles) acoustic mixing in a chamber at (a) time 0; (b) 28 sec; (c) 1 min 7 sec; (d) 1 min 46 sec

3.2 Microvalves

Microvalve is an essential component in microfluidic device/system. Most conventional active microvalves couple a flexible diaphragm to a thermopneumatic, piezoelectric, electrostatic, electromagnetic, bimetallic, or other actuators [42]. The scaling of these actuation forces to the micro scale is often unfavorable [43]. Some actuator components are made heavy and large (mm size) in order to produce enough force to regulate a microscale fluid. The displacement of conventional diaphragms (e.g., silicon, silicon nitride) is typically limited to tens of microns or less. Fabrication of many active microvalves involves bulk processes (e.g., bulk etching of silicon) and surface processes (e.g., thin film process) [42]. Disparate materials (e.g., thin film resistive heaters in thermopneumatic valves, piezo material in piezoelectric valves, TiNi alloy in shape memory alloy valves, etc.) are often used in addition to the substrate material (e.g., Si). Actuators that use resistive heating (thermopneumatic), high voltage (electrostatic), and magnets (electromagnetic) often require high power consumption. Not only is the fabrication process of such microvalves composed of electronic components complicated, but the integration into complex microfluidic systems has also proven to be non-trivial using traditional approaches, such as modular or lithographic methods [44]. In this section, we will present a hydrogel microvalve that has many advantages over most conventional microvalves.

An unconventional approach to fabricate microvalves using functional hydrogels as actuators has been recently developed [45]. The microvalves consist of a pH-responsive hydrogel material that undergoes a volume change in response to changes in local pH. Stimuli-responsive hydrogels have a significant advantage over conventional microfluidic actuators due to their ability to undergo abrupt volume changes in response to the surrounding environment (via direct chemical to mechanical energy conversion) without the requirement of an external power source [46].

The fabrication of hydrogel microvalves consists of two steps: fabrication of microchannels and *in situ* polymerisation. Microchannels can be fabricated using thick SU-8 (EPON) or PDMS process. After the microchannel fabrication, an *in situ* polymerisation method that introduces hydrogel components into microchannels via direct photo patterning of a liquid phase precursor mixture is used to create actuators for microvalves. The precursor mixture consists of monomers (acrylic acid (AA) and 2-hydroxyethyl methacrylate (HEMA) in a 1:4 mol ratio, a crosslinker (ethylene glycol dimethacrylate) and a photoinitiator . Hydrogel components were fabricated by irradiating the precursor mixture through a photomask (as shown in Fig. 17).

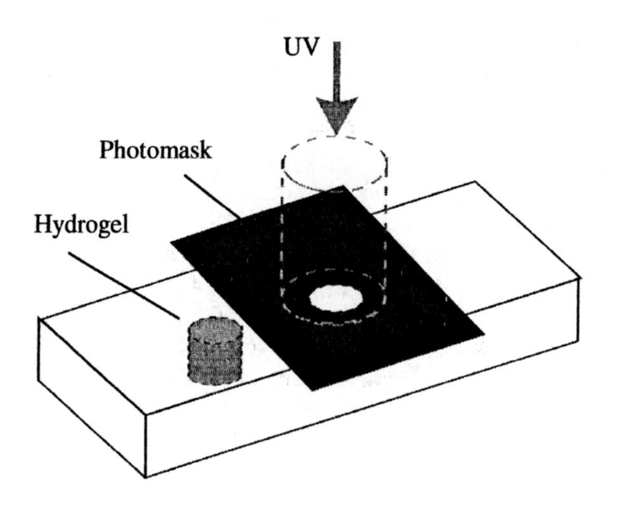

Figure 17. Schematic of the "mask mode" photopolymerization process

Using this *in situ* polymerisation technique, a hydrogel post valve was developed. In order to fabricate stable hydrogel actuators with fast response times, a hydrogel jacket (50 μm thick) was polymerised around a prefabricated circular EPON post (50 μm diameter) in the microchannel using the "mask mode" photo-polymerisation as shown in Fig. 18. When a pH 11 solution flowed into a side branch at a flow rate of 0.01 ml/min as shown in Fig. 18(b), the hydrogel jacket expanded and closed the regulated channel (pH 7). If a pH 2 solution flowed into the other side branch at a flow rate of 0.01 ml/min instead as shown in Fig 18(c), the contracted hydrogel allowed the fluid in the regulated channel to flow down the waste channel. The pressure drop measurement of the array valve indicates that the response time to completely close the regulated channel is 12 seconds. The maximum differential pressure the valve can sustain is 43.5 psi when the valve is closed. At least 200 operating cycles were achieved over a pH range of 2 to 11 without failure.

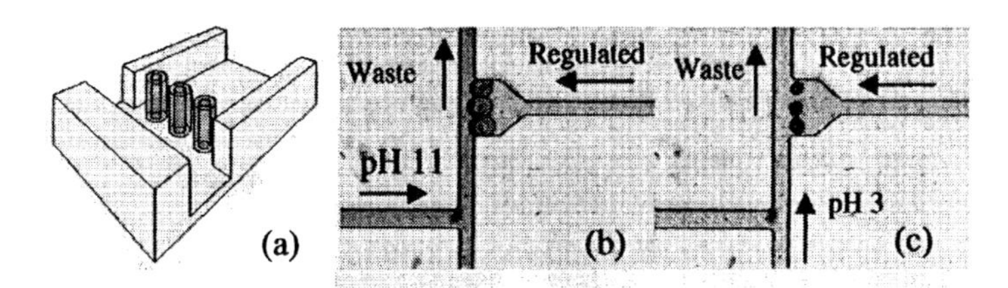

Figure 18. (a) Schematic of a 2D shut-off microvalve consisting of hydrogel "jackets" (50 μm thick) around three prefabricated EPON posts. (b) Micrograph of the hydrogel jackets blocking the regulated channel (pH 7) in their expanded state in a pH 11 solution (dyed). (c) Micrograph showing the contracted hydrogels allowing fluid to flow down the side branch

For the above 2-D shut-off microvalve, the sensing pH solution is directly or indirectly (through the porous hydrogel) in contact with the regulated solution, resulting in a cross talk problem between the two fluids. In order to address this cross talk problem, a 3D hybrid microvalve that couples a flexible PDMS membrane to a hydrogel actuator was designed

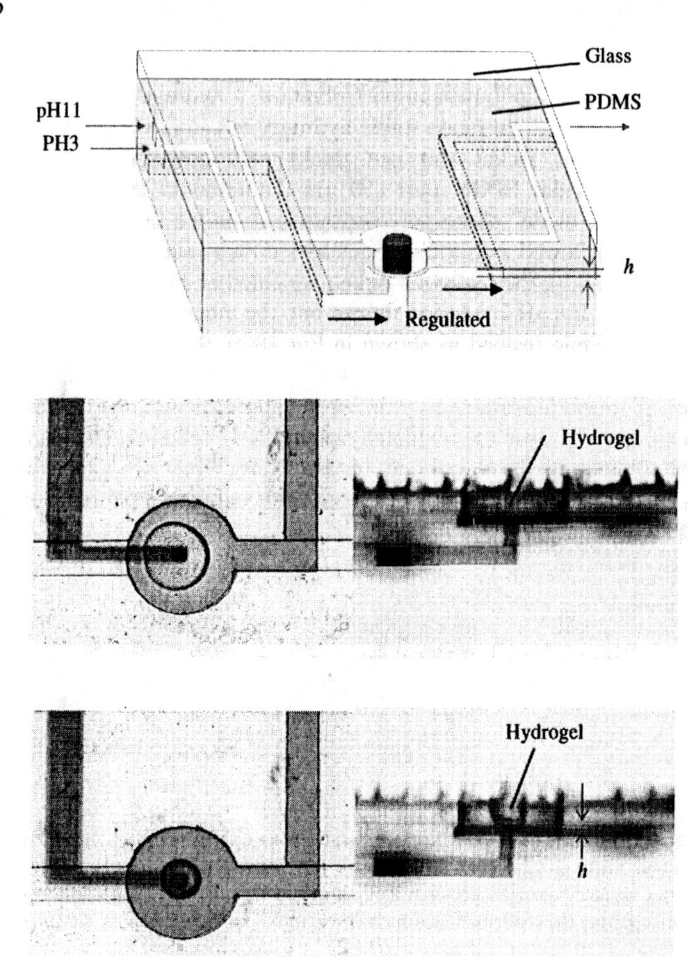

Figure 19. (a) Schematic of the 3D hybrid microvalve. (b), (c) The top view and the cross section, respectively, of the valve when a pH 11 solution is flowing in the upper channel at a flow rate of 0.01 ml/min. The hydrogel actuator expands and deforms the PDMS membrane to seal the orifice in the regulated channel. (d), (e) show that the hydrogel contracts at a pH=3 buffer (with a flow rate of 0.01 ml/min) and reopens the regulated channel

(Fig. 19(a)) and fabricated (Fig. 19(b-e)). The membrane acts as a physical barrier between the sensing and regulated flows. The hydrogel actuator can sense the pH in the upper channel and expands or contracts isotropically as the surrounding pH is changed. The force associated with these volumetric changes is sufficient to deform the membrane and consequently control the flow in the lower channel. The membrane can deform to completely block the orifice with a displacement of up to 185 μm.

No leakage was observed. A maximum differential pressure P_{max} was measured to be 26.7 psi with a height *h* of 75 μm. Simple fabrication, no power consumption and no integrated electronics make hydrogel microvalves attractive in many microfluidic applications.

3.3 Cell Capture Device

The separation and isolation of specific cells from the initial sample mixture is considered the first step in genetic sample preparation and can be performed using a number of different techniques [47]. Among all, immuno-magnetic method using magnetic beads that are coated with specific antibody to bind with target cells is commonly used. To ensure maximum capture efficiency, fully mixing between the magnetic capture beads and the sample solution containing target cells is required. The chaotic mixer discussed in Section 3.1.1 was used to achieve homogeneous mixing before the cell capture step. In the experiments, bacteria cell (*Esherichia coli* K-12) capture by mixing a blood/bacteria sample solution with magnetic capture beads was performed in the "L-shaped" micromixer with five mixing cycles. Two streams, one with bacteria and the other with colloid magnetic bead labels were mixed using this mixer. The bacteria stream consists of *Esherichia coli* K-12, DH5 alpha in PBS/5mM EDTA that was labelled with biotinylated *E. coli* antiserum. This solution was pumped through one inlet using a syringe pump. The total volume of the bacteria solution pumped through the mixer is 1 ml. Simultaneously, streptavidin coated microbeads (Miltenyi Biotech) in PBS/EDTA were pumped through the other inlet. Two flowrates were used: 0.1 ml/min and 0.2 ml/min. The outlet was directed into a miniMACS separation column (Miltenyi Biotech) where magnetically labelled bacteria were captured and subsequently plated onto LB agar. Capture rates were compared to those using a conventional rotary mixer with and without initial vortexing. Results shown in Table 1.1 indicate that the chaotic mixer is highly efficient and comparable to conventional bench-top mixing despite its much short mixing time. The result also showed that high flow rate results in more mixing (thus higher capture rate) than low flow rate, which is consistent with the dye intensity measurement results.

Table 1. *E.coli* capture experiments by mixing bacteria solution with colloid magnetic beads, followed by column separation and plating (counting)

	Micromixer ml/min (0.3 sec)	Micromixer ml/min (0.15 sec)	Rotary mixer (25 min)	Rotary mixer with 5 sec vortex (25 min)
Cell capture	69%	99%	93%	99%

	Micromixer ml/min (0.3 sec)	Micromixer ml/min (0.15 sec)	Rotary mixer (25 min)	Rotary mixer with 5 sec vortex (25 min)
rate				

The bacteria capture experiments also demonstrated that at least 99% of *E. coli* (~ 1 μm size) survived after the mixing process (note that they flowed through a number of 50 x 50 μm orifices in the Si micromixer (see Fig. 10(b)), indicating that the shear strain rate inside the mixer is low. Another experiment was performed to test for possible shear strain field that could lyse blood cells in the "L-shaped" Si micromixer. Blood diluted 1:10 in PBS/EDTA was pumped through the micromixer at a flow rate of 0.1 ml/min or 0.2 ml/min. The absorbance of the sample at 414 nm was compared before and after mixing with no significant difference for either flow rate. This shows no detectable increase in free hemoglobin, an indication that red blood cells stayed intact through the mixing process. During all the experiments, no clogging of the mixer due to flowing large particles (i.e., bacteria and blood cells) was observed. It is because of the high flow rates and relatively low surface-to-volume ratio of the device that the chance for blood cells or bacteria to stick on the channel surface is minimized. The experiments show the "L-shaped" micromixer not only gives high cell capture efficiency but also has low shear strain field.

Following the mixing and incubation, the magnetically labeled bacteria cells were separated and isolated from the initial sample mixture using an on-chip capture channel. The channel device (Fig. 20) was fabricated by compression moulding polycarbonate against a Si stamper. The channel is 3.3 mm wide, 300 μm deep and 20 mm long, giving a total volume of 20 μL. The channel was enclosed using a biocompatible pressure sensitive adhesive tape (Adhesive Research Co., IN). Since the channel structures are relatively shallow, the thin cover tape can easily sag into the channel and in some cases adhere to the bottom of the channel. To prevent this problem, a number of plastic posters (300 μm dia.) were also constructed inside the channel (Fig. 20(b)).

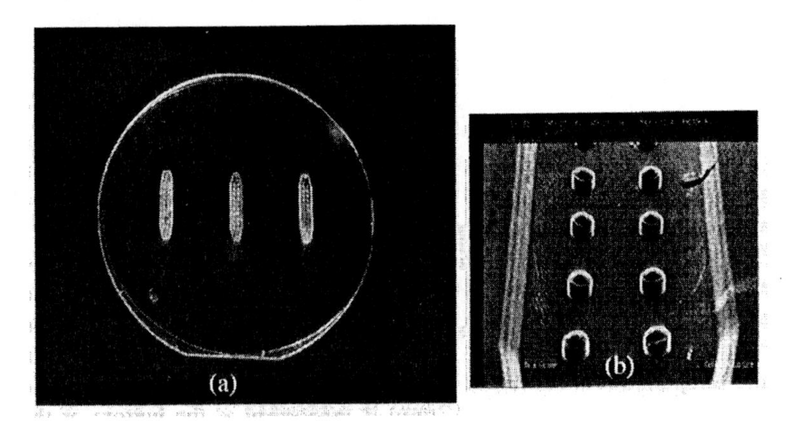

Figure 20. Images showing the plastic capture channels. (a) Three channels in parallel; (b) Close-view of the channel

During the capture experiments, the biological sample mixture solution, which consists of *E. coli* in a cultural medium, cell capture magnetic beads (Protein A coated Paramagnetic beads, Bangs Laboratories) and DNA capture magnetic beads (carboxylate-modified super-paramagnetic beads, Bangs Laboratories), was continuously flowed through the channel at a flow rate of 100 μl/min. The magnetic particles, including the cell-bound beads and DNA beads, were immobilized on the bottom of the channel with a permanent magnet. In the next step, the beads were washed to eliminate unbound sample matrix material and other components. Following a thermal lysis (95 oC for 5 min), bound E.coli cells were disrupted. The released DNA of interest was captured on the carboxylate-modified super-paramagnetic beads that were also held in the channel by the magnet. Cellular debris and other components (e.g., small ions, lipids and proteins) were flushed out of the channel with a high salt wash buffer solution. Under the high salt wash conditions (e.g., 3-4 M Na^+) the DNA remains associated with the beads surfaces. After the high-salt wash step, the immobilized micro-beads containing the adsorbed sample DNA, were washed with a low salt (e.g., 10 mM Na^+) elution buffer, resulting a purified and concentrated DNA solution that is ready for further steps, such as PCR.

3.4 On-chip Micro-PCR

In conventional bench-top PCR thermal cyclers, samples are mixed in stationary vessels to about 100 μL range and undergo a series of temperature shifts programmed to optimize the efficiency of each of the PCR steps. The time at a set temperature is the most critical component for each step. Reduction of the sample volume and improvement of the ramp times between temperature steps makes micro-PCR devices desirable.

Using casting, combined with embodiment of other functional elements, we were able to demonstrate plastic micro-PCR devices with integrated heaters and coolers [25]. As shown in Fig. 21, a glass capillary cylindrical reaction chamber is surrounded by a resistive heater coil. The reaction mixture temperature is sensed through a thermocouple. A thermal electric device (Melcor, 4mm x 4mm x 2.2mm) is positioned at the bottom of the chamber to facilitate cooling and to assist uniform heating. The entire assembly is embedded in a transparent polymer matrix (Epotek, 301-2F1) with the normal heating side of the TE device exposed for better heat condition. We have measured the heating rate of ~2.4°C /s and the cooling rate of ~2.0°C /s for devices tested under active heating/cooling control. *E. coli* cell lysis and the subsequent amplification of the released genomic DNA segments were carried out using the micro-PCR chamber. The cell and PCR mixture (containing 10 mM Tris-HCl, 50 mM KCl, 1.5 mM $MgCl_2$, 0.001% gelatin, 250 g/ml bovine serum albumin, 2.5 units/100 μL AmpliTaq DNA polymerase, etc.) of 1.5 μL was loaded in the micro reaction chamber and then taken through a temperature cycle program. The PCR product was evaluated and compared with control reaction using gel electrophoresis technique (Fig. 21(c))

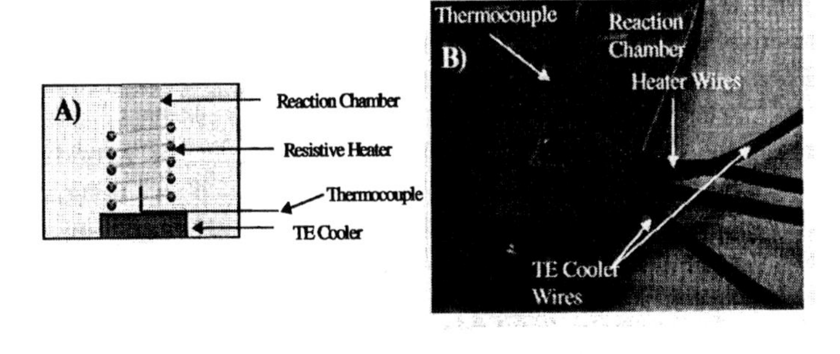

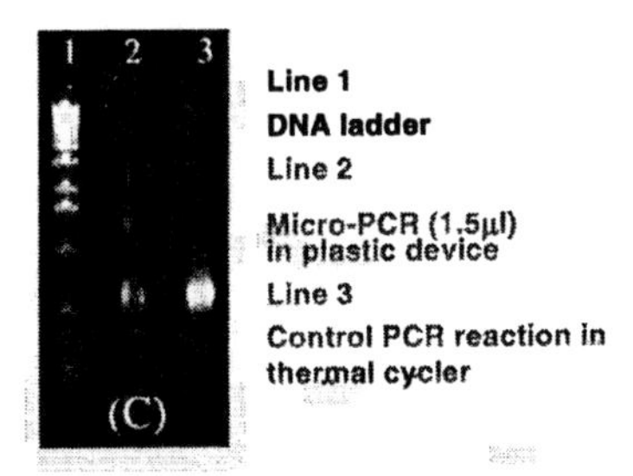

Figure 21. (a) and (b) Plastic micro-PCR device with integrated heaters and coolers. (c) Gel electrophoresis scanning image of the PCR product

3.5 Hybridisation Enhancement Using Microfluidics

Following PCR, the amplified DNA can be detected using either microarray hybridisation or capillary electrophoresis separation. Most prevalent method of on-chip detection is through capillary electrophoresis separation [48]. However, when higher specificity and highly parallel detection schemes are needed, the use of hybridisation arrays is appropriate. In the following sections, two techniques to use microfluidics to enhance conventional DNA microarray will be presented. One uses acoustic mixing that has been discussed in Section 3.1.2 to accelerate hybridisation, and the other one uses microchannels to enhance hybridisation kinetics and reduce sample/reagent consumption.

3.5.1 Hybridisation Enhancement Using Acoustic Mixing

The efficacy of acoustic mixing on hybridisation kinetics enhancement was tested in a Motorola eSensor™ chip [49]. An eSensor™ device is composed of a printed circuit board (PCB) that consists of an array of gold electrodes modified with a multi-component self-assembled monolayer (SAM) that includes pre-synthesized oligonucleotide (DNA) capture probes that are covalently attached to the electrode through an alkyl thiol linker [49]. When a sample solution containing target DNA is introduced into the e-Sensor™, capture probes on an electrode surface encounter complementary DNA from the sample. Then binding, or hybridisation, occurs. The system also contains DNA sequences, called signalling probes, attached with ferrocene-containing electronic labels. These signalling probes also bind to the target DNA sequence. When a slight voltage is applied to the sample following hybridisation, the ferrocene labels release electrons, producing a characteristic signal that can be detected through the electrode. This indicates the presence of the target DNA. As shown in Fig. 22(a), a plastic cover layer of the eSensor™ device was drilled with 4x4 air pockets (500 μm in diameter and 500 μm deep). The cover layer was then assembled with the PCB chip that has 4x4 detection electrodes (Fig. 22(b)). A PZT disk was then glued on the top of the plastic cover layer (Fig. 22(c)) to induce acoustic mixing during the hybridisation (Fig. 22(d)).

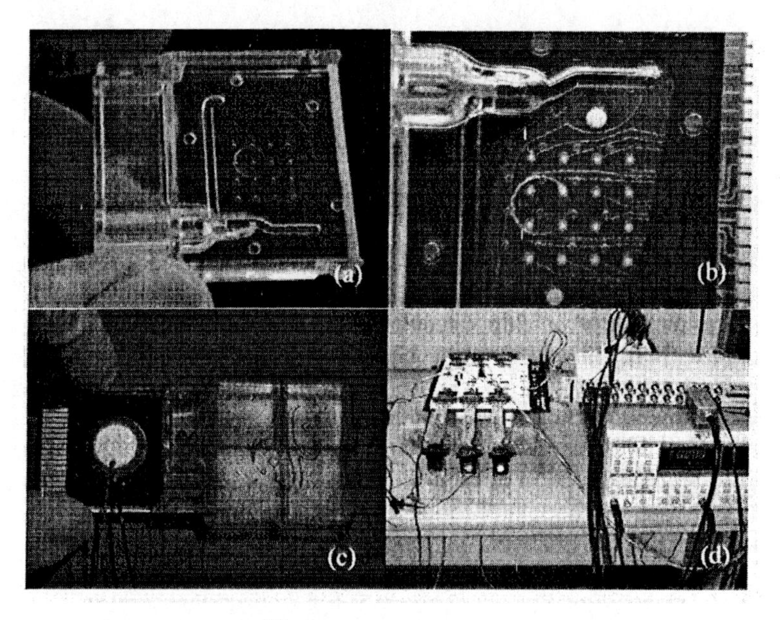

Figure 22. Images of an eSensor™ device coupled with a PZT disk to induce acoustic mixing in hybridisation assay

An assay for single nucleotide polymorphisms (SNP's) associated with hematochromatosis (HFE-H) was performed in the device with acoustic mixing. The DNA target solution containing the HFE-H polymorphism was amplified from human genomic DNAs characterized for HFE genotype. To genotype the HFE samples, 50-200 ng of human genomic DNA was asymmetrically amplified by PCR (A-PCR) with the forward primer in 5-fold excess over the reverse primer. The resulted DNA amplicon is 200 bp. After PCR, the amplicon solution was mixed with the hybridisation solution and then pipetted into the chamber. The chip was then plugged into an instrument that continuously scanned the electrodes/DNA probes during the hybridisation that occurs at 35 °C. For comparison purpose, we also performed hybridisation reaction in a conventional diffusion-based chamber (70 μL) using the same amplicon mixture. To demonstrate the capability of acoustic mixing to achieve bulk fluid-to-fluid mixing, we also performed hybridisation in another acoustic mixing chip, in which the PCR solution and the hybridisation buffer were injected into the chamber without prior mixing. This is also compared to an identical experiment having no acoustic mixer integrated into the cartridge.

Kinetic data of target binding to sensor electrodes were collected by measuring electrochemical signal as a function of time. Figure 23 summarizes the kinetics results from an acoustic mixing based hybridisation device and a pure diffusion based device. Note that the y-axis in the figure is the measurement of the Faradaic current from the electrodes. The Faradaic current is directly proportional to the number of ferrocene moieties immobilized at the electrode surface that in turn is proportional to the number of target nucleic acid molecules [49]. The results show that the device with acoustic mixing reached saturation signal within a short time compared to the diffusion-based one. This indicates that acoustic mixing provides convectional flow that enhances mixing between the target DNA in the solution and DNA capture probes on the surface, which in turn accelerates the hybridisation kinetics. When considering the quality and efficiency of hybridisation, of fundamental importance is the mixing/binding of the target DNA with the capture probe. It is believed that mixing enhancement can decrease depletion effect and reduce the thickness of the diffusion layer above the capture probe, and thus bring more DNA targets to the capture probes in a much faster manner.

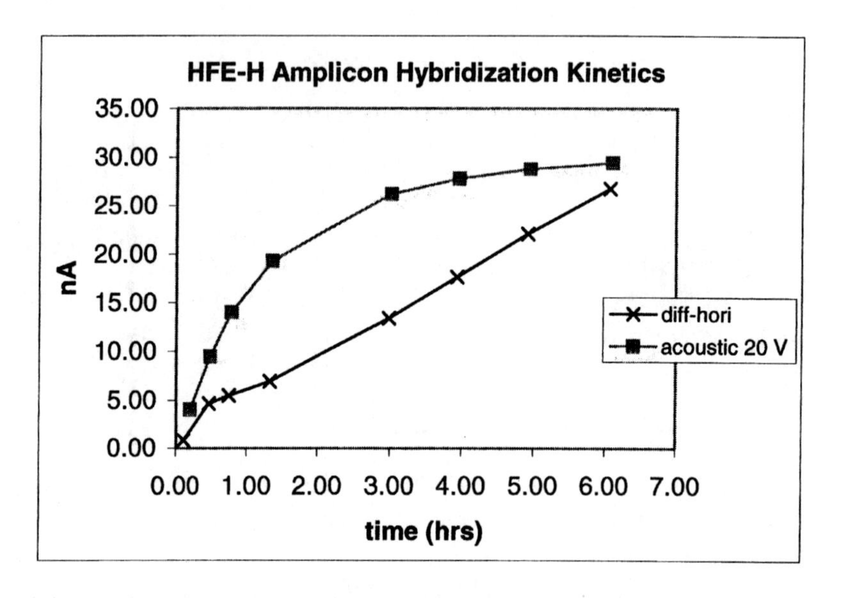

Figure 23. Single nucleotide polymorphism detection assay (HFE-H) comparing different mixing procedures

3.5.2 Hybridisation arrays in channel networks

Conventional array chips use ultra-high-density arrays (>10,000 probes/array), which are powerful in expression analysis. Lower or medium density, but highly parallel arrays will be useful in many other applications such as clinical diagnostics. For conventional DNA arrays, the sample volume is usually fairly large (tens of microlitres). In majority of the cases sample preparation is carried out on the bench, purified target is brought to the array for further analysis. The integration of these arrays with front-end microfluidic devices is not straightforward and prevents the system from developing increased functionality and automation.

We have demonstrated that channel networks made of PDMS can be attached to the DNA array devices [50]. The device is composed of two components as shown in Fig. 24: 1) a disposable, microfabricated PDMS device that consists of tens or hundreds of open microchannels (~200 μm wide, ~50 μm deep) as short as a few centimetres; 2) a microarray glass-based chip containing a number of DNA oligonucleotide probes that allow specific or non-specific capture of DNA via hybridisation. The PDMS channels were fabricated using a micromoulding technique and then bonded on a microarray glass slide using a selective plasma surface treatment. With channel volumes in the range of nanolitres (~ 100 nl), the overall volume for a reaction and the reagent cost are reduced by a factor of 100.

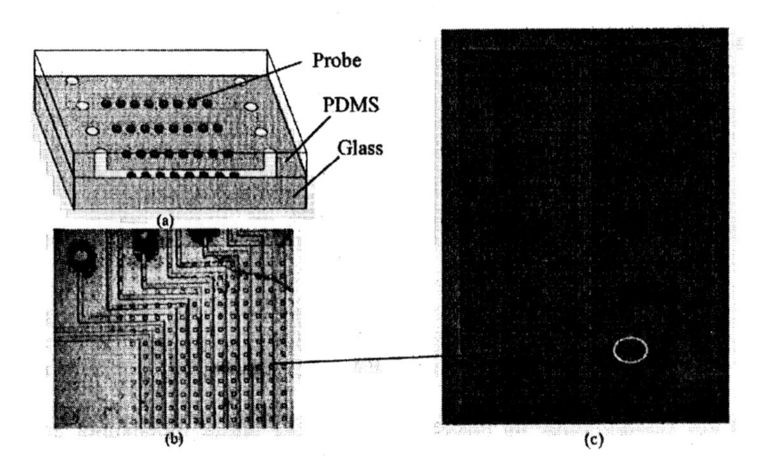

Figure 24. Schematic of a highly parallel biochannel array chip consisting of a PDMS microchannel layer and a microarray glass chip. (b) A close view of the high-density PDMS channel array running across gel pad based DNA array. The microchannels are 250 μm wide and 50 μm deep. (c) Micrograph showing two biochannel array chips, each has various channel array designs. The chip on the right-hand side consists of 52 channels in parallel

Single base extension (SBE) enzymatic reaction was performed in the biochannels with single nucleotide polymorphism (SNP) DNA arrays. PCR was used to amplify potential SNP targets of human genomic DNA before the hybridisation reaction. 100nL of DNA hybridisation reaction mixture contains 10ng human genomic DNA targets. Hybridisation reactions were completed within minutes (~ 3 min) during a thermal denaturation process (i.e., 2 min at 85°C, 1 min at 60°C). Figure 25 shows some of scanned fluorescent images of biochannel arrays, where high intensity of the spots indicates positive signals ("hits"). Note that in Fig. 25(a) a thermal bubble micropump was used to induce oscillation flow inside the channel (i.e., to move the DNA sample plug back and forth) during the hybridisation reaction, which in turn improves the mixing and hybridisation. 95% accuracy was achieved for SNP assays in the biochannels.

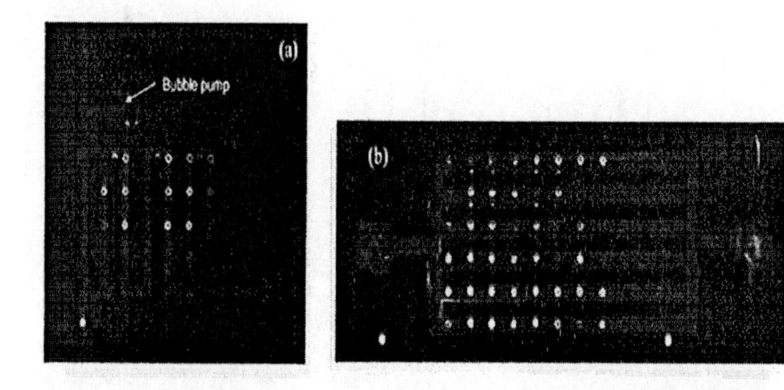

Figure 25. (a) Fluorescent image of single base extension (SBE) single nucleotide polymorphism (SNP) DNA hybridisation assay in a PDMS biochannel coupled with a thermal bubble micropump. (b) Fluorescent image of SBE assays in parallel PDMS biochannel arrays

In the further work of fluorescent detection based biochannel arrays, electrochemical detection based biochannels were also investigated. A microchannel structure was integrated with Motorola eSensor™ (Fig. 26). A channel pattern (500 μm wide and 200 μm deep) was cut out of the double-sided adhesive film and placed between the PCB chip and a plastic cover plate. A micro bubble pump was also integrated with the biochannel to provide oscillation flow in the channel.

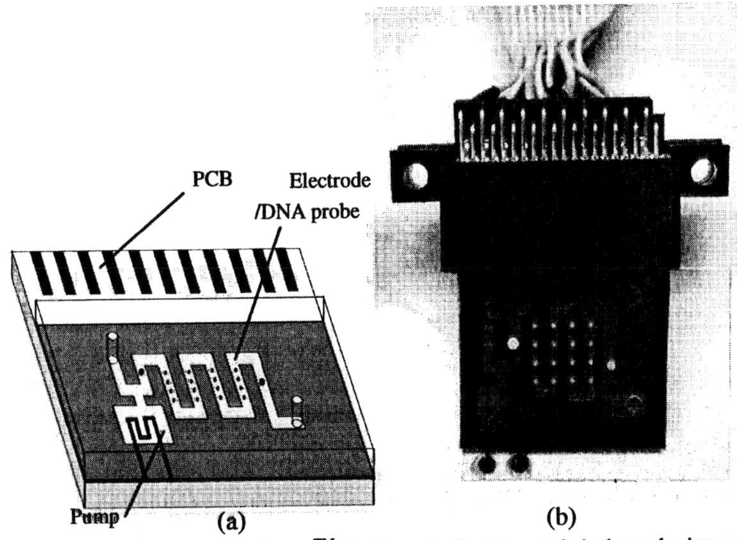

Figure 26. (a) Schematic of an eSensor™ biochannel with integrated air thermal micropump. (b) Photograph of eSensor™ biochannel with integrated micropump

Hybridisation kinetics was studied in the biochannel and compared with conventional diffusion-based eSensor™ chambers. Kinetic data of target binding to sensor electrodes can be collected by measuring electrochemical signal as a function of time. Figure 27 summarizes the hybridisation kinetics results for biochannels with pumps, conventional diffusion-based chamber, and PZT acoustic mixing chambers. Note that the y-axis in the figure is the measurement of the faradaic current from the electrodes. It is shown from Fig. 27 that the biochannels with micropumps has significantly hybridisation kinetics improvement compared to the diffusion-based chamber and the acoustic mixing chamber. The hybridisation in the biochannels (with pumps) reaches saturation within 3-4 hrs, while other methods took much longer This new platform is not only suitable for overcoming the inferior performance of conventional diffusion-based hybridisation (large sample volume consumption and lengthy hybridisation process), but also easy to integrate with the front-end sample preparation and PCR microfluidic components. Due to the mixing enhancement in z direction and focusing effect in y direction, the biochannel with integrated micro pump has been demonstrated a significant hybridisation kinetics acceleration over many other systems.

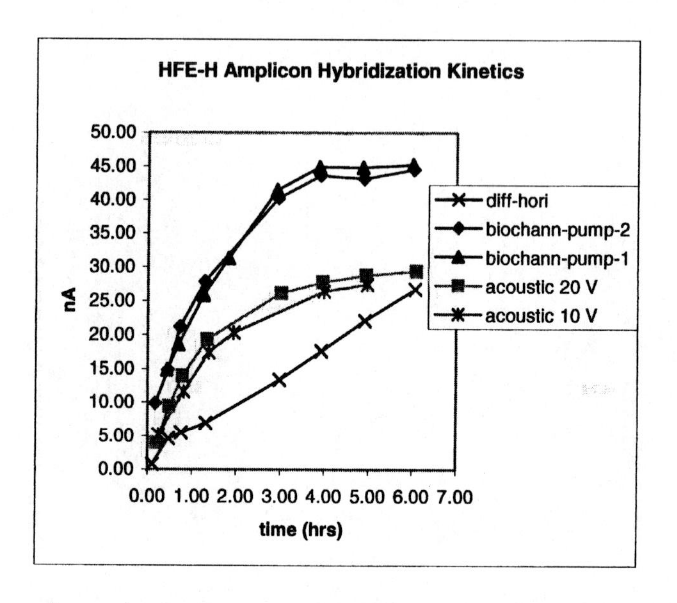

Figure 27. Hybridisation kinetics study of eSensor™ biochannels with pumps, conventional diffusion-based eSensor™ chamber, and PZT acoustic mixing eSensor™ chambers. HFE-H amplicon (diluted 1:4) was used as target DNA in all these chips

3.6 System Integration

Final integration of individual microfluidic components into systems has proven to be non-trivial. A universal protocol for the integration of these components on a large scale is not currently available. Although monolithic approaches have been successfully used in IC industry, many disparate materials and fabrication processes required in the fabrication of microfluidic devices hinder the fabrication of complex systems [44]. The modular approach shows promising but most current modular approaches do not have standard fluidic and electrical connectors and suffer certain drawbacks in their difficulty for chip reconfiguration and application-specific designs [51], [52], [53]. We are currently developing a unique solution relying on multi-chip-module docking board (MCM-DB) approach with individual device components performing single analytical functions. These component units are designed with standardized fluidic and electric interfaces. They will be assembled together into the complete system using common interfaces for the purpose of inter-chip communication. The schematic design of proposed approach is shown in Fig. 28. This design offers great flexibility in the system development compared to the monolithic device integration approach. It allows for selection of the optimum material platform and fabrication processes for given device component. It provides ease of system reconfiguration, increases system functionality through addition of other docking terminals on the board. It also standardizes the microfluidic interface through the development of individual electric and fluidic interfaces on each component.

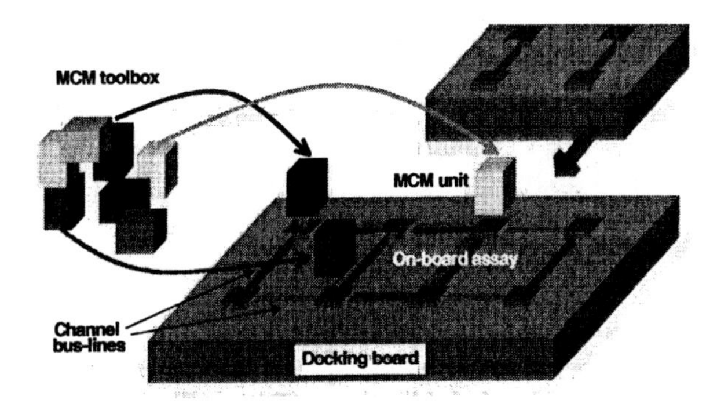

Figure 28. Schematic of the multi-chip-module docking board (MCM-DB)

Besides MCM-DB, another integration approach "microfluidic integrated circuit (μFluidic IC)" was also developed. μFluidic IC incorporates a series of functional modules by a set of elements/chips that are stacked vertically to form a series of internal flow streams. This modular approach combines multi-chip stacked module with standard electrical circuit board, and confers similar advantages in terms of simplifying system interconnects. The μFluidic IC system (Fig. 29) is composed of a series of functional chips that are plugged into a docking board. The docking board comprises a desired number of sockets and an electrical circuit board. Only those connectors between chip modules provide fluidic passageways. Therefore, the docking board can be reused while the chip modules can be disposed after a single use. An example of the system using barbed fitting connectors is shown in Fig. 29(b). The barbed fitting connectors showed easy connection, excellent reusability and reliable seal. All these fluidic/electric connections are also releasable by simply unplugging thc connectors from the bores and unplugging the chips from the sockets. This allows quick and easy module installation, or replacement of defective modules. The μFluidic IC approach, analogous to standard electronic IC, provides a module and assembly standard for integrated microfluidic systems for biological applications.

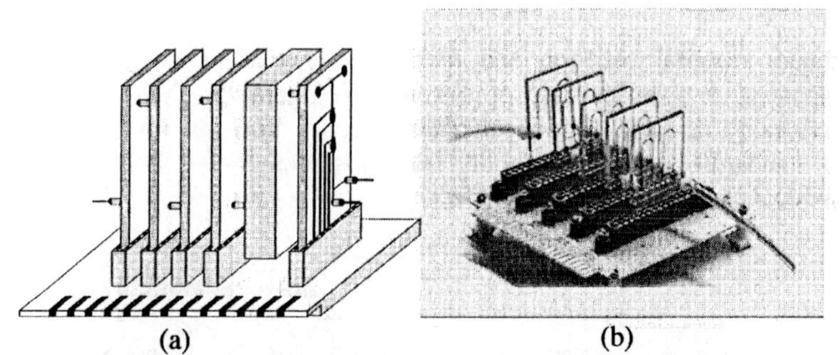
(a) (b)

Figure 29. (a) Schematic of a μFluidic IC system. (b) Image showing a μFluidic IC that uses barbed fitting connectors

4. CONCLUSIONS

In this chapter we have discussed the device developments towards the front-end functions in the area of sample preparation as well as the back-end detection using on-chip hybridisation. This will allow for further consolidation of on-chip functionalities and further assay automation. We have shown that disposable plastic devices can become a viable platform for such device components. In particular, we have discussed here development of micromixers, microvalves, cell capture device, integrated micro-PCR device, and micro-channel DNA hybridisation arrays. We have also provided a new concept leading to common integration platform of single-function microfluidic components through multi-chip-module (MCM) and μFluidic IC scheme.

ACKNOWLEDGEMENT

The authors would like to express their thanks to Justin Bonanno, Dale Ganser, Dr. Ralf Lenigk, Dr. Jianing Yang, Thomas Smekal, David Rhine, Dr. Huinan Yu, Dr. Betty Chen, Dr. Kenneth Luehrsen, Dr. Pankaj Singhal, and Cory Rauch at Motorola Inc. for their works in plastic device fabrication and testing. Thanks also go to Prof. David Beebe, Prof. Jeffery Moore, and Qing Yu at UIUC for their helps in hydrogel microvalves. We would also like to thank Prof. Ronald Adrian and Kendra Sharp at UIUC for their helps in the chaotic micromixer. Most of the work presented in this chapter has been sponsored in part by NIST ATP contract #1999011104A, DARPA contract #MDA972-01-3-0001 and DARPA-MTO contract # F33615-98-1-2853.

REFERENCES

[1] R. G. Crystal, "Transfer of genes to humans: early lessons and obstacles to success," Science, Vol. **270**, 404 (1995).

[2] D. E. J. Bassett, M. B. Eisen, and M. S. Boguski, "Gene expression informatics - It's all in Your Mine," Nat. Genet., Vol. **21**, 51 (1999).

[3] D. J. Graves, "Powerful Tools for Genetic Analysis Come of Age," Tibtech, Vol. **17**, 127 (1999).

[4] C. R. Cantor and C. L. Smith, *Genomics: The Science and Technology Behind the Human Genome Project* (Wiley-Interscience, 1999).

[5] A. T. Woolley and R. A. Mathies, "Ultra-high-speed DNA Sequencing Using Capillary Electrophoresis Chips," Anal. Chem., Vol. **67**, 3676 (1995).

[6] P. O. Brown and D. Botstein, "Exploring the New World of the Genome with DNA Microarrays," Nat. Genet., Vol. **21**, 33 (1999).

[7] H. Koster *et al.*, "A strategy for rapid and efficient DNA sequencing by mass spectrometry," Nature Biotechnology, Vol. **14**, 1123 (1996).

[8] P. Mitchell, "Microfluidics - Downsizing Large-scale Biology," Nature Biotechnology, Vol. **19**, 717 (2001).

[9] S. C. Terry, J. H. Jerman, and J. B. Angell, "A Gas Chromatographic Air Analyzer Fabricated on a Silicon Wafer," IEEE Transactions on Electron Devices, ED-**26**, No. 12, 1180 (1979).

[10] E. Bassous, E., H. H. Taud, and L. Kuhn, "Ink jet printing nozzle arrays etched in silicon," Appl. Phys. Lett., Vol. **31**, 135 (1986).

[11] P. Gravesen, J. Branebjerg, and O. S. Jensen, "Microfluidics-a review," J. Micromech. Microeng., Vol. **3**, 168 (1993).

[12] K. Petersen, "Biomedical Applications of MEMS," in *IEDM*, San Jose, CA (1996).

[13] N. P. Chereminisof, *Fluid Flow* (Ann ARbor Science, 1981).

[14] D. J. Harrison *et al.*, "Capillary Electrophoresis and Sample Injection Systems Integrated on a Planar Glass Chip," Anal. Chem., Vol. **64**, 1926 (1992).

[15] A. G. Hadd *et al.*, "Microchip Device for Performing Enzyme Assays," Anal. Chem., Vol. **69**, 3407 (1997).

[16] S. C. Jacobson, and J. M. Ramsey, "Integrated Microdevice for DNA Restriction Fragment Analysis," Anal. Chem., Vol. **68**, 720 (1996).

[17] I. Kheterpal and R.A. Mathies, "Capillary array electrophoresis DNA sequencing," Anal. Chem., Vol. **71**, 31A (1999).

[18] L. C. Waters *et al.*, "Microchip Device for Cell Lysis, Multiplex PCR Amplification and Electrophoretic Sizing," Anal. Chem., Vol. **70**, 158 (1998).

[19] M. A. Northrup *et al.*, "A Miniature Analytical Instrument for Nucleic Acids Based on Micromachined Silicon Reaction Chambers," Anal. Chem., Vol. **70**, 918 (1998).

[20] M. Kopp, A. D. Mello, and A. Manz, "Chemical amplification: Continuous-Flow PCR on a chip," Science, Vol. **280**, 1046 (1998).

[21] T. Woolley *et al.*, "Functional Integration of PCR Amplification and Capillary Electrophoresis in a Microfabricated DNA Analysis Device," Anal. Chem., Vol. **68**, 4081 (1996).

[22] C. H. Mastrangelo and D. T. Burke, "Integrated Microfabricated Devices for Genetic Assays," in *Microprocesses and Nanotechnology '99*, Yokohama, Japan (1999).

[23] O. Larsson *et al.* "Silicon Based Replication Technology of 3D Microstructures by Conventional CD-Injection Moulding Techniques," in *Transducers '97* (1997).

[24] H. Becker, W. Dietz, and P. Dannberg, "Microfluidic Manifolds by Polymer Hot Embossing for Micro Total Analysis System Applications," in *uTas 98*, Banff, Canada (Kluwer Academic Publishers, 1998).

[25] H. Yu *et al.* "A Miniaturized and Integrated Plastic Thermal Chemical Reactor for Genetic Analysis," in *uTAS 2000*, The Netherlands (Kluwer Academic Publishers, 2000).

[26] R. M. McCormick *et al.*, "Microchannel Electrophoretic Separations of DNA in Injection-Moulded Plastic Substrates," Anal. Chem., Vol. **29**, 2626 (1997).

[27] Zhao, M., *et al.* "Functional and Efficient Electrode-Integrated Microfluidic Plastic Devices," in *uTas 2001*. 2001. Monterey, CA (Kluwer Academic Publishers, 2001).

[28] J. P. Brody *et al.*, "Biotechnology at Low Reynolds Numbers," Biophysical Journal, Vol. **71**, 3430 (1996).

[29] H. T. Evensen, D. R. Meldrum, and D. L. Cunningham, "Automated fluid mixing in glass capillaries," Rev. Scien. Inst., Vol. **69**, 519 (1998).

[30] J. Evans, D. Liepmann, and A. Pisano, "Planar Laminar Mixer," IEEE, 96 (1997).

[31] R. M. Moroney, R. M. White, and R. T. Howe. "Ultrasonically induced microtransport," in MEMS '95, The Netherlands (1991).

[32] J. Branebjerg *et al.*, "Fast mixing by lamination," in MEMS '96, San Diego, CA (1996).

[33] H. Mesinger *et al.* "Microreactor with integrated static mixer and analysis system," in uTAS '94, The Netherlands (1994).

[34] R. Miyake *et al.*, "Micro Mixer with Fast Diffusion," IEEE, p. 248 (1993).

[35] D. Leckband and G. Hammes, "Interactions between nucleotide binding sites on chloroplast coupling factor one during ATP hydrolysis," Biochemistry, Vol. **26**, 2306 (1997).

[36] R. H. Liu *et al.* "Plastic In-line Chaotic Micromixer for Biological Applications," in *uTas 2001*, Monterey, CA (Kluwer Academic Publishers, 2001).

[37] H. Aref, "Stirring by Chaotic Advection," J. Fluid Mech., Vol. **145**, 1 (1984).

[38] W. Arter, "Ergodic Stream-lines in Steady Convection," Phys. Lett. A, Vol. **97**, 171 (1983).

[39] D. R. Sawyers, M. Sen, and H.-C. Chang, "Effect of Chaotic Interfacial Stretching on Biomolecular Chemical Reaction in Helical-Coil Reactors," The Chemical Engineering Journal, Vol. **64**, 129 (1996).

[40] S. W. Jones, O. M. Thomas, and H. Aref, "Chaotic Advection by Laminar flow in a Twisted Pipe," J. Fluid Mech, Vol. **209**, 335 (1989).

[41] S. A. Elder, "Cavitation Microstreaming," J. Acoust. Soc. Am., Vol. **31**, 54 (1959).

[42] G. T. A. Kovacs, *Micromachined Transducers Sourcebook* (WCB McGraw-Hill, Boston, 1998).

[43] W. S. N.Trimmer, "Microrobots and micromechanical systems," Sensors and Actuators, Vol. **19**, 267 (1988).

[44] M. Madou, *Fundamentals of Microfabrication* (CRC Press, 1997).

[45] D. J. Beebe *et al.*, "Functional Structures For Autonomous Flow Control Inside Microfluidic Channels," Nature, Vol. **404**, 588 (2000).

[46] N. A. Peppas, *Hydrogels in Medicine and Pharmacy, Vol. I* (CRC Press, 1986).

[47] K. Wilson and J. M. Walker, *Principles and Techniques of Practical Biochemistry, 4th ed.* (Cambridge University Press, UK, 1997).

[48] D. J. Harrison *et al.*, "Micromachining a Miniaturized Capillary Electrophoresis-Based Chemical Analysis System on a Chip," in Science, Vol. **261**, 895 (1993).

[49] D. H. Farkas, "Bioelectronic DNA chips for the clinical laboratory," Clinical Chemistry, Vol. **47**, 1871 (2001).

[50] R. H. Liu *et al.* "Highly Parallel Integrated Microfluidic Biochannel Arrays," in Technical Digest of the 14th International Conference on Micro Electro Mechanical Systems, Interlaken, Switzerland (2001).

[51] J. C. Fettinger, "Stacked Modules for Micro Flow Systems in Chemical Analysis: Concept and Studies using an enlarged model," Sensors and Actuators B, Vol. **17**, 19 (1993).

[52] Shikida, *Valve and Semiconductor Fabricating Equipment Using the Same*, U.S. Pat. No. 5,284,179.

[53] Stoll, *Electrofluidic Circuit Board Assembly with Fluid Ducts and Electrical Connections*, U.S. Pat. No. 4,549,248.

Chapter 7

MICROFLUIDIC DEVICES ON PRINTED CIRCUIT BOARD

Stefan Richter,[1] Nam-Trung Nguyen,[2] Ansgar Wego,[1] and Lienhard Pagel[1]
University of Rostock, Germany;[1] Nanyang Technological University, Singapore[2]

Abstract: This chapter discusses a new approach for fabrication of microfluidic devices based on printed circuit board (PCB) technology. The chapter shortly describes the basic process steps in PCB fabrication considering the special needs of fluidic components to make it easier to understand the technological approach. Design, fabrication and characteristics of a number of sensors and actuators are presented. Results of active component such as diffuser/nozzle pumps, peristaltic pumps, check-valve pumps and sensors such as pressure sensors, flow sensors, bubble sensors, and pH sensors prove the feasibility of this new fabrication concept.

Key words: PCB, PCB-Technology, micropumps, thermopneumatic actuators, piezoelectric actuators, pressure sensors, flow sensors, bubble sensors, pH-sensors.

1. INTRODUCTION

The development of microsystems for sensing, analysing and actuating is necessary and demanded by the needs of the future society. Only small and powerful systems have the capability to decrease consumption of resources. Microsystems combine and integrate miniaturized sensors, actuators and electronics in a single device. That is why system integration is essential for a powerful device. Regarding reliability, performance, volume and cost, the system approach can improve the quality of a commercial product [1].

Microfluidics has been established as a new engineering discipline with a huge scientific and commercial potential. In the last decade, research on microfluidic devices and fluidic phenomena in micro scale became a

F.E.H. Tay (ed.), Microfluidics and BioMEMS Applications, 185–217.

strategic topic of international research communities [2][3]. Numerous fluidic components have been realized in silicon technology. The most common components are micropumps, valves, flow sensors, separators and mixers [4-6].

There is a huge interest for the widespread introduction of cheap fluidic microsystems in laboratory devices to minimize the consumption of resources and probe-materials, and in some cases to reduce the analysis time. In biological, medical, and chemical analysis equipment, minimum amounts of fluids have to be detected, analysed and actuated [7][8].

Despite the research effort and the promisingly huge market, microfluidic systems still cannot establish widely in commercial applications. One of the reasons is silicon technology, which requires relatively high development and fabrication costs. Complex silicon based microfluidic systems (lab-on-a-chip, micro total analysis system) are expensive in development and fabrication. High expenses for mask fabrication and clean room processes are the major burdens. Silicon technology is only cost-effective for multi-purpose systems, which have a large market.

Moreover, the problem of expensive packaging is still not solved. Nowadays the cost of packaging can make up to 75% of the costs of the whole microfluidic system [2]. The cost calculations discussed below are derived from those of classical microelectronics [9]. The cost of a silicon-based microfluidic system c_{system} is calculated from the cost of a die c_{die}, the testing cost $c_{testing}$, the packaging cost $c_{packaging}$ and the final test yield α_{final}:

$$c_{system} = \frac{c_{die} + c_{testing} + c_{packaging}}{\alpha_{final}} \tag{1}$$

While the packaging cost $c_{packaging}$ of a microfluidic system is critically high due to the lack of micro-macro interconnection standards, the cost of a die c_{die} is also high because of the relatively large size of the system and consequently the small number N of dies per wafer:

$$c_{die} = \frac{c_{wafer}}{N \cdot \alpha_{die}} \tag{2}$$

The die yield per wafer α_{die} is proportional to the complexity of the whole system β_{system}:

$$\alpha_{die} \sim \frac{1}{\beta_{system}} \tag{3}$$

The Eqs. (1) to (3) show that the cost of a silicon-based microfluidic system could be much higher than a traditional microelectronic chip. Using alternative fabrication methods such as plastic moulding and hot embossing may solve these cost-related problems.

Another approach is the use of printed circuit boards (PCBs), not only as substrates for the wiring of electronic components but also as a carrier of fluidic elements. The fabrication of PCBs is an established technology that provides lower costs than silicon and LIGA processes [10]. However, this comparatively low-cost technology still allows the development of complex integrated microfluidic systems.

Table 1 compares some common parameters of MEMS-technology (silicon and LIGA), PCB-technology and those required by a microfluidic system. The comparison shows that PCB can be one of the solutions for systems with no critically small structures.

Lammerink et. al. [11] presented a demonstrator of a microfluidic system with channels milled in PCB, they introduced the term of MCB (mixed circuit board). Silicon components are glued to the fluidic board. Recently, research efforts have been made in order to integrate passive fluidic elements (channels, orifices), sensors and actuators into the PCB [12-15].

First, the following sections shortly describe the basic process steps in PCB fabrication considering the special needs of fluidic components to make it easier to understand the technological approach. Second, design, fabrication and characteristics of a number of sensors and actuators are presented and followed by a conclusion section and a list of references.

Table 1. Typical parameters of MEMS, PCB and microfluidic devices

Parameters	MEMS	PCB	Microfluidics
Resolution	5 μm	100 μm min.	50 μm – 10 mm
Clean room class	100 – 1000	1000 – 10000	Typ. 1000
Electronic compatibility	Typ. 4 metal layers	4 – 20 metal layers	Typ. 1 metal layer
Bonding techniques	Anodic bonding, direct bonding, adhesive bonding	Eutectic bonding, adhesive bonding (lamination)	Anodic bonding, direct bonding, adhesive bonding

2. PCB-TECHNOLOGIES

There is a huge variety of materials and processes in PCB fabrication [16][17]. First, PCB-substrates can be categorized as rigid and flexible materials. Second, PCB-technology can be divided into two main classes: additive and subtractive techniques. In fabrication processes of fluidic components based on PCB-technology, copper cladding of rigid substrates is usually structured by subtractive techniques [12][15]. The following sections focus on the major PCB-technologies used for microfluidic devices.

Etching

The common subtractive technique is wet etching [16]. In this process up to 90% of the copper area on the substrate surface is resolved by an etchant. A mask protects the later wiring image of the PCB. The mask is resistant to the etchant. Such a mask is a resist film, which can be patterned by photolithography or screen-printing.

Adhesive bonding and lamination

Adhesives are widely used for mechanical interconnections. In the field of PCB technology adhesive bonding is used e.g. to fix surface mounted devices (SMD)- or chip-on-board devices in their places on the board, before soldering respective wire bonding [18].

Lamination is a special adhesive bonding technique. For instance, copper cladding can be laminated onto a substrate. One or both parts are coated with adhesive and then bonded together at high pressures and elevated temperatures.

Lamination plays an important role in the production process of multi-layer boards. The different layers are coated with semi-cured epoxy resin sheets (prepregs). After lamination, heating, and compression, stable mechanical bonds are achieved between the layers [16].

Eutectic bonding (Soldering)

The most important technique for interconnecting the electronic components on a PCB is soldering [17][18]. The contact region is heated up to a temperature higher than the melting point of the bonding material called solder, which is usually an eutectic tin-lead-alloy. During melting a small amount of the structural materials and the PCB's materials are dispersed in the molten alloy. Thus cooling down the contact leads to an intimate electrical and mechanical connection.

Hybrid assembly with other MEMS-components

Besides the approach of direct fabrication of MEMS components on a PCB, the hybrid integration of MEMS devices is also possible. Adhesive or eutectic bonding can be used for assembling these parts. Electrical interconnects are created by soldering or wire bonding [19][20].

3. MOVABLE STRUCTURES WITHIN PCB'S

3.1 Circular Membrane Structures

The insertion of a flexible layer between the rigid laminates of a PCB makes the design of movable elements with one degree of freedom possible. Utilizing these functional elements, one can design membranes for actuators, valves and sensors. Some of the device examples presented in the following sections use pre-strained polymer foils as membranes of circular shape. In order to handle the membranes, the foils are mounted in a carrying system (clamping rings, see Fig. 1), which enables the machining such as drilling, cleaning, coating as well as exact positioning for assembling with the PCBs. By clamping the membrane in the carrying rings, the membrane is stretched. The tension on the membrane σ_0 can be estimated by Eq. (4).

$$\sigma_0 = \varepsilon \cdot \frac{E}{1-\nu} \tag{4}$$

with

$$\varepsilon = \frac{2 \cdot 7 \cdot (\sqrt{h^2 + a^2} - a)}{d} \tag{5}$$

where ε is the membrane's radial strain, ν is the Poisson's ratio, E is the Young's modulus, d is the membrane diameter, h and a are geometrical data of the clamping rings, Fig. 1. A tension of 18.5 MPa is calculated for Kapton with the handling system describe in Fig. 1.

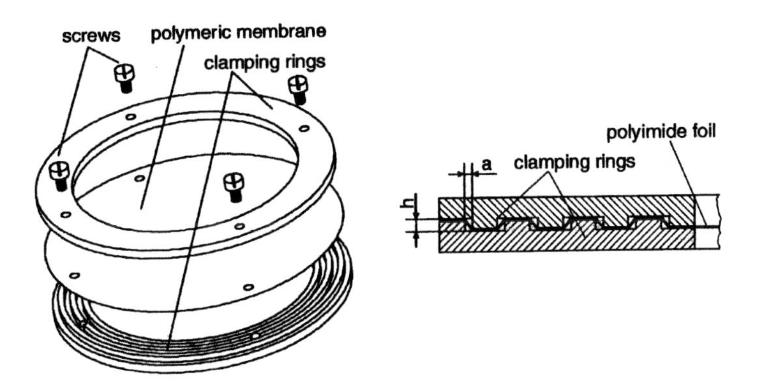

Fig. 1. Schematically view of the membrane mounting system used

A sufficient approximation for the relation between pressure load p and deflection w of a radial symmetric membrane's centre is given as [21]:

$$p = \frac{4d\sigma_0}{R^2} \cdot w + \frac{8d}{3R^4} \cdot \frac{E}{1-\nu} \cdot w^3 \tag{6}$$

where R is the radius of the membrane, d the thickness of the foil and σ_0 the intrinsic tension of the membrane caused by the clamping process. Eq. (6) is valid for the assumption of negligible bending forces. This assumption is sufficiently complied upon the used membranes. This tension consists of two components: the intrinsic tension σ_0 and the tension caused by the deflection of the membrane. Equation (7) shows this relation:

$$\sigma(w) = \sigma_0 + \frac{2}{3} \cdot \frac{E}{1-\nu} \cdot \frac{w^2}{R^2} \tag{7}$$

where ν is the Poisson's ratio of the foil, and E is the Young's modulus.

Figure 2 confirms the validity of Eq. (6) for a membrane with 5-mm radius. A good agreement of the measured and theoretical deflection curve can be clearly seen. The knowledge of the intrinsic tension σ_0 is important for designing valves and actuators. Measuring the membrane deflection using a microscope and the pressure load allows the tension σ to be determined.

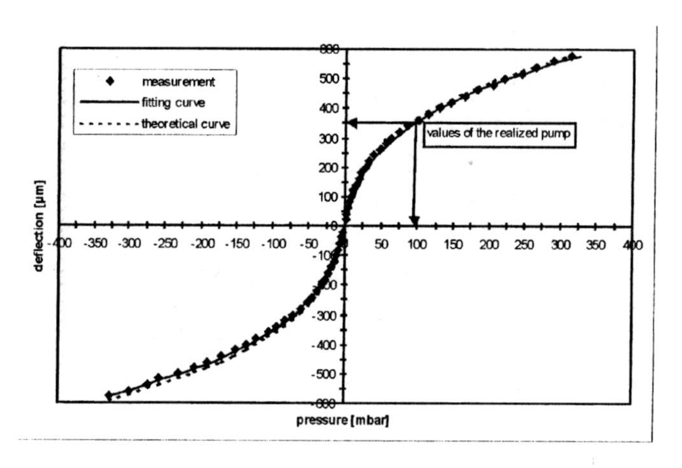

Fig. 2.Measured and calculated deflection curves of a polyimide membrane (thickness 7.8 µm, diameter 10 mm)

The intrinsic tension σ_0 can be calculated [22] by fitting the measured data to Eq. (7). Utilizing this method, an average intrinsic tension of about 9 MPa is measured for a number of fabricated membranes. The difference between the theoretical tension value and the experimental one can be explained by relaxing processes (creeping) of the polyimide foil after stretching [23].

Long-term tests are carried out to examine the influence of the number of actuating cycles on the deflection behaviour. During a period of 18 months, the membrane is periodically deflected by a thermopneumatic actuator at atmospheric pressure and room temperature. The corresponding cycle number is approximately 70 million. No change in the mechanical behaviour is observed.

3.2 Actuators

3.2.1 Thermopneumatic Volume Actuator

Basic Structure

Figure 3 schematically illustrates the realized actuator with measurement set-up. The actuator chamber encloses air as the driving medium. The chamber forms a closed thermodynamic system.

Assuming air as an ideal gas, a relation between the chamber's increasing mean temperature ΔT and the deflection w of the membrane is given by Eq. (8):

$$\Delta T = T_0\left[\frac{p_a}{p_0}\left(1+\frac{1}{2}\frac{w}{h_a}\right)-1\right] \tag{8}$$

where p_a is the pressure in the actuator chamber, Eq. (6). T_0 and p_0 are determined by the ambient temperature and pressure of the air during the fabrication process. h_a names the height of the actuator chamber.

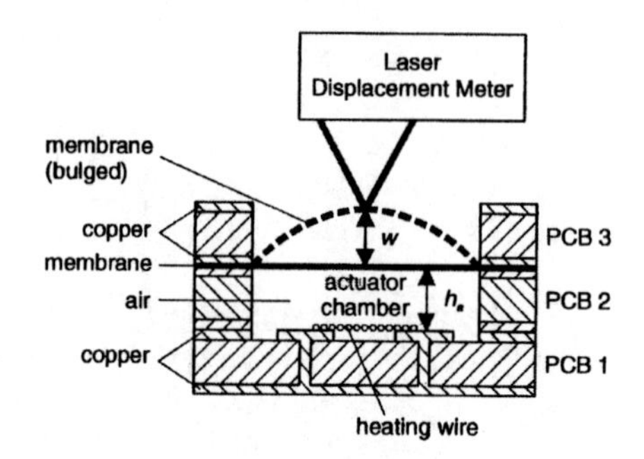

*Fig. 3.*Scheme of dynamical membrane deflection measurement

Figure 4(a) depicts the above relations in a diagram. The measured data are obtained at thermal equilibrium. The difference to the theoretical curve can be explained by the decreasing Young's modulus of the polymeric membrane material as a result of the rising temperature, and by thermal strain. A temperature rise of 100 K (i.e. ambient temperature plus 100 K) results to a deflection of 350 μm and a chamber pressure of 100 mbar (arrows in Fig. 4(a)). The temperature should be kept below 100°C over ambient temperature to avoid damages of the PCBs related to thermal stresses between the different laminated layers.

Closing the actuator chamber during the fabrication at a higher temperature (T_0 of Eq. (8)) yields a negative pre-deflection of the membrane, when the actuating chamber is cooled down to ambient temperature. This effect could be used to take advantage of the high slope in the p-w curve for the small pressure region. E.g. a symmetrical deflection of $\Delta w = \pm 250$ μm requires a pressure difference of $\Delta p = \pm 50$ mbar. These are more convenient values compared to $p = +100$ mbar and $w = +350$ μm, if the membrane is only deflected in positive directions, Fig. 2. However, this option has not been used yet for the actuators developed.

The membrane's centre above the pressure chamber is placed under a laser displacement meter (KEYENCE LK-031) to observe the dynamic behaviour of the actuator, Fig. 3. This experimental set up allows the determination of the system's time constants. Figure 4(b) shows a measured deflection curve for a frequency of 1 Hz and a duty cycle of 10 %.

The average consumed electrical power is adjusted to 1 W. A deflection of 350 μm is obtained. Therefore, the mean chamber temperature does not exceed the allowed value (arrows in Fig. 4(a)).

For the parabolic shape of the deflected membrane the cyclic volume displacement can be estimated as 10 μl:

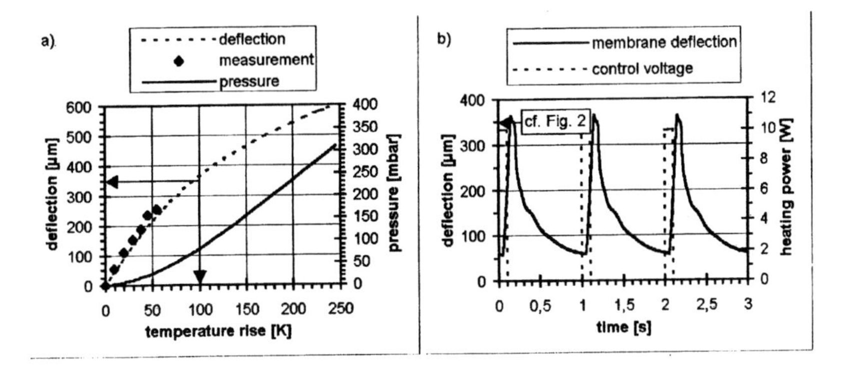

*Fig. 4.*Behaviour of a polyimide membrane 10 mm in diameter. (a) theoretical relation of deflection, temperature and pressure, (b) dynamically measured membrane deflection (f = 1 Hz, P = 1 W)

$$V = \frac{1}{2}\pi \cdot R^2 \cdot \Delta w \qquad (9)$$

An estimation of the maximum actuator pressure leads to values on the order 100 mbar.

Different Heater Concepts

The behaviour of varying designed thermopneumatic actuators has been studied to optimise the ratio of the membrane deflection versus the electrical power consumption. The structures of these different concepts are shown in Fig. 5.

The actuator shown in Fig. 5(a) uses the membrane as the carrier of a copper heater fabricated by thin film technologies. It works well as a single actuator. But for pumps, which are designed to deliver liquids, the cooling effect induced by the fluid leads to an insufficient membrane deflection.

In the second concept a constantan wire 70 µm in diameter manually threaded through small holes serves as the heater. Hermetic sealing of the drills is reached using epoxy resin (see Fig. 5(b)). Though a relatively small efficiency can be reached due to a largc part of the wire length lying outside the actuator chamber the good mechanical strength of the concept is proved by the long-term test mentioned above and promises a high working reliability.

Patterning the copper layer on a PCB-laminae as a heating resistor is the simplest way to produce an actuator because no additional process steps are required (see Fig. 5(c)). But due to the very small specific resistance and the relatively high thickness of the copper layer large structures are required for a sufficient heating effect.

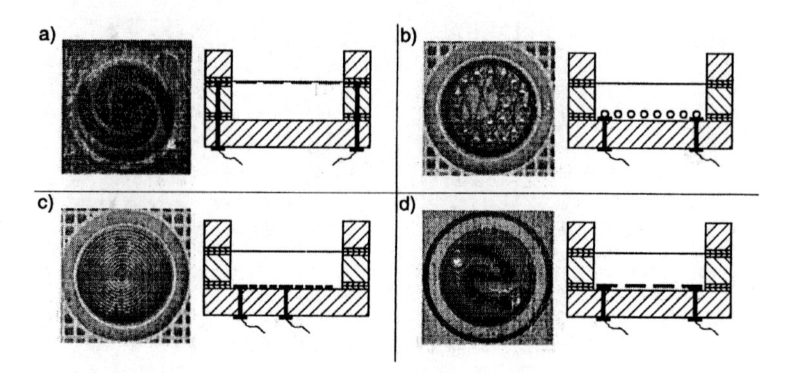

*Fig. 5.*Photographs and schemes of thermopneumatic volume actuators with different heating concepts: (a) thin film copper heater on the membrane, (b) Constantan heater wire, (c) heater structured in the PCBs copper cladding, (d) Constantan cantilever heater on a polymeric carrier

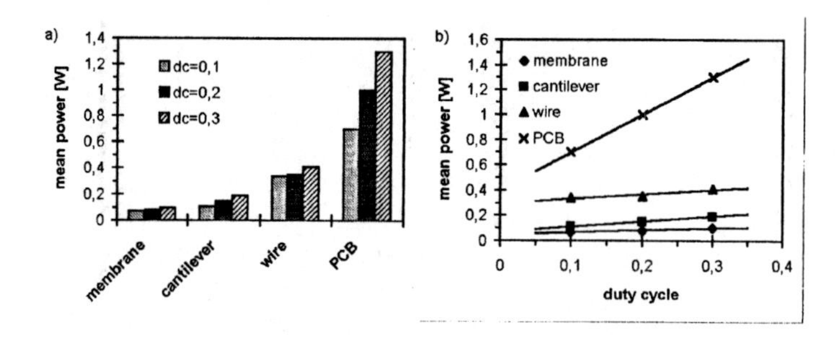

*Fig. 6.*Comparison of the mean power consumption of different heater types for the same volume displacement rate (app. 240 μl/min)

Additionally, the tight contact between the base laminate and the copper leads to a considerable heat loss.

The concept depicted in Fig. 5(d) uses a polymeric foil perforated by plasma etching serving as a carrier for the heater structure made of a thin constantan film. A cantilever heater is manufactured by soldering the heater onto copper bumps inside the actuator chamber. This concept promises the highest efficiency, but the technology is more complex. There is no test data regarding the long-term stability of these heaters.

Characteristics

The diagram in Fig. 6 shows the energy required by each design variant described to reach the same volume displacement. It can be seen clearly, that thin film membrane heaters and cantilever heaters consume the least amount of heating power. The unexpected high effectiveness of the membrane concept is due to the buckling effect of a bimorph.

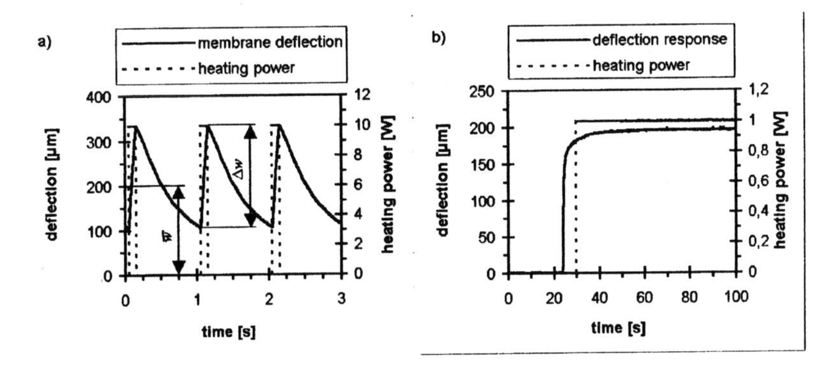

*Fig. 7.*The thermopneumatic actuators dynamic behaviour: (a) Typical movement of an excited actuator membrane, (b) deflection response on a heating power step

Taking the costly fabrication process of the small quantities of thin film structures into account, the wire concept seems to be a good compromise between manufacturing expense and energetic efficiency.

Figure 7 shows the dynamic behaviour of an actuator with the concept of constantan wire heater. The mean heating power dissipated an amount of 1 W and leads to a ±100 µm peak-to-peak deflection, or a mean deflection $\bar{w}$ of approximately 200 µm.

3.2.2 Piezoelectric Disc Actuators

Structure

Commercially available piezo discs can be selected as actuators to keep designs simple. The piezo disc consists of a 175 µm thick piezoelectric ceramic layer 12 mm in diameter glued on a brass disc. The dimensions of the brass disc are 95 µm (thickness) and 15 mm (diameter). With a maximum driving voltage of 200V, the maximum electrical field strength is 1.1 KV/mm, which is lower than the break down field of most common piezoelectric materials (more than 2 KV/mm).

In this section, we define the top electrode of the piezo disc as the voltage terminal and the brass disc as the ground terminal. The square wave signal at the voltage terminal has the high potential level V+ and a low potential V-.

Working Principle

The electric field applied on the piezoelectric layer induces an expanding strain in the disc perpendicular to the electric field and a contracting strain in the direction of the disc thickness (assuming that the piezoelectric coefficient d_{31} is negative and d_{33} is positive). Since the piezoelectric layer is tightly glued to the brass disc, there are reacting forces from the brass disc opposing the expansion of the piezoelectric layer. This motional restriction causes the deflection of the disc.

Characteristics

The characterization of the piezo disc is carried out using a laser vibrometer (Polytech). The piezo disc described above is investigated. It is soldered onto a copper ring of a PCB. Figure 8 shows the dynamic characteristics of the piezo disc's centre. Since the vibrometer only measured the vibration velocity (Fig. 8(a)), the disc deflection is calculated based on the velocity and time data. Using this measurement system, the absolute deflection cannot be measured because of the unknown initial condition. The results of the disc deflection are therefore corrected by a first order fitting function, which eliminates the offset and centres the deflection on zero (Fig. 8(b)). Based on the deflection data and the excitation signal, the transfer function of the disc is calculated (Fig. 8(c)). The transfer function identifies the piezo disc as a typical second order system with a resonant frequency of about 4 kHz. Figure 9 shows the membrane deflection versus the applied voltage. The circles in this figure represent the measured results; the solid line is a polynomial fitting function of second order. The deflection-voltage characteristics show a typical parabolic behaviour.

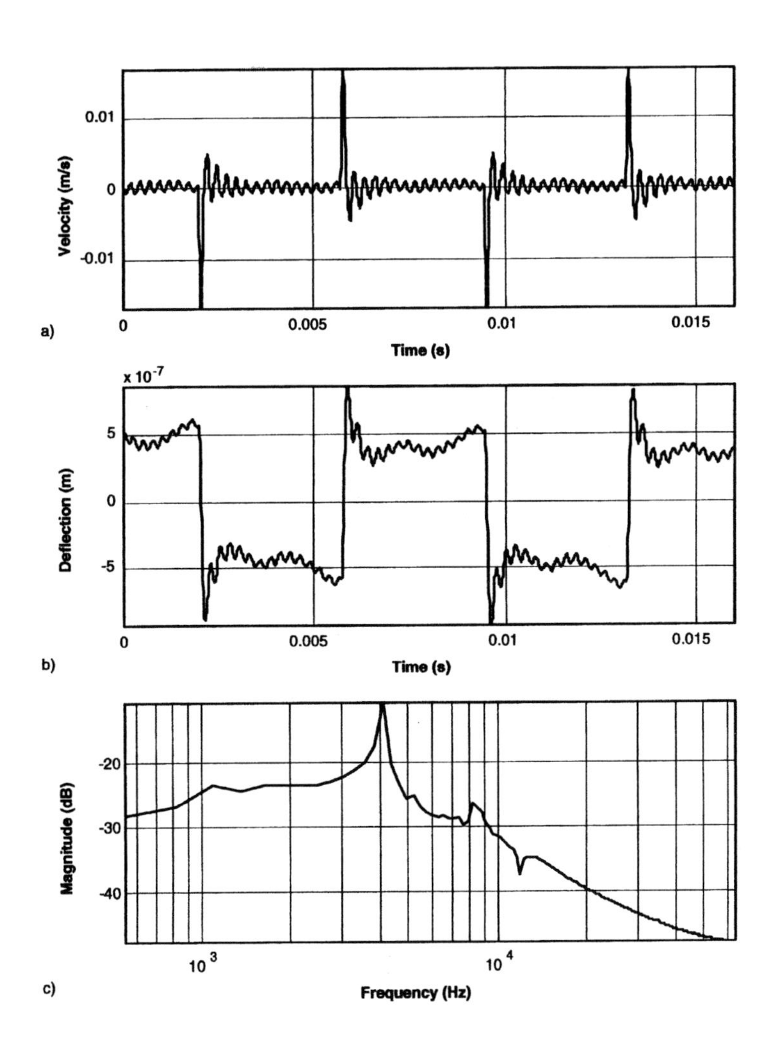

*Fig. 8.*Results of the vibration measurement (V=±30V, pump chamber filled with air): (a) Velocity, (b) Displacement, (c) Transfer function of the piezo disc

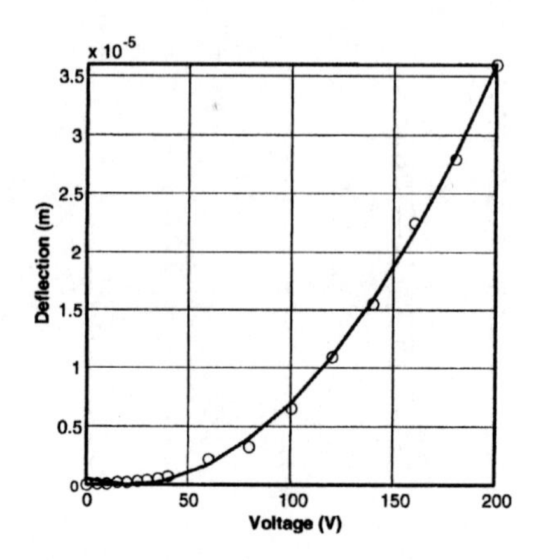

Fig. 9. The deflection as function of voltage (o: measured data, solid line: fitting function $dz=0.0011 \cdot V^2+0.0477 \cdot V+0.5495$)

4. MICROFLUIDIC DEVICES WITH PCB

4.1 Microvalve (Passive membrane check valve)

Figure 10(a) schematically shows the cross section with geometrical data of a single passive check valve. A pressure difference between inlet and outlet controls the flow through the valve.

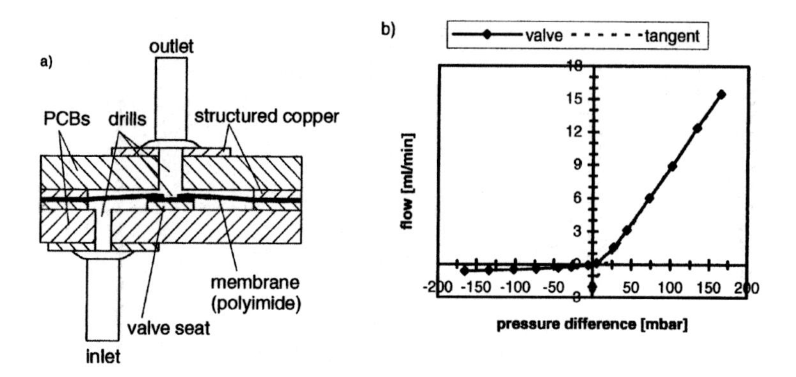

Fig. 10. Passive membrane check valve: (a) Scheme, (b) Characteristics

A higher pressure at the valve's inlet forces the membrane to bulge. The valve seat is free and the fluid can pass the valve. In the reverse case (higher pressure at the valve's outlet) the membrane is pressed against the valve seat, so that the flow is inhibited.

Figure 10(b) depicts the measured flow rate versus the pressure difference. A leakage ratio of over 40 can be reached between fluidic resistance of reverse to forward direction. The characteristic curve is similar to a semiconductor diode. After overcoming a pressure threshold of approximately 15 mbar (see tangent of Fig. 10(b)) the flow linearly rises to the pressure difference. The obtained forward resistance is approximately 9.5 $mbar \cdot (ml/min)^{-1}$.

4.2 Micropumps

Micropumps are important components for liquid and sample transport in microfluidic systems. The role of the pumps in those systems is similar to the power supply in an electronic system. Because of its robust and reliable design, miniature pumps also have their place in aerospace applications [24]. Therefore, miniature pumps are selected as first objects of our study on PCB-based microfluidic systems.

Micromachined pumps are categorized by actuating principles (piezoelectric, pneumatic, thermopneumatic, thermomechanic, electrostatic), pump principles (reciprocating, peristaltic, electro-hydrodynamic, electroosmotic, ultrasonic) or principles with check-valves and without them (valveless) [4-6].

4.2.1 Diffuser/Nozzle Pump

In recent years, numerous micropumps with a wide spectrum of operation principles were reported. Diffuser/nozzle pumps are valveless pumps. Valveless pumps don't require check valves. Generally, they are divided into the following categories:

- Peristaltic pumps, which can have a maximum flow rate of 100 μl/min was reported in [25] for water,
- diffuser/nozzle- and valvular conduits pumps, which can generate a maximum flow rate of 2000 to 3000 μl/min water [26][27],
- ultrasonic pumps, which induce acoustically a flow rate on the order of 10 μl/min [28][29],
- electroosmotic and electrophoresis pumps, which can move liquids with a velocity less than 1mm/s. Depending on the capillary diameter and the applied voltage, a flow rate of several hundred nano litre per minute can be achieved [30].

Design

The development of the diffuser/nozzle pump is our first effort to sink the packaging cost by combining functional elements (diffuser/nozzle) with packaging materials (PCB, inlet/outlet tubes) (Fig. 11). Figure 11(a) illustrates the developed diffuser/nozzle pump.

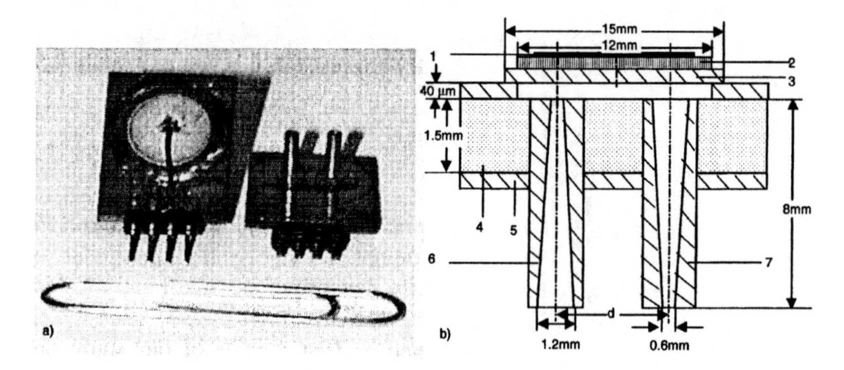

Fig. 11. The diffuser/nozzle pump: (a) Assembled pump, (b) Pump structure (1: Conductive layer, 2: Piezoelectric disc, 3: Brass plate, 4: PCB-substrate, 5: Copper layer, 6: Outlet tube, 7: Inlet tube)

Fabrication

Commercial PCB with coated positive photo resist is used as substrate material. The copper layer on both sides had a thickness of about 40 microns. The pump chamber is etched in the copper layer by iron chloride solution. Afterward, inlet and outlet holes are drilled into the substrate. The piezo disc is soldered directly to the copper layer of the PCB. The inlet and outlet copper tubes are also diffuser/nozzle elements. The diffuser structure is fabricated using classical cutting techniques. The two tubes are then soldered to the backside of the PCB (cf. Fig. 11(b)). For large-scale production, inlet and outlet tube can be made of plastic and pressed into the holes in the PCB. The self-fitting tubes will not require adhesive sealing or soldering.

Characterization

The pump is tested with water. Figure 12(a) shows experimental results of the pump performance in terms of flow rate versus drive frequency. During the test, the performance of the pump at high drive frequencies is found not as good as that at low frequencies, due to the inertia of the system. Besides, a high temperature is detected in the pump chamber when the pump operates at high frequencies, which is caused by the mechanical work of the piezo disc dissipated into the thermal energy.

Figure 12(b) shows the characteristics of the flow rate versus disc deflection at a constant drive frequency of 100 Hz. The results of two pumps with different distances between inlet and outlet are illustrated. We can see that the pump with inlet and outlet closer to disc centre have a better pump performance.

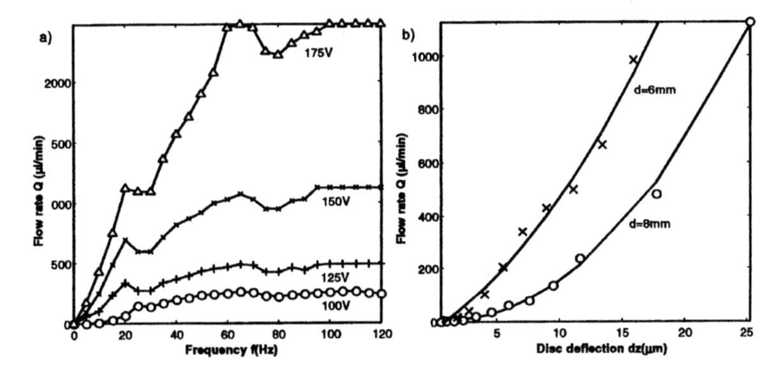

Fig. 12. Pump characteristics: (a) Pump performance with different frequencies and voltages of drive signals (Water, 8 mm inlet/outlet distance), (b) Flow rate versus drive disc deflection (water, f=100 Hz, inlet/outlet distances of d=6 mm and d=8 mm)

The deflection is calculated from the drive voltages and the fitting function described in Fig. 9. The influence of the back pressure on the flow rate is investigated by changing the height between the measurement tube and the water level in the reservoir.

4.2.2 Peristaltic Pump

Design

In contrast to diffuser/nozzle pumps, peristaltic pumps synchronize several piezo discs in a wave-like motion. This peristaltic motion transports the fluid in one direction and requires no diffuser/ nozzle. Classical peristaltic pumps generate the wave-like motion by a wheel with roles along the circumference. The roles press on a flexible silicon rubber tube and cause the wave-like motion when the wheel rotates. The micromachined peristaltic pump reported in [24] used active valves at the outlet and the inlet, the signalling scheme did not describe a true wave-like motion.

In our design, a minimum number of 3 piezo discs is required for a wave-like motion. Figure 13(a) illustrates the signalling scheme for the three piezo discs in our design. The piezo discs are actuated by square-wave signals. The second and the third signals have a phase shift of 90° and 180° to the first signal. In Fig. 13(b) the wave motion goes from left to right, while the fluid flows from right to left. When the first signal is switched to high level, the first disc sucks the fluid from the second disc into the first pump chamber while the third disc is closed and pushes the fluid to the second chamber (stage 1, Fig. 13(b)). With a 90° phase shift, the second disc is opened and sucks the fluid from the third chamber into the second chamber. Due to its inertia the fluid continues to flow to the outlet.

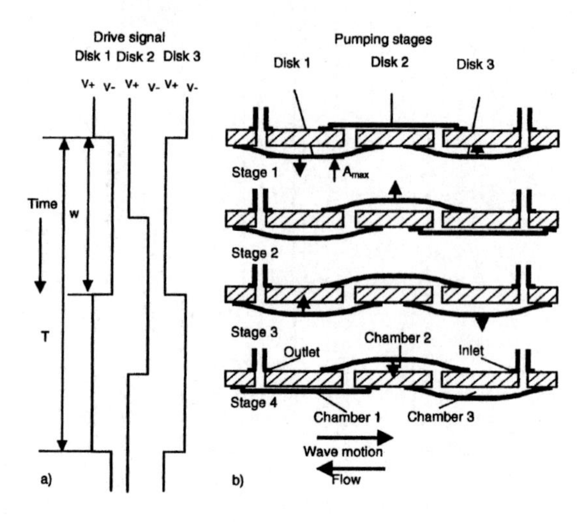

Fig. 13. Operation principle of the peristaltic pump: (a) Signalling scheme, (b) The corresponding pumping stages.

In the third stage, disc 1 is closed and pushes the fluid to the outlet. While the second chamber is still opened, disc 3 sucks the fluid from the inlet into the third chamber. In the last stage of the pumping period, disc 2 pushes the fluid out of the second chamber. The fluid is pushed through the first chamber to the outlet.

Fabrication

The PCB-pump is fabricated with the same technology described in section 4.2.1. The pump chambers are sealed and electrically connected by soldering. The assembly parts are illustrated in Fig. 14(a). Figure 14(b) shows the photographic view of the finished peristaltic pump.

Characterization

In the experiments, the lower voltage level is kept at ground ($V_- = 0$ V), while the piezo discs are driven by the high voltage level V_+. Figure 15 illustrates the circuitry implementing the signalling scheme described above.

The universal timer LM555 generates the square wave signal for disc 1. A simple phase shifter based on 2 logical NANDs and a RC-bridge shifts the second signal. A logical negator implements the third signal, which is shifted 180° from the first signal. The three signals are used for switching 3 electromechanical relays. The piezo discs are switched between two voltage levels: V+ and V-. These two drive voltages are supplied from an external high voltage source. For high frequency switching and avoiding noises, solid-state relays with optoelectronic coupling can be used instead of electromechanical relays.

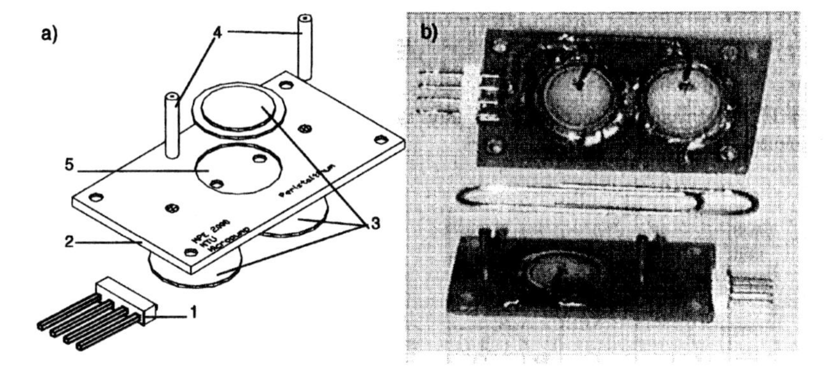

Fig. 14. The peristaltic pump: (a) Structure (1: Header for electric interconnection, 2: PCB, 3: Piezo disc, 4: Inlet/outlet tubes, 5: Pump chamber etched in the copperlayer), (b) Photograph of the finished pump

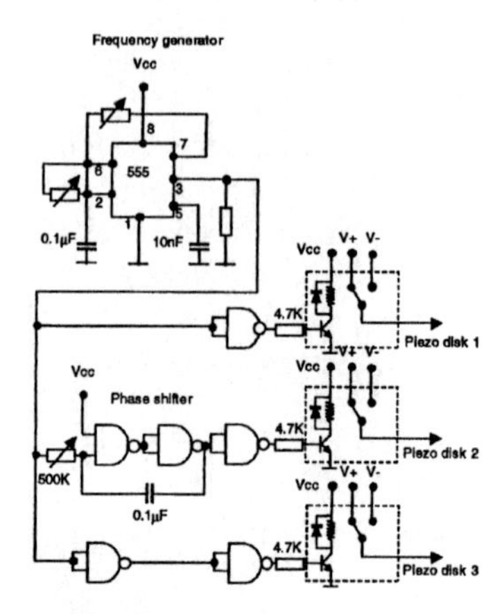

Fig. 15. Schematic of the control circuitry for the peristaltic pump (dotted boxes are commercial relays)

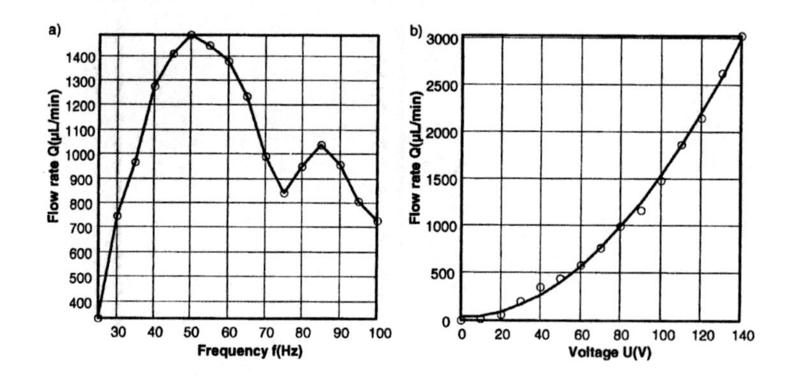

Fig. 16. Characteristics of the peristaltic pump: (a) Flow rate vs. drive Frequency ($V_- = 0$, $V_+ = 100$ V), (b) Flow rate vs. drive voltage ($V_- = 0$, f = 50 Hz)

Figure 16(a) shows the measured flow rate versus the drive frequency at a constant drive voltage (V+ = 100V, V_- = 0 V). For frequencies less than 50 Hz, the flow rate increases proportionally with the frequency.

The flow rate reaches its maximum at 50 Hz. The flow rate decreases at higher frequencies because of the large inertia of the fluid in the pump. If the flow rate is controlled by the drive frequency, the operation range should be between 0 and 50 Hz. Smaller pumps with smaller fluid volume will give a wider flow range.

Figure 16(b) illustrates the flow rate versus the drive voltage V_+. The circles are measured results. The line is the parabolic fitting curve. A relatively high flow rate of 3 ml/min can be achieved with 140V. A number of this pump working in parallel is able to generate a flow rate for active cooling applications.

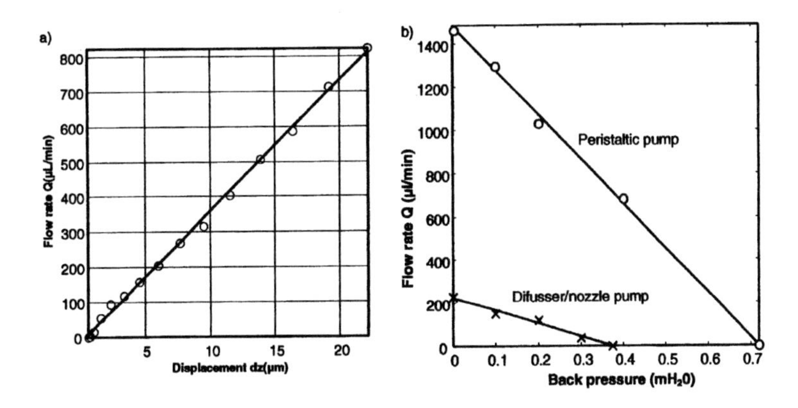

Fig. 17. (a) Flow rate of the peristaltic pump vs. displacement of the pump membrane, (b) Comparison of the flow rates reached by the diffuser/nozzle and the peristaltic pumps

Figure 17(a) illustrates the flow rate versus the disc deflection. The circles are measured results. The line is the fitting curve. Comparing the characteristics in Fig. 12(b) and Fig. 17(a), it is seen that with the same drive voltage, the peristaltic pump delivers higher flow rate. Figure 17(b) gives a comparison between the back pressure – flow rate characteristics of the two pumps. At the same drive frequency of 50 Hz and the same drive voltage $V_+ = 100$ V the head loss of the peristaltic pump is 72 cm, while the diffuser/nozzle pump can only deliver 38cm.

4.2.3 Check Valve Pump

Structure

The pump consists of two passive check valves and a thermopneumatically driven volume actuator as shown the cross section in Fig. 18 (see sections 3.2.1 and 4.1). Without electronics the pump measures 14 mm × 17,5 mm × 3,2 mm in size.

Four PCB layers and one membrane layer are used for the structure. The valves and the actuator use the same membrane for their function. The membrane consists of a thin polymeric foil (Kapton® or Mylar®) which measures 7,8 μm (Kapton) or 6 μm (Mylar) in thickness.

Fabrication

A special adhesive technique is used to connect the structured PCBs and the membrane [12][31]. The valve seat is coated with adhesive. A slice of Kapton is punched and manually placed on the valve seat before connecting with the membrane to prevent the membrane from sticking on the valve seat.

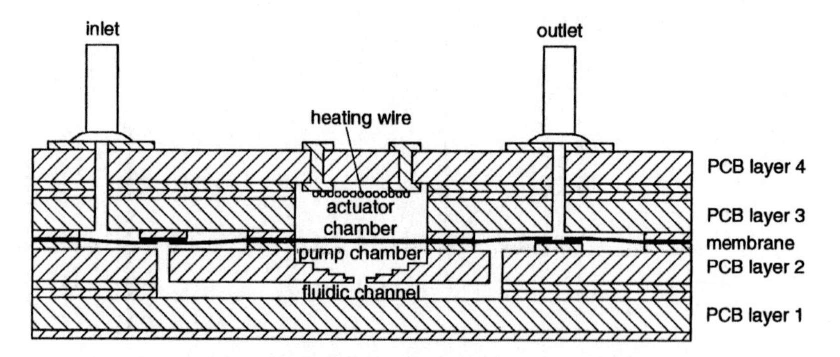

Fig. 18. Structure of a thermopneumatically actuated pump with passive check valves

Fig. 19. Photograph of the check valve pump with integrated electronics

The heater for the volume actuator is an isolated Constantan® wire (70 μm in diameter). The wire is threaded through small holes (300 μm in diameter) and mounted by soldering. The actuator chamber has a diameter of 10 mm containing air as the working medium.

In order to reach a self-filling capability the dead volume of the pump has to be minimized [32][33]. The shape of the pump chamber is adapted to the form of the bulged membrane by use of the milling tool of a Circuit Board Plotter (LPKF). This optimised design minimizes the dead volume of the pump. Hence, a maximum ratio of the volume compression $\Delta V/V_0$ of approximately 30 % (corresponding to a theoretical pressure rise of ca. 390 mbar) could be reached. A photograph of a delivering pump with integrated electronics is shown in Fig. 19.

Characterization

The flow rate versus backpressure is measured (medium: water) to characterize the pump. Typical characteristics of the pump are depicted in Fig. 20(a). The results show a linear relation between the flow on the applied backpressure. A maximum flow rate of app. 470 µl/min is achieved at zero backpressure. The flow stops at a backpressure of 135 hPa. The driving electrical power is on the order of 1 W. The drive frequency is adjusted to 1 Hz with a duty cycle of 0,1. The maximum backpressure correlates well with the measured and calculated chamber pressure (see arrows in Fig. 2 and Fig. 4).

Measuring the flow rate versus the exciting frequency at constant pulse width yields the diagram of Fig. 20(b). A peak flow of 0.66 ml/min is measured at a frequency of 1.8 Hz (duty cycle 0.18; average power consumption 1.8 W). First duration tests over 540,000 pump cycles (more than 150 hours) without any failure and loss of capability lead to the assumption of a high working reliability.

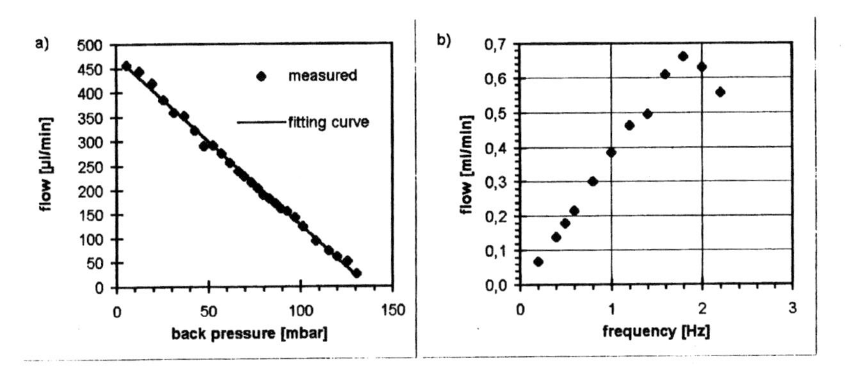

Fig. 20. Characteristics of the check valve pump: (a) Flow rate vs. back pressure, (b) Flow rate vs. driving frequency

4.3 Flow sensor

4.3.1 Design

The flow sensor is based on the classical electrocaloric principle. Due to the forced convection, the temperature field around a heater is shifted into flow direction. The asymmetrical temperature profile around the heater can be used for measuring the flow velocities. Two temperature sensors upstream and downstream detect the temperature difference. The difference signal is used as the sensor signal output [34].

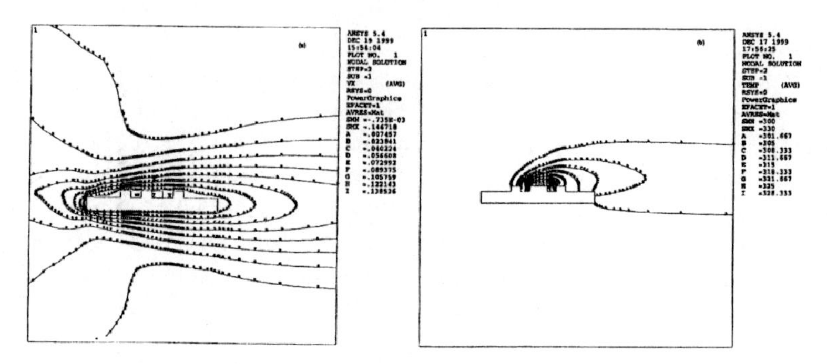

Fig. 21. Simulation results (air, u=0.1 m/s): (a) Contour plot of x-component of the flow velocity, (b) Contour plot of the temperature field

Numerical Model

A simple two-dimensional finite element analysis (FEA) model with ANSYS is also used for optimising the flow sensor. This model only considers the governing equations for the fluid, and in contrast to the analytical model it neglected the heat conduction in the PCB. Figure 21 shows the velocity field around the sensor and the corresponding temperature field. The asymmetrical temperature caused by the forced convection can be seen clearly.

With this model two operational modes are simulated: constant heating power mode and constant heater temperature mode. Figure 22 illustrates the temperature profile along the sensor in flow direction. In Fig. 22(a), the cooling effect can be seen clearly. The solid lines in Fig. 22 represent the temperature profile of the sensor with a distance of 1.2mm between the heater and the temperature sensor. The dashed lines are profiles of the sensor with a distance of 2mm. The temperature profiles in constant heater temperature mode are illustrated in Fig. 22(b).

With the FEA model, different sensor designs can be investigated. Similar to the discussion with the analytical model, we concentrate on the influence of the distances between the temperature sensors and the heater on the sensitivities. Figure 23(a) shows the difference between sensor characteristics of the investigated designs ($l = 1.2$ mm and $l = 2$ mm).

Figure 23(b) illustrates the dependency of the sensitivity on the distance between temperature sensors and the heater. The sensitivity is calculated as:

$$S = \frac{\Delta T|u}{u} \tag{10}$$

Results of the two operating modes at a velocity u=20 mm/s give the same behaviour as expected in the analytical model. The highest sensitivity can be achieved with a distance between 2 mm and 3 mm. This high sensitivity will allow the detection of very low air velocities.

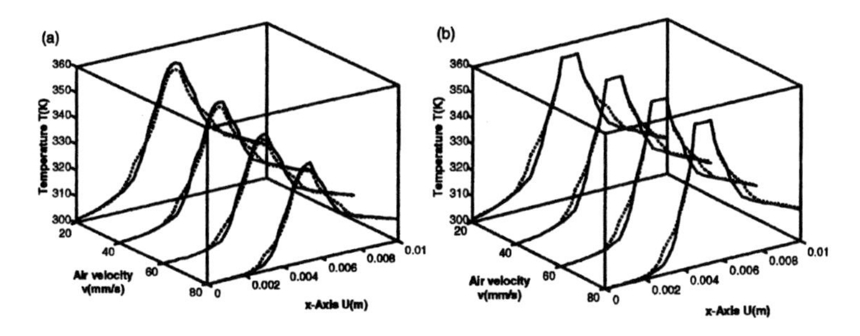

Fig. 22. Simulation results for the temperature profile along flow direction: (a) Constant heating power, (b) Constant heater temperature (solid line: l=1.2 mm, dashed line: l=2.0 mm)

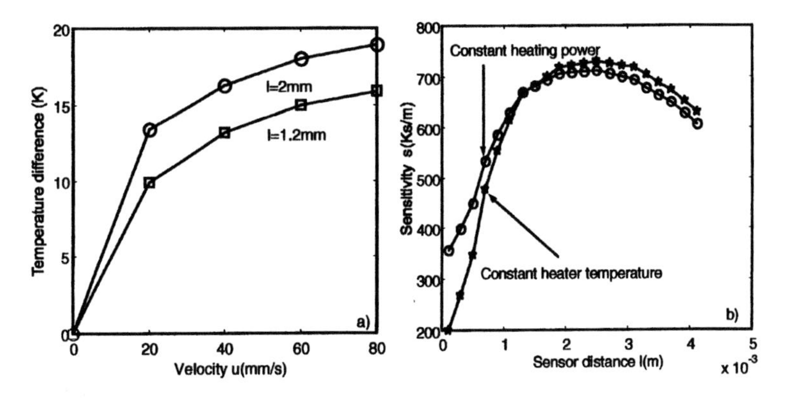

Fig. 23. Simulation results: (a) Sensor characteristics in constant heating power mode, (b) Sensitivity versus distance between temperature sensors and heater

Fabrication

Figure 24 illustrates the prototypes of the thermal flow sensor. The heater and temperature sensors are conventional surface mounted devices. An isolation trench is milled in the substrate material (ceramic or hard epoxy).

The electrical connections and solder pads are printed on a mask transparency, which is then transferred onto the substrate using common technology (photoresist development and copper etching with iron chloride). The heater resistor and temperature sensors are then soldered on the substrate by using solder paste and subsequent annealing. In the last step, interconnecting headers are soldered on the PCB in form of a dual-in-line (Fig. 24(a)) or single-in-line package (Fig. 24(b)).

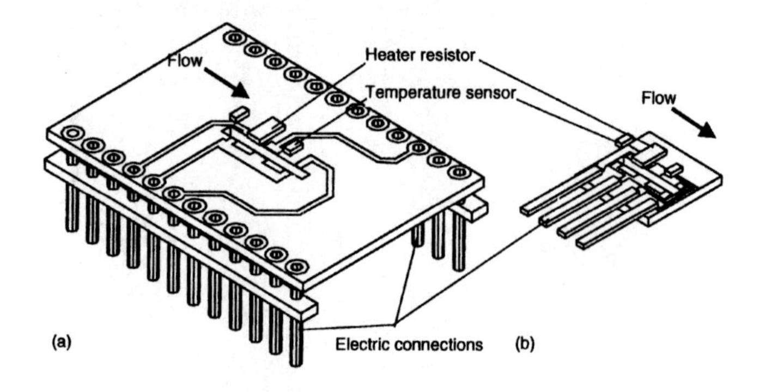

Fig. 24. The thermal flow sensor based on PCB-technology: (a) Dual in line module, (b) Single in line module

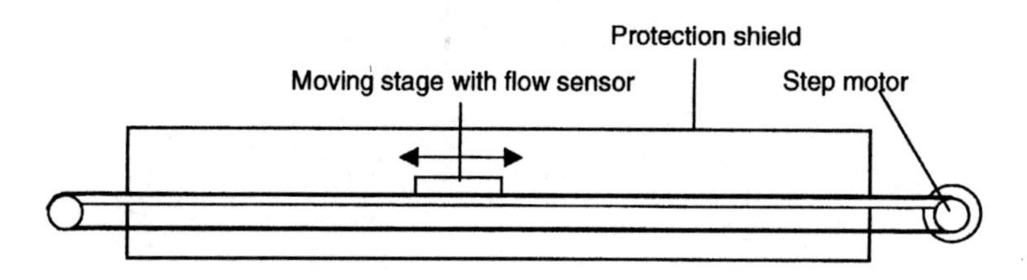

Fig. 25. Experimental set-up for calibration of ultra-low velocity sensors

Characterization

Because of the extremely high sensitivities, the flow sensor is able to detect air velocities in the mm/s and cm/s range. Since the most wind tunnel facilities are only able to generate flow velocities in the range of m/s, wind tunnel are not suitable for characterization of the flow sensors reported in this chapter. Figure 25 shows the experimental set-up for the calibration of ultra-low velocity flow sensors.

The sensor is mounted on a stage, which is moved by a step motor. The step motor is programmable and can control stage velocities with a resolution of 1 mm/s. Since the length of the movement is limited (2.2 m) and the time response of the flow sensor is on the order of few seconds, the set-up allowed a maximum velocity of 80 mm/s in order to make an accurate measurement.

To start with, the temperature profile around the heater is measured by using two thermocouples located upstream and downstream. The flow sensor is glued on a large aluminium plate, in order to generate a stable heat sink. The two thermocouples are calibrated to have the same output at the same temperature.

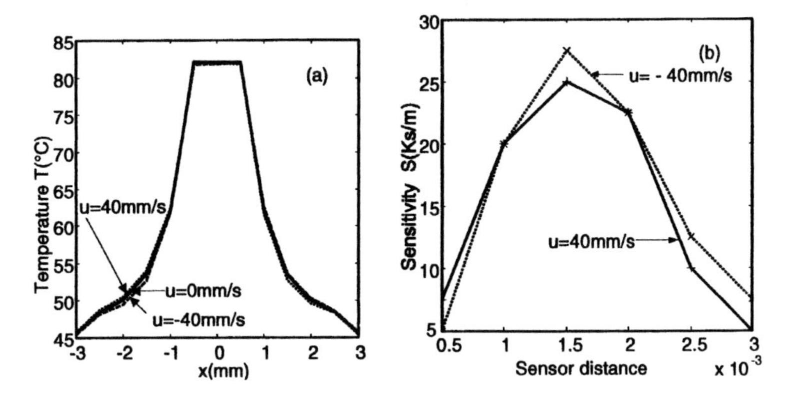

Fig. 26. Experimental results: (a) Temperature profile along flow direction for 0 mm/s (solid line), 40 mm/s (dotted line), and –40 mm/s (dashed line), (b) Sensitivities vs. heater to sensor distances for 40 mm/s (solid line) and –40 mm/s (dotted line)

The first thermocouple is positioned at a fixed distance to the heater (0.5 mm, 1 mm, 1.5 mm, 2 mm, 2.5 mm, and 3 mm). The heater (47 Ohm) is supplied with a constant voltage of 6 V or a constant heating power of 766 mW. The position of the second thermocouple is then adjusted for the same output. The whole set-up is then placed into the stage illustrated in Fig. 25. This procedure warrants that the temperature difference between upstream and downstream temperature sensors is always set to zero.

Figure 26(a) shows measured temperature profile around the heater at 0 mm/s, 40 mm/s, and –40 mm/s. The measurements show that asymmetrical effect caused by the flow is not that strong like the simulation results. One of the reasons may be the large aluminium heat spreader. However, the sensitivity resulting from this measurement (cf. Fig. 26(b)) agrees well with the simulation. The sensitivity is calculated with Eq. (10). The measurement gives an optimal distance of 1.5 mm.

In the next step, flow sensors with temperature sensing resistors are characterized. Two resistors are used as temperature sensors (1000 Ω, temperature coefficient 3000 ppm, Farnell's code 732-310). The two sensing resistors are connected with two resistors of the same type in a Wheatstone bridge. The two external sensors are kept at ambient temperature. Figure 27 shows the measured characteristics of the investigated flow sensors. Both curves are corrected with the offset at zero flow. Airflow in both directions is considered. The results agree with the analytical and numerical results. The sensor with a distance of 2 mm has a better sensitivity. Both sensors are able to measure flow velocities less than 80 mm/s in two directions.

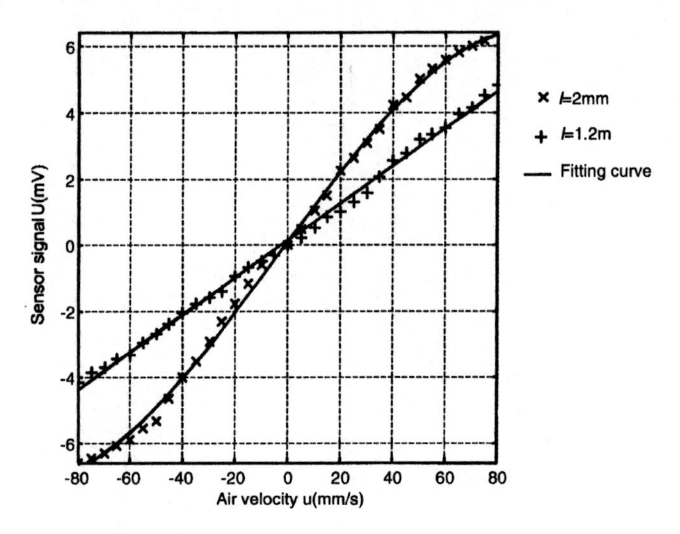

Fig. 27. Measured sensor characteristics

4.4 Pressure sensor

4.4.1 Structure

The structure of a membrane type capacitive pressure sensor is schematically shown in Fig. 28(a). The sensor consists off three rigid PCB-layers and an intermediate flexible foil as the membrane. The first rigid layer carries a horizontal channel required for the fluidic connection between the reference pressure tube and the sensor chamber, which is placed in the next rigid layer. The metal layer of the latter board is structured to form the fixed electrode of the sensing capacitance. The surrounding copper layer of a greater height serves as seals for the chamber and defining the nominal electrode spacing.

This gap between the electrodes is determined either by milling the fixed electrode area or by selective electroplating the surrounding copper layer. The deformable electrode is a 20 nm thin film of aluminium, which is sputtered on the 8 μm polyimide membrane. The last rigid layer of the PCB stack carries the pressure chamber, and on its front side the pressure connection tubes and the electric connectors of the transducer.

4.4.2 Fabrication

Similar to the check valve pump described in section 4.2.3 the sensor is fabricated using a special adhesive bonding technique. In addition, the fabrication process needs a pre-strained membrane, which is coated with metals by sputtering. Figure 28(b) shows a prototype of the sensor described above.

4.4.3 Working principle

The metal layer on the membrane surface and the fixed electrode in the sensor chamber form the sensing capacitor. A pressure difference between the inlet and the sensor chamber represents a mechanical load on the membrane leading to the deformation of the membrane electrode. The deformation in turn changes the gap between the electrodes, and consequently the capacitance.

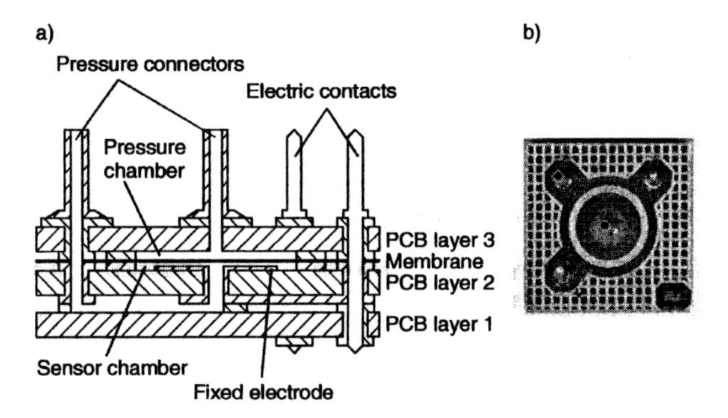

Fig. 28. Capacitive working differential pressure sensor: (a) Structure, (b) Photograph

4.4.4 Characteristics

A syringe pump generates the pressure for deformation of the membrane to measure the characteristics of the sensor. The actual values of the pressure and the capacitance are captured and evaluated using LabView® instrumentation environment. The resulting characteristics for a pressure range of 350 mbar are shown in Fig. 29(a).

The curve shows a non-linear behaviour due to the inversely proportional correlation between the electrode's distance and the sensor's capacitance, even in the case of parallel deflection of rigid electrodes. The non-linear deformation of the membrane adds another source of non-linearity. Only in the small range from –5 to +15 mbar the sensor characteristics can be assumed to be linear, in Fig. 29(b). Thus the sensor is characterized by a nominal capacitance of 22 pF and an averaged sensitivity of 1 pF/mbar. Compared to similar sensors fabricated by other technologies these parameters are relatively high.

By variation of the membranes thickness and of the initial strain the characteristics can be adapted to a wide range of pressure measurement tasks. More linear characteristics can be expected if the sensor is operated in touch mode, if the electrode area of the membrane is stiffened or if a push-pull structure is built up using a pair of identical sensors [35-37].

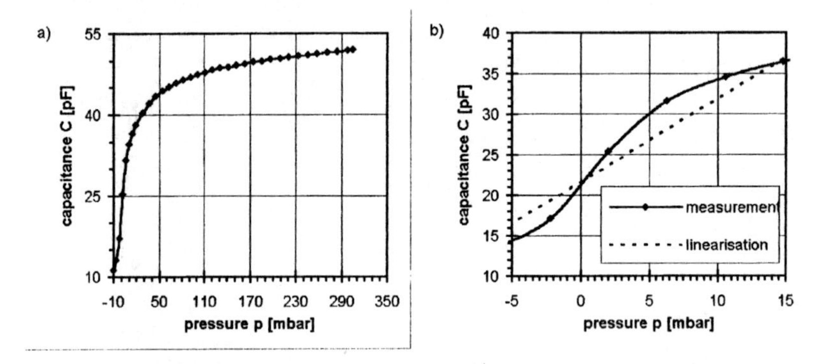

Fig. 29. Characteristics of the capacitive pressure sensor: (a) Full range, (b) With linear approximation for a small pressure range

4.5 Other sensors

Capacitive bubble detector

Many technical, medical and biological systems respond sensitively to gas bubbles in fluids. It is common to obtain incorrect experimental results because of bubbles. In the worst case, the systems fail to function. A bubble detector can give a warning if gas bubbles are detected.

The first bubble detector fabricated with PCB technology [38] consists of two capacitance channels in a row. If a bubble appears in the flow, the first value of capacitance changes while the second value remains constant. When the bubble leaves the first channel and arrives at the second channel, the signals are reversed. The electronic circuit of the bubble detector compares both values of capacitance. If they are different, a LED indicates a detected bubble. Figure 30(a) shows a photograph of the bubble detector.

Microfluidic pH-regulation system

Microfluidic components for a pH-regulation system based on PCB technology are presented in [39]. Such a system can be used for biological cell culture systems. It basically consists of a fluidic part and an electronic part. A chamber contains cultured cells for optical and electrical monitoring. A micropump delivers the nutrition liquid to the cultured cells. For long-term cultivation, a stable physiologic environment is necessary. Therefore, the pH-value of the nutrition medium must be controlled. The sensing concept is realised by an optical sensor and a CO_2-diffuser (see Fig. 30(b)). The signal analysis is done by a microcontroller in the electronic part of the system.

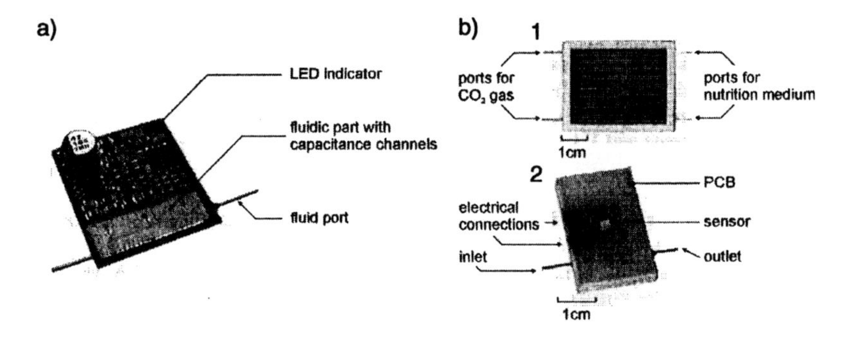

Fig. 30. (a) Capacitive bubble detector, (b) Fluidic components of a pH-regulation system (1: CO_2-diffuser, 2: Optical pH-sensor)

5. CONCLUSION

This chapter describes the basic technology steps to fabricate fluidic components on printed circuit boards. The design, fabrication, and the characterization of a number of device examples are described and discussed. These examples show that different micropumps for handling fluids can be implemented in PCB technology. The liquid flow rates can have a range from some microlitres per minute up to several millilitres per minute. Furthermore, an airflow sensor and a differential pressure transducer for gases are described. These sensors can be used for a closed control loop of the pumps. The measured characteristics of these components impressively show the capability of PCB-technologies for the manufacturing of microfluidic components.

REFERENCES

[1] Walter Lang, "Reflections on the future of microsystems," Sensors and Actuators A, Vol. **72**, 1 (1999).

[2] *MEMS 1999 – Emerging Applications and Markets* (System Planning Corporation, 1999).

[3] *Microfluidic systems – new products*, MST-News, No. 17 (October 1996).

[4] P. Gravesen, J. Branebjerg, and O. S. Jensen, "Microfluidics – a review," J. Micromech. Microeng., Vol. **3**, 168 (1993).

[5] M. Elwenspoek, T. S. Lammerink, R. Miyake, and J. H. J. Fluitman, "Towards integrated microliquid handling system," J. Micromech. Microeng., Vol. **4**, 227 (1994).

[6] G. Stemme, "Micro fluid sensors and actuators," Proc. 6th International Symposium on Micro Machine and Human Science, Vol. **4**, 45 (1994).

[7] B. L. Gray, D. Jaeggi, N. J. Mourlas, B. P. van Drieenhuizen, K. R. Williams, N. I. Maluf, and G. T. A. Kovacs, "Novel interconnection technologies for integrated microfluidic system," Sensors and Actuators A, Vol. **77**, 57 (1999).

[8] J. Jang and S. S. Lee, "Theoretical and experimental study of MHD (magnetohydrodynamic) micropump," Sensors and Actuators A, Vol. **80**, 84 (2000).

[9] D. A. Patterson and J. L. Hennessy, *Cost and trends in cost, Chapter 1.4 in Computer architecture - a quantitative approach, pp. 18-28* (Morgan Kaufmann Publishers, San Francisco, 1997).

[10] G. Herrmann (Ed.), *Handbuch der Leiterplattentechnik, Vol. 3, Eugen G. Leuze Verlag,* (Saulgau, Germany, 1993).

[11] T. S. J. Lammerink, V. L. Spiering, M. Elwenspoek, J. H. J. Fluitman, and A. van den Berg, "Modular concept for fluid handling systems: a demonstrator micro analysis system," Proc. MEMS 96, 389 (1996).

[12] T. Merkel, M. Graeber, and L.Pagel, "A new technology for fluidic microsystems based on PCB technology," Sensors and Actuators A, Vol. **77**, 98 (1999).

[13] T. Merkel, L.Pagel, and H.-W. Glock, "Electric fields in fluidic channels and sensors applications with capacitance," Sensors and actuators A, Vol. **80**, 1 (2000).

[14] S. Richter, A. Wego, and L.Pagel, "Fabrication of micro-fluidic devices using PCB-technology," Proc. Int. MEMS Workshop 2001, 4-6 July, NUS Singapore, 106 (2001).

[15] N.-T. Nguyen, X. Huang, "Microfluidic devices based on PCB-technology," Proc. Int. MEMS Workshop 2001, 4-6 July, NUS Singapore, 722 (2001).

[16] H.-J. Hanke, *Baugruppentechnologie der Elektonik – Leiterplatten*, 1st Ed. (Verlag Technik GmbH, Berlin, 1994).

[17] G. Herrmann and K. Egerer (Eds.), *Handbuch der Leiterplattentechnik – Band.: Neue Verfahren, Neue Technologien*, 1st Ed., Eugen G. Leuze (Verlag Saulgau, Germany, 1991).

[18] W. Scheel (Ed.), *Baugruppentechnologie der Elektronik – Montage*, 1st Ed., (Verlag Technik GmbH, Berlin, 1997).

[19] J. Wissink, A. Prak, M. Wehrmeijer, and R. Mateman, "Novel low cost modulas assembly technology for μTAS using PCB-technology," Proc. MICRO.tec 2000, VDE Verlag Berlin, Vol. **2**, 51 (2000).

[20] N. T. Nguyen, S. Richter, S. Schubert, J. Mehner, W. Doetzel, and T. Gessner, "Micro dosing system," Proc. of SPIE Vol. **3514**, 415 (1998).
[21] J. W. Beams, *Structure and properties of thin films*, in: C. A. Neugebaur, J. B. Newkirk, and D. A. Vermilyea (Eds.) (Wiley, New York, 1959).
[22] E. I. Bromley, J. N. Randall, D. C. Flanders, and R. W. Mountain, "A technique for the determination of stress in thin films," J. of Vac. Sc. & Technology B, Vol. **1**, 1364 (1983).
[23] *Kapton Polyimidfolie – Zusammenfassung der Eigenschaften*, Product description H-38492-2, DuPont (1996).
[24] L. Lencioni, M. Carrozza, A. Menciassi, D. Accoto, N. Croce, and P.Dario, "A micromechatronic system for oil supply to momentum wheels bearing in space satellites," Artificial and natural perception: proceedings of the 2nd Italian Conference on Sensors and Microsystems, p. 338 (1997).
[25] J. G. Smits, "Piezoelectric micropump with three valves working peristaltically," Sensors Actuators A, Vol. **40**, 203 (1990).
[26] Aders Olsson, "Valveless diffuser micropumps," Ph. D. Thesis, Royal Institue of Technology, Sweden (1998).
[27] F. K. Forster, L. Bardell, M. A. Afromowitz, N. R. Sharma, and A. Blanchard, "Design, fabrication and testing of fixed-valve micropumps," Proc. of ASME Fluids Engineering Division, IMECE, Vol. **234**, 39 (1995).
[28] N. T. Nguyen and R. M. White, "Design and optimization of an ultrasonic flexural plate wave micropump using numerical simulation," Sensors Actuators A, Vol. **77**, 229 (1999).
[29] N. T. Nguyen, A. H. Meng, J. Black, and R. M. White, "Integrated flow sensor for in situ measurement and control of acoustic streaming in flexural plate wave micro pumps," Sensors Actuators A, Vol. **79**, 115 (1999).
[30] A. E. Herr, J. I. Molho, T. W. Kenny, J. G. Santiago, M. G. Mungal, and M. G. Garguilo, "Variation of capillary wall potential in electrokinetic flow," Proc. of Tranducers 99, 710 (1999).
[31] M. Graeber, "Entwicklung einer Technologie fuer fluidische Mikrosysteme auf Basis der Leiterplattentechnologie," Ph. D. Thesis, University of Rostock, Germany (1999).
[32] K.-P. Kaemper, J. Doepper, W. Ehrfeld and S. Oberbeck, "A self-filling low cost membrane micropump," IEEE Workshop on MEMS 1998.
[33] R. Linneman et al., "A self-priming and bubble-tolerant piezoelectric silicon micropump for liquid and gases," IEEE Workshop on MEMS 1998.
[34] N. T. Nguyen, *Micromachined flow sensors: a review, Flow Measurement and Instrumentation, p. 7-16* (Elsevier, 1997).
[35] L. Rosengren, J. Soederkvist and L. Smith, "Micromachined sensor structures with linear capacitive response," Sensors Actuators A, Vol. **31**, 200 (1992).
[36] G. Meng and W. H. Ko, "Modeling of circular diaphragm and spreadsheet solution programming for touch mode capacitive sensors," Sensors Actuators A, Vol. **75**, 45 (1999).
[37] X. Li, M. Bao and S. Shen, "Study on linearization of silicon capacitive pressure sensors," Sensors Actuators A, Vol. **63**, 1 (1997).
[38] T. Merkel, "Fluidische Mikrosysteme auf der Basis der Leiterplattentechnologie," Ph. D. Thesis, University of Rostock, Germany (1999).
[39] C. Laeritz, "Konzeption, Konstruktion und Realisierung eines mikrofluidischen Supportsystems auf Basis der Leiterplattentechnologie," Ph. D. Thesis, University of Rostock, Germany (1999).

Chapter 8

NANO AND MICRO CHANNEL FLOWS OF BIOMOLECULAR SUSPENSION

Fan Xijun,[1] Nhan Phan-Thien,[2] Ng Teng Yong,[1] Wu Xuhong,[1] and Xu Diao[1]
Institiute of High Performance Computing, Singapore;[1] Bioengineering Division, National University of Singapore, Singapore[2]

Abstract: In this chapter, we described the two particle methods for simulating flow problems: molecular dynamics (MD) and dissipative particle dynamics (DPD). MD simulates the motion of individual molecules or atoms in a system and provides information on phenomena occurring in atomic scales. DPD is a mesoscaled method dealing with the motion of fluid particles and is feasible to cope with the complex system involving inherently disparate length and time scales, such as polymer suspensions and colloids. The principles, their governing equations and their numerical implementations of these two methods were outlined in this chapter. The molecular models used in polymer rheology were introduced to model bio-macromolecules. Some of the simulated results were presented on the flow of a simple liquid through nano periodic nozzles, using MD simulation, and on the flow of a bio-macromolecular suspension flow a micro channel, using DPD simulation. Finally, some remarks and findings based on MD and DPD simulations were discussed.

Key words: Molecular dynamics simulation, Dissipative dynamics simulation, Molecular models, Nano channel flow, Micro channel flow, Bio-macromolecular suspensions

1. INTRODUCTION

There has been a considerable interest in flow phenomena in and around MEMS (Micro-Electro-Mechanical Systems) devices. An excellent review has been provided by Gad-el-Hak [1]. The mechanics of such a device operating in a gaseous medium have been studied by the Direct Monte Carlo Simulation method based on the well-developed kinetic theory of gases [2], or by using continuum models incorporating slip boundary conditions. The

F.E.H. Tay (ed.), Microfluidics and BioMEMS Applications, 219–264.

situation is much more complicated in liquids, because of molecules are more closely packed. The flow is often characterised by the Knudsen number *Kn* (ratio of the mean free path to the characteristic length of the flow). When $Kn \leq 10^{-3}$, continuum models are supposedly valid, and when $10^{-3} \leq Kn \leq 10^{-1}$, slip flow regime is supposed to take place. When $Kn \geq 10$, rarified flow occurs, and the region corresponding to $10^{-1} \leq Kn \leq 10$ is termed transition flow. Because of the less developed state of micro/nano fluid mechanics, flow verification still is an important part of the research.

Experiments on liquid flow through micro tubes showed that the tube friction coefficient can be quite different from that predicted by the Navier-Stokes equations, even though the flow is laminar, and that the critical transition Reynolds number (from laminar to turbulence) is much less than the accepted value of 2300. However, there is a considerable scatter in the data from various studies. For example, some reported a smaller friction coefficient than the laminar value 64/Re (Re is the Reynolds number); but larger values have also been reported, see, for example, Weilin *et al* [3], Harley and Bau [4] and Pfahler [5]. Some suggested a slippage at the solid wall, but others believe that strong surface force acting on the fluid molecules in the vicinity of the solid wall would prevent slippage. Clearly, the flow behaviour is not very well understood.

Microfabrication technology has made it possible to design and manufacture micro devices which are capable of processing, analysing and delivering biochemical materials, and can be used for a broad range of biomedical and biological applications, such as in clinical assays, diagnostics and injections. Comparing with the conventional medical techniques, these devices promise more efficient and accurate analysis and treatment. One interesting application is the micro needle, used for drug and DNA delivery into cells, local tissue and skin regions. Arrays of hollow microcapillaries have been developed to achieve a high efficiency in cell transportation as well as a precise control over the amount of material delivered [7]. A multichannel silicon probe has been fabricated to deliver a very small and precise amount of drugs into neural tissue of guinea pigs *in vivo*, while simultaneously monitoring and stimulating neuronal activity [8]. Various microneedles have been developed for hypodermic injection and transdermal drug delivery [9][10][11], without pain and tissue trauma caused by using conventional needles.

The delivery of drug and DNA is carried out by the flow of bio-macromolecular suspensions through micro channels. Hence, it is important to understand the effects of macromolecular conformation on micro channel flow in designing micro devices. The characteristic size of microcapillaries is usually in the same order of magnitude as the length of a typical DNA molecule. For example, the microchannel size is about 9-40 μm [6] and the uncoiled length of a λ-DNA is about 25 μm [14]. The ratio of the macromolecular length to the characteristic length of flow field can be taken as the equivalent Knudsen number in bio-molecular flow. When $Kn << 1$, the suspension can be treated as a continuum and the standard viscoelastic constitutive modelling is valid. When $Kn = O(1)$, the suspension may not be treated as a continuum. The conformation of DNA molecules has such a strong influence on the flow and even the Brownian motion of DNA segments may not be neglected. It may be misleading to use continuum rheological constitutive equations derived for polymer solutions with the usual balance equations of fluid mechanics to model the flow.

Some experimental studies have been conducted to understand the mechanics of micro channel flow of bio-molecular suspensions. Perkins *et al* [12] and Smith *et al* [13][14] reported their experimental investigation of DNA conformations both in simple elongation and simple shear flows recently. The flow through microchannels is more complicated, being a combination of inhomogeneous elongation and shear flows. Shrewsbury *et al* [15] investigated the effect of flow on complex biological macromolecules in microfluidic devices in which the channel had a rectangular section and connected two large reservoirs. The fluid accelerated when entering the channel from a reservoir and decelerated when exiting from channel into another reservoir downstream. DNA molecules were observed undergoing elongation, non-uniform shear and compressing while passing through the devices.

Numerical simulation can provide more detailed information on flow and molecular conformations, and it should be an efficient way to explore the flow phenomena in fluidic devices. However, only simple flows have been numerically simulated successfully to date. Some recent reports include, for example, the simulation of freely-draining flexible polymers in steady linear flows [16], bead-rod chains in start-up of extensional flow [17], relaxation of dilute polymer solutions following extensional flow [18], a single DNA molecule in shear flow [19]. The common numerical method used in the above-mentioned articles is the Brownian Dynamics Simulation (BDS). In the method, the bulk flow field is specified *a-priori*, the effects of polymer molecules on the flow field cannot be taken into account without coupling the BDS with the conservation equations (through the stress tensor). When the flow domain size is of the same order of magnitude as the size of the macromolecules, the application of continuum mechanics may be questionable. Hence, the BDS may not be a suitable mean for simulating micro/nano channel flows of bio-macromolecular suspensions.

In this chapter, we are describing two numerical methods, Molecular Dynamics (MD) and Dissipative Particle Dynamics (DPD) simulations, which are based on discrete molecule or particle methods and simulate the flow directly from Newton's second law. The conservation equations of continuum mechanics are not required at all in determining the bulk flow properties. MD is basically a simulation technique at molecular scales; it can accurately account for inter- and intra-molecular interactions that exist in real materials and therefore is especially suitable for simulating molecular processes, i.e., those with nano length and time scales. But in complex structured fluids, such as macromolecular suspensions and colloids, there is a spectrum of characteristic length and time scales for various suspended components, which makes it difficult to deal with using MD, since in this technique the time step should be as small as the lowest vibration period. However, MD may be useful in shedding light on flow phenomena that occur in nanodevices. DPD is a mesoscale technique, in which coarse-grained fluid particles are simulated, and may be more suitable for complex fluids. This method has been widely applied to predict rheological properties of polymer suspensions and colloids.

In this chapter, we first introduce the MD and DPD methods and then describe the molecular models used to mimic bio-macromolecules and report some simulation results.

2. SIMULATION TECHNIQUES

2.1 Molecular Dynamics Simulation

Molecular Dynamics simulation is based on statistical mechanics of non-equilibrium liquids (Evans and Morriss [20]). In essence, it replaces the continuum description with a discrete number of particles, and can provide detailed information on the flow field. This method has been attracting more and more attention in the last two decades. As computer power rapidly increases, doubling in performance every two years, MD simulation has found more and more applications in complex flow areas, including molecular hydrodynamics, Koplik and Banavar [21]. It has been successfully used in studying the lubrication processes and properties of thin liquid films for both Newtonian and non-Newtonian lubricants, see for example, Todd *et al.* [28], Jabbarzadeh *et al* [33][34], Hu *et al* [37][38]. The flow behaviour near the solid surface has been simulated for simple liquids [30][24], and for polymers [35]. Corner flows in the sliding plate and re-entrant problems have been reported by Koplik and Banavar [31][32] for simple liquids and polymers.

The principle, numerical implementation, and code fragments of MD have been described in some monographs, e.g., [29][57][58][59]. Hence, only a brief description is required for completeness here.

In MD, both the liquid and walls are modelled using simple spherical molecules undergoing motions governed by Newton's second law. For a *simple* liquid made up of a collection of particles (called *molecules*), the velocity $\dot{\mathbf{r}}_i$ and the momentum $\dot{\mathbf{p}}_i$ of the molecule i are given by

$$\dot{\mathbf{r}}_i = \frac{\dot{\mathbf{p}}_i}{m}, \quad \dot{\mathbf{p}}_i = \sum_{j \neq i} \mathbf{f}_{ij} + \mathbf{F}_e, \tag{1}$$

where m is the mass of the fluid molecule, $\mathbf{f}_{ij}$ is the internal force on molecule i by molecule j, and $\mathbf{F}_e$ is the force due to external field.

The interaction forces between fluid/fluid and fluid/wall molecules depend on the physical properties and molecular structures of the fluid and wall materials. For simplicity, we adopt the Weeks-Chandler-Anderson (WCA) potential [23], which is a modification of the Lennard-Jones potential:

$$\phi(r_{ij}) = 4\varepsilon\left[\left(\frac{\sigma}{r_{ij}}\right)^{12} - \left(\frac{\sigma}{r_{ij}}\right)^{6}\right] + \varepsilon, \quad r_{ij} < r_c = 2^{\frac{1}{6}}\sigma. \tag{2}$$

Here, $r_{ij} = |\mathbf{r}_{ij}|, \mathbf{r}_{ij} = \mathbf{r}_i - \mathbf{r}_j$, ε is a parameter characterising the strength of the interaction and σ is a molecular length scale, r_c is the cut-off distance, i.e., $\phi(r) = 0$, when $r \geq r_c$. The potential of interaction forces between fluid and wall molecules can be specified in terms of the WCA potential with ε_{fw} and σ_{fw}, which may be different from ε and σ. In the simulation, we choose $\sigma_{fw}/\sigma = 1$ but allow $\varepsilon_{fw}/\varepsilon$ to vary. This potential has been used in the simulation of soft spheres [29].

The external force field is used to drive the fluid in the simulation, and is treated as a gravity field, $\mathbf{F}_e = mg\boldsymbol{\delta}_x$, where $\boldsymbol{\delta}_x$ is a unit vector in the x-direction, and g is the "gravitational" constant. The work done by the external force would partly be converted into heat and increases the temperature of a system. To keep the temperature at a constant, it is necessary to use a thermostat. One of the thermostats is the so-called Gaussian thermostat, which generates the isokinetic ensemble by a differential feedback mechanics [20]. It introduce a non-holonomic constraint into the equations of motion, Eq. (1), to fix the temperature of the system at a constant value,

$$T = \frac{1}{3(N_a - 1)k_B} \sum_{i=1}^{N_a} \left(\frac{\mathbf{p}_i}{m} - \mathbf{v}(\mathbf{r}_i) \right)^2, \tag{3}$$

where N_a is the total number of molecules in the system and $\mathbf{v}(\mathbf{r}_i)$ the mean velocity of the flow at $\mathbf{r}_i$. If we use $\mathbf{F}_i$ to denote the total force acting on molecule i, the constrained equation of motion for molecule i can be written as

$$\dot{\mathbf{p}}_i = \mathbf{F}_i + \alpha\left[\mathbf{P}_i / m - \mathbf{v}(\mathbf{r}_i)\right], \tag{4}$$

here α is a Lagrange multiplier as a control of the feedback. The constant temperature constraint requires $\dot{T} = 0$, i.e.,

$$\sum_{i=1}^{N_a} (\dot{\mathbf{p}}_i / m - \dot{\mathbf{v}}) \cdot (\mathbf{p}_i / m - \mathbf{v}(\mathbf{r}_i)) = 0, \tag{5}$$

where $\dot{\mathbf{v}}$ is acceleration the molecule experienced at $\mathbf{r}_i$ due to the inhomogeneous mean velocity field. From Eqs. (4) and (5), the Lagrange multiplier is determined as

$$\alpha = \frac{\sum_{i=1}^{N_a} (\mathbf{F}_i - m\dot{\mathbf{v}}) \cdot (\mathbf{p}_i / m - \mathbf{v})}{\sum_{i=1}^{N_a} (\mathbf{p}_i / m - \mathbf{v})^2}. \tag{6}$$

Another widely used thermostat is the Nosé-Hoover thermostat, which generates the canonical ensemble by an integral feedback mechanism [20]. The thermostats work well for the system in equilibrium, where $\mathbf{v}$ and $\dot{\mathbf{v}} = 0$, and for simple flow where $\mathbf{v}$ and $\dot{\mathbf{v}}$ are known. In complex flow, the mean velocity of flow is not known *a priori*, and thermostatting may not work efficiently. We feel that thermostats impose additional constraints on the equations of motion, which may or may not be physical. A more physically meaningful method is to adjust the wall temperature to a given constant at each time step, as described by Heinbuch and Fischer [24]. It is equivalent to connecting the wall to an infinite thermal bath. Heat generated by the external force field is removed from the wall into the bath. The flow may not be therefore isothermal.

To render the equations dimensionless, molecular units are used in scaling. The unit of mass is the mass of a molecule, m; the unit of length is σ; the unit of energy is ε; the unit of time is $\sigma\sqrt{m/\varepsilon}$; the unit of density is $m\sigma^{-3}$; the unit of viscosity is $\sqrt{m\varepsilon}\sigma^{-2}$; and the unit of stress is $\varepsilon\sigma^{-3}$.

2.2 Dissipative Particle Dynamics

There are some of coarse-grained simulation techniques. Lattice-Gas Automata (LGA) [60] is one of such techniques, where the time step used can be orders of magnitude larger than that used in MD. However, there have been fundamental problems: isotropy and Galilean invariance are not observed, since LGA restricts particles to move on a regular lattice. For complex fluids, these problems cannot be easily overcome. It is also quite cumbersome to satisfy no slip boundary conditions on a 3D complex boundary. This is a major disadvantage of LGA from a practical point of view. Likewise, another coarse-grained technique is the Brownian Dynamics simulation (BDS). It has been used extensively in Rheology and requires the kinematics to be specified *a priori*. It conserves only the number of particles in the system and its macroscopic behaviour is diffusive only.

The Dissipative Particle Dynamics (DPD) is a mesoscale fluid simulation method introduced by Hoogerbrugge and Koelman [43], and has its basis in statistical mechanics (Español and Warren [44], Marsh [45]). DPD facilitates the simulations of static and dynamics complex fluid systems on physically interesting length and time scales. Unlike LGA, the DPD system exists in continuous space, rather than on lattices. Hence it removes the problems in LGA, and isotropy and Galilean invariance can both be observed. However, it is of a coarse-grained scale simulation and retains the computationally efficiency of the LGA. The DPD conserves not only the number of particles but also the total momentum of the system. This implies that the macroscopic behaviour will not only be diffusive like the BDS, but also hydrodynamic in nature, i.e., the transport equations of the system will be of the form of mass and momentum conservation. Thus, the flow kinematics can be found as a part of the solution procedure.

A DPD particle is not regarded as a single molecule in the MD simulation, but a group of "fluid particles". The interaction between DPD particles is much softer than that between atoms; consequently much larger time steps can be used in the simulation. Dissipation can be introduced into DPD by means of Brownian dashpots. The fluid particles no longer conserve the energy in each "collision" but conserve the momentum of system. Since the system is described at a coarse-grained level, its hydrodynamic behaviour can be observed with a considerably smaller number of particles.

DPD has been applied to various complex systems, such as polymer suspensions [47][48][49][50], colloids [51][52][53], and multi-phase fluids [54][55][56]. Although it is a promising technique for complex fluids, there has been no micro-fluidic application of DPD reported.

DPD simulates the motion of a set of interacting particles; for a simple DPD particle *i*, we have

$$\frac{d\mathbf{r}_i}{dt} = \mathbf{v}_i, \quad \frac{d\mathbf{v}_i}{dt} = \sum_{j \neq i} \mathbf{f}_{ij} + \mathbf{F}_e, \tag{7}$$

where $\mathbf{r}_i$ and $\mathbf{v}_i$ are the position and velocity vectors of particle *i*, and similar to molecular units adopted in MD, the unit of mass is taken to be the mass of particles; and $\mathbf{f}_{ij}$ is the interparticle force acting on particle *i* by particle *j*, which is assumed to be pairwise additive and consists of three parts: conservative force, $\mathbf{F}_{ij}^C$, dissipative force, $\mathbf{F}_{ij}^D$, and random force, $\mathbf{F}_{ij}^R$:

$$\mathbf{f}_{ij} = \mathbf{F}_{ij}^C + \mathbf{F}_{ij}^D + \mathbf{F}_{ij}^R. \tag{8}$$

In Eq. (7), the sum runs over all other particles within a certain cutoff radius r_c, taken as the unit of length, i.e., $r_c = 1$. The conservative force, $\mathbf{F}_{ij}^C$, is a soft repulsion and given by

$$\mathbf{F}_{ij}^C = \begin{cases} a_{ij}\left(1 - r_{ij}\right)\hat{\mathbf{r}}_{ij}, & r_{ij} < r_c \\ 0, & r_{ij} \geq r_c \end{cases}, \tag{9}$$

where a_{ij} is a maximum repulsion between particles i and j; and $\mathbf{r}_{ij} = \mathbf{r}_i - \mathbf{r}_j$, $r_{ij} = \left|\mathbf{r}_{ij}\right|$ and $\hat{\mathbf{r}}_{ij} = \mathbf{r}_{ij} / r_{ij}$ is the unit vector directed from particle j to particle i. The dissipative force and random force are

$$\mathbf{F}_{ij}^D = -\gamma\omega^D\left(r_{ij}\right)\left(\hat{\mathbf{r}}_{ij} \cdot \mathbf{v}_{ij}\right)\hat{\mathbf{r}}_{ij}, \tag{10}$$

and

$$\mathbf{F}_{ij}^R = \sigma\omega^R\left(r_{ij}\right)\theta_{ij}\hat{\mathbf{r}}_{ij}, \tag{11}$$

where γ and σ are coefficients characterising the strengths of the dissipative and random forces, $\omega^D\left(r\right)$ and $\omega^R\left(r\right)$ are r-dependent weight functions vanishing for $r \ge r_c = 1$, $\mathbf{v}_{ij} = \mathbf{v}_i - \mathbf{v}_j$, and θ_{ij} is a random function with the properties

$$\langle \theta_{ij}(t) \rangle = 0 \text{ and } \langle \theta_{ij}(t)\theta_{kl}(t') \rangle = (\delta_{ik}\delta_{jl} + \delta_{il}\delta_{jk})\delta(t-t'). \tag{12}$$

The strength of the random force is the integral correlation over a time scale considerably larger than its correlation time scale:

$$\left[\sigma\omega^R(r_{ij})\right]^2 = \int_{-\infty}^{\infty} \langle \mathbf{F}_{ij}^R(t) \cdot \mathbf{F}_{ij}^R(t+\tau) d\tau \rangle. \tag{13}$$

This strength may not be prescribed arbitrarily; in fact a detailed balance requires that

$$\omega^D(r) = \left[\omega^R(r)\right]^2 \text{ and } \gamma = \frac{\sigma^2}{2k_B T}, \tag{14}$$

where k_B is the Boltzmann constant and T is the temperature of the system. This is seen as a *fluctuation-dissipation* theorem for the system (see Huilgol and Phan-Thien [27]), constraining the mean square velocity to be proportional to the temperature (i.e., a thermostat).

A simple choice for the weight function may be

$$\omega^D(r) = \left[\omega^R(r)\right]^2 = \begin{cases} (1-r)^2, & r < r_c, \\ 0, & r \ge r_c. \end{cases} \tag{15}$$

The random force between i and j, $\mathbf{F}_{ij}^R$, represents the results of thermal motions of all molecules contained in particles i and j. It tends to ``heat up" the system. While the dissipative force, $\mathbf{F}_{ij}^D$, reduces the relative velocity of two particles and removes kinetic energy from their mass centre to cool down the system. When the detailed balance is satisfied, the system temperature will approach the given value. The dissipative and random forces act as thermostats here. Comparing Eqs.(7) and (8) to Eq.(4), one can see that DPD is a version of MD, with a softened repulsive force and Brownian-dashpot thermostatting. However, the mathematical frameworks are quite different: the governing equations for MD are deterministic, but they are a set of stochastic differential equations for DPD. The corresponding evolution equation for the distribution function is a parabolic Fokker-Planck equation for DPD, but a hyperbolic equation, without the diffusion terms, for MD. The reader interested in the kinetic origins of these methods is referred to [20] and [45].

Solid walls are usually represented by frozen DPD particles, consisting of rigid objects moving in a rigid body motion, which is determined by the total forces and torques acting on the objects. When DPD particles represent solid walls, they are simply fixed at the lattice sites and move with the velocity of the wall. Various coarse-grained molecular models in polymer rheology can be employed to model bio-macromolecules. Beads in molecular models will be replaced by DPD particles. The intermolecular forces will act on these particles and should be added to the right hand side of Eq.(7). These models will be outlined in next section.

2.3 Molecular Models

Various molecular models have been used in MD and DPD simulations. We may classify the models into three categories: detailed models, simplified models and coarse-grained models. A detailed model is designed to reproduce the behaviour of a real polymer. It takes into account detailed features of a molecule, such as the mass and charge of each atoms, the length and strength of covalent bonds, the angles between chemical bonds, the potential energies of bonding, non-bonding, angle, torsion, improper torsion and dihedrals, etc. This kind of models is used in MD for simulating the properties and structures of complex biopolymers, such as proteins and DNA molecules [61]. In the second category are much simpler models, aiming at capturing some general aspects of molecular chains rather than all the myriad quantitative details. As far as the conformations and trajectories of molecular chains are concerned, some of the degrees of freedom can be frozen and only those relevant to molecular motion are studied [29]. The most useful model in this category is the chain model, e.g., alkane chains. Each chain segment represents a CH_2 or CH_3 group. The internal degrees of freedom in the group are frozen and neglected. In the last category are the coarse-grained models, which have commonly been used in polymer physics and rheology. The models are similar to those in the second category, but are different in their coarse-grained levels. A chain segment may represent many chemical groups, i.e., a part of the molecule. The most coarse-grained model is the dumbbell, which represents a whole molecular chain. Coarse graining also allows for more flexibility in incorporating artificial force potentials into the model.

Here we briefly describe some coarse-grained models, since they are closely related to DPD simulations in bio-MEMS. Readers interested in detailed models are referred to [61]. The coarse-grained models are usually composed of beads and connectors. The mass of the molecule is lumped into the beads, which is distributed along the backbone of the molecule, and the connectors connect neighbouring beads to form chains, rings or stars. The feature of models is determined by the mechanical models of connect forces and the number of segments. Some of models commonly used in rheology are [46]:

1. Bead-rod models, such as rigid dumbbells, Kramers freely jointed bead-rod chains and Kirkwood-Riseman freely rotating chains;
2. Linear bead-spring models, such as Hookean dumbbells and Rose bead-spring chains;
3. Non-linear bead-spring model, such as finitely extendable non-linear elastic (FENE) dumbbells and FENE and worm-like chains.

The bead-rod models are geometrically constrained molecular models, in which the connector force is indeterminate. The distance between two neighbouring beads is constrained to be the length of the rod all the time. The number of constrains is equal to that of rods. A constrain imposed on the system can eliminate one degree of freedom. This constrained problem is usually converted into a unconstrained one by introducing Lagrange multipliers and then solved by a matrix or a relaxation method [29].

Bead-spring models are unconstrained and are relatively easier to implement than constrained models. The spring force is model-dependent; various springs can be employed to construct chains. The simplest one is the Hookean spring,

$$\mathbf{F}_{ij}^{S} = -H\mathbf{r}_{ij}, \tag{16}$$

where H is the spring constant and subscripts i and j denote the numbering of beads connected by the spring. In the absence of tension, the equilibrium distance between the beads is zero. A slight modification is the Fraenkel spring, whose force law is

$$\mathbf{F}_{ij}^{S} = -H\left(1-\frac{r_0}{r_{ij}}\right)\mathbf{r}_{ij}, \tag{17}$$

where r_0 denotes the segment length when the chain is free of tension. When r_0 approaches to zero, the Hookean spring is recovered, and when H to infinity, we get a rigid rod of length r_0. Since the length of linear springs can be stretched to infinity, linear-spring models can be extended without bounds, and together with this, an infinite elongational viscosity at a finite elongation rate.

Non-linear bead-spring models are more realistic in their extension behaviours. A simple and widely used non-linear model is the Finitely Extendable Non-linear Elastic (FENE) spring:

$$\mathbf{F}_{ij}^{S} = -\frac{H\mathbf{r}_{ij}}{1-\left(r_{ij}/r_m\right)^2}, \tag{18}$$

where r_m is the maximum length of one segment. When r_{ij}/r_m approaches zero, the model reduces to Hookean spring. The spring force increases drastically with r_{ij}/r_m, and becomes infinite as $r_{ij}/r_m = 1$. This model can predict the shear-rate dependent viscosity and finite elongational viscosity. The Fraekel spring results in a simple modification of the FENE spring:

$$\mathbf{F}_{ij}^{S} = -\frac{H\left(r_{ij}-r_0\right)\hat{\mathbf{r}}_{ij}}{1-\left[\left(r_{ij}-r_0\right)/r_m\right]^2}, \tag{19}$$

When no tension applies to the chain segment, its equilibrium length is r_0. Chain, ring and star models can be constructed with the above-mentioned spring laws.

Worm-like chains are thought to be more realistic models for DNA molecules in an aqueous solution. The spring force law is given by [69]

$$\frac{F_{ij}\lambda_p}{k_B T} = \frac{1}{4}\left(1-\frac{r_{ij}}{l}\right)^{-2} - \frac{1}{4} + \frac{r_{ij}}{l}, \tag{20}$$

where λ_p is the effective persistence length and l is the length of a chain segment.

The relaxation time is an important parameter to characterise the molecular motion. The time constant for various models can be outlined as follows [46]. The basic time constants are defined in terms of the parameters of dumbbell models. For Hookean dumbbells we can define a time constant, λ_H, to be

$$\lambda_H = \frac{\zeta}{4H}, \tag{21}$$

where ζ is the friction coefficient of a bead. For rigid dumbbells, we have another time constant

$$\lambda_Q = \frac{\zeta r_m^2}{12k_B T}, \tag{22}$$

where r_m is the length of the rod. For FENE dumbbells, we can obtain the above two time constants. The ratio of these two constants is useful to characterize the behaviour of the FENE dumbbell. We have

$$b = \frac{3\lambda_Q}{\lambda_H} = \frac{H r_m^2}{k_B T}. \tag{23}$$

Chain molecules usually have a spectrum of relaxation times. The Rouse chain, consisting of beads and Hookean springs, has a discrete spectrum. The j-th relaxation time an be expressed as

$$\lambda_j = \frac{\zeta / 2H}{4\sin^2\left(j\pi / 2N_b\right)} = \frac{\lambda_H}{2\sin^2\left(j\pi / 2N_b\right)}, \; j = 1, \ldots, N_b - 1, \tag{24}$$

where N_b is the number of beads in the chain. For FENE chains, there is no expression found for the spectrum of relaxation times. However, a modification of FENE chains, called the FENE-PM chain by Wedgewood [66], has the same spectrum as the Rouse chain, Eq. (24). They defined a time constant for the FENE-PM chain to be

$$\lambda_{fene} = \frac{b}{b+3} \lambda_H \frac{N_b^2 - 1}{3}. \tag{25}$$

The geometry of macromolecules in solution can be characterised by the gyration radius, end-to-end or contour lengths in equilibrium [46]. The contour length is more appropriate to represent the molecular size in simulation and equal to $\left(N_b - 1\right) L_c$, here L_c is the length of one segment of the molecular chain. For the bead-rod models, L_c is just the length of the rod. For the bead-spring models, we may use the equilibrium length of a segment as L_c. For Hookean springs, we have

$$L_c = \sqrt{\frac{3k_B T}{H}}, \tag{26}$$

and for FENE springs,

$$L_c = r_m \sqrt{\frac{3}{b+5}}. \tag{27}$$

2.4 Numerical details

Some of simulation details on MD and DPD are briefly given here. Since the numerical schemes employed in both simulations are basically the same, we will use the term "particle" to denote molecules in MD and particle for DPD and most of schemes are applicable to both simulations.

The initial particle configurations for the fluid and the wall are generated separately by a pre-processing program and read in as input data. The total number of particles depends on the size and geometry of the flow domain and the densities of the fluid and wall materials. All fluid particles are initially located at the sites of a face-centred cubic (FCC) lattice. The initial velocities of fluid particles are set randomly according to the given temperature. In MD, the initial velocities of wall molecules are set in the same way as those of fluid molecules but in DPD the wall particles are frozen. At the beginning of simulation the particles are allowed to move without applying external forces until a thermodynamic equilibrium state is reached. Then the external force field is applied to fluid particles and the non-equilibrium simulation starts.

In MD the equations of motion are solved by the Leap-Frog method, which is most efficient and with the accuracy of $O\left(\Delta t^2\right)$. It also conserves the system energy well [29]. In DPD, the dissipative force is dependent on the velocity and a modified version of the velocity-Verlet algorithm is used [62]. This algorithm can be described as follows:

$$\mathbf{r}_i(t+\Delta t) = \mathbf{r}_i(t) + \Delta t \mathbf{v}_i(t) + \frac{1}{2}\Delta t^2 \dot{\mathbf{v}}_i(t), \tag{28}$$

$$\tilde{\mathbf{v}}_i(t+\Delta t) = \mathbf{v}_i(t) + \lambda \Delta t \dot{\mathbf{v}}_i(t), \tag{29}$$

$$\dot{\mathbf{v}}_i(t+\Delta t) = \dot{\mathbf{v}}_i\left(\mathbf{r}_i(t+\Delta t), \tilde{\mathbf{v}}_i(t+\Delta t)\right), \tag{30}$$

$$\mathbf{v}_i(t+\Delta t) = \mathbf{v}_i(t) + \frac{1}{2}\Delta t\left[\dot{\mathbf{v}}_i(t) + \dot{\mathbf{v}}_i(t+\Delta t)\right]. \tag{31}$$

where $\dot{\mathbf{v}}_i(t)$ denotes the acceleration of (i.e. the total force on) particle i at instant t and position $\mathbf{r}_i(t)$, and $\tilde{\mathbf{v}}_i(t+\Delta t)$ is the prediction of the velocity of particle i at instant t+Δt, and λ is an empirically introduced parameter, which accounts for some additional effects of the stochastic interactions. If the total force were independent of the velocity, the standard velocity-Verlet algorithm would be recovered for $\lambda = 0.5$. Groot and Warren found the optimum value of λ is 0.65. For this value of λ, the time step can be increased to $\Delta t = 0.06$ without loss of temperature control in simulating an equilibrium system with $\rho = 3$ and $\sigma = 3$ [62].

Periodic boundary conditions are applied on the fluid boundaries of the computational domain. The solid boundaries are represented by wall particles, which are located at the sites of a planar face-centered lattice. The mass of wall particles is usually assumed to be identical to that of fluid particles. In MD each wall molecule is anchored at its lattice site by a non-linear spring. The equations of motion for wall molecules are similar to Eq. (1), except the external force is now replaced by the spring force. Hookean spring is commonly used to keep a wall molecule oscillating about its lattice site. Here we propose to use a non-linear spring, with potential

$$\phi_{spring} = \frac{H_1}{2} R^2 + \frac{H_2}{4d_{cr}^2} R^4 + \frac{H_3}{6d_{cr}^2} R^6, \tag{32}$$

where H_1, H_2 and H_3 are the spring constants, R is the displacement of the wall molecule from its lattice site and d_{cr}^2 is the critical mean-square displacement of wall molecules. This critical value is determined according to Lindemann criterion (McDonald [22]) for melting:

$$\langle R^2 \rangle / d_{\min}^2 < 0.023, \tag{33}$$

in which $d_{\min}^2$ is the distance of a wall molecule to its nearest neighbour. This criterion requires the mean displacement of wall molecules be small enough to keep the wall in solid state. Hence the critical mean-square displacement, d_{cr}^2, should be smaller than $0.023\,d_{\min}^2$ and is assumed to be $0.01\,d_{\min}^2$ here. The spring constants should be adjusted to keep the mean-squared displacement of wall molecules to be about d_{cr}^2. With a complex geometry, the local force acting on the different part of the wall may vary drastically and the linear spring forces may not be able to guarantee a small enough molecular displacement everywhere, even when the criterion has been satisfied globally. This is due to the fact that only two layers of molecules are employed to represent a solid wall. In reality, the solid wall consists of a large number of molecules and the movement of molecules on the wall surface would be resisted by other molecules inside. Hence the large displacement of surface molecules would not occur. The linear force is a reasonable model for wall molecules under small displacement.

In DPD, the solid wall is usually represented using frozen particles; there is no real physical need to allow for random motion and dissipation in the wall particles. Due to the soft repulsion between DPD particles, it is difficult to prevent fluid particles from penetrating wall particles, and to efficiently slow down particles near the wall surface to yield no-slip boundary conditions. A higher density of wall material and larger repulsive forces are necessary to strengthen the wall effect. These, however, yield large density distortions in the flow field, similar to MD results. Special treatments have been proposed to implement no slip boundary condition in DPD simulation without using frozen wall particles. Revenga *et al* [64][65] used effective forces to represent the effect of wall on fluid particles instead of using wall particles. For planar wall, the effective forces can be obtained analytically. But these forces are not sufficient to prevent fluid particles from penetrating the wall. When particles cross the wall, they used a wall reflection law to reflect particles back to the fluid side. Willemsen *et al* [63] added an extra layer of particles outside of the simulation domain. The particle positions and velocities in this layer are determined by the particles inside the simulation domain near the wall, such that the mean velocity of a pair of particles inside and outside the wall satisfies the given boundary conditions. We still use frozen particles to represent the wall. The wall particles are fixed at their lattice sites. Near the wall we assume that there is a thin layer, in which the no slip boundary condition holds. We enforce a random velocity distribution in this layer with zero mean and corresponding to the given temperature. Similar to Revenga et al's reflection law, we further require that particles in this layer are always to leave the wall. The velocity of particle i should be

$$\mathbf{v}_i = \mathbf{v}_R + \mathbf{n}\left(\sqrt{(\mathbf{n}\cdot\mathbf{v}_R)^2} - \mathbf{n}\cdot\mathbf{v}_R\right), \tag{34}$$

where $\mathbf{v}_R$ is the random vector and $\mathbf{n}$ the unit vector normal to the wall and pointing to the fluid. The thickness of the layer and strength of repulsion between wall and fluid particles are chosen artificially to reduce the velocity and density distortion. In the numerical examples reported here, the layer thickness is chosen to be the minimum between 1% of channel width and 0.5 of the cutoff radius. A thin layer is necessary to prevent the frozen wall to cool down the fluid and this method is more flexible when dealing with a complex geometry.

The main effort in solving the equations of motion is to compute the interaction forces between the particles. A cell sub-division and linked-list approach as described in Rapaport [29] are employed to reduce the computation time in the force calculation. The linked list associates a pointer with each data item and to provide a non-sequential path through the data. In the cell algorithm, linked lists associate particles with the cells in which they reside at any given instance. It needs only a one dimension array to store the list and makes the search of neighbouring molecules more efficiently. The flow domain is divided into grids and local data are collected in each bin. The flow properties are calculated by averaging over all sampled data in each bin.

The stress tensor is calculate by using the Irving-Kirkwood model [25]:

$$\begin{aligned}\mathbf{S} &= -\frac{1}{V}\left[\sum_i m\mathbf{u}_i\mathbf{u}_i + \sum_i \sum_{j>i} \mathbf{r}_{ij}\mathbf{f}_{ij}\right] \\ &= -n\left\langle \sum_i m\mathbf{u}_i\mathbf{u}_i + \sum_i \sum_{j>i} \mathbf{r}_{ij}\mathbf{f}_{ij} \right\rangle,\end{aligned} \tag{35}$$

where $\mathbf{u}_i = \dot{\mathbf{r}}_i - \mathbf{v}(\mathbf{x})$ is the peculiar velocity of particle i, with $\mathbf{v}(\mathbf{x})$ being the stream velocity at position **x**, n the number density of particles, and $\langle ... \rangle$ denotes the ensemble average. The first term on the right side in the above equation describes the contribution to the stress from the micro momentum transfer and the second term from the interparticle forces. The forces $\mathbf{f}_{ij}$ are calculated from Eq. (2) or Eqs. (8) – (11) for a simple particle i. If particle i denotes a bead on molecular chains, $\mathbf{f}_{ij}$ should include the total spring force on the bead. There is an obvious similarity to the expression for the average stress tensor in a suspension [26][27]. Non-Newtonian properties arise naturally from the interparticle interaction terms. The constitutive pressure can be determined from the trace of the stress tensor:

$$p = -\frac{1}{3} tr\mathbf{S}. \tag{36}$$

Note that neither the stress nor the pressure participate in determining the motion of the particles - they are used in a post-processing to obtain rheological properties of the liquid

3. NANO CHANNEL FLOW OF A SIMPLE LIQUID

In this section we employ MD to simulate a complicated nano channel flow of a simple liquid (size ~10 nm), and attempt to shed some light on the mechanisms of micro channel flow. This flow is more complicated than Couette (simple shear) or Poiseuille flows, which have been extensively investigated so far. Poiseuille flow of Lennard-Jones-type liquids in narrow pores of a few molecular sizes has recently been attracting much research interest. In such a simple flow, the departure of flow behaviour from what predicted by the Navier-Stokes (NS) solution can be revealed clearly. It is usually assumed that the liquid density is a function of temperature alone, which does not vary appreciably over molecular length and molecular relaxation time scales. The liquid transport coefficients are also assumed independent of position and time. Hence the classical theory predicts the quadratic velocity profiles in Poiseuille flow. However, MD simulations show that the density is not uniform across the channel, especially near the solid wall: significant variation in the density exists. Travis and Gubbins further showed that for channels of about 4 molecular-diameter width, the density oscillates along the flow direction with a wavelength of the order a molecular diameter [39]. It is remarkable that in planar Poiseuille flow an approximate quadratic velocity profile is still obtained for channels of 10 molecular-diameter width [39], or larger channel [28], despite the observed variations in the density. For channels of less than 10 molecular-diameter width, the velocity profile is no longer quadratic [39][40]. Travis and Gubbins further found that the NS equations seem to break down for channel of less than 5 molecular-diameter width [40]. It is believed that, for Couette and Poiseuille flows, when the size of channels is larger than about 10 molecular diameters, the velocity and stress fields are in complete agreement with the NS solutions with no-slip boundary conditions [21].

The flow occurs in practical micro channels may be more complicated than Poiseuille or Couette flow, since the wall surfaces of real channels are more than molecular rough. The roughness of the wall surface would have a large influence on the flow due to the significantly high ratio of surface-area to volume. Hence, there may be no simple flow existing in real nano channels. The flow through a nozzle can be regarded as a typical complicated flow. The area of flow passage is changing abruptly along the flow direction due to sudden contraction and diverging geometry. Molecules would be undergoing acceleration and deceleration as they move though the channel. When the size of the surface roughness is in the same order of magnitude as the size of channel, abrupt contraction and diverging would exist in nano channels. Hence, we suppose that this kind of flow may be close to that in real nano channels in some respects. It is thus interesting to find out what kind of deviations from the NS predictions in a complicated nano channel flow.

To compare the results of the MD simulation with those of the classical NS theory, we assume the density and temperature of the fluid are uniform in the flow field and the viscosity of fluid is constant, and write the NS equation in dimensionless form as

$$\mathrm{Re}\frac{d\mathbf{u}}{dt} = -\nabla p + \nabla^2 \mathbf{u} + \tilde{g}\boldsymbol{\delta}_x, \quad \nabla \cdot \mathbf{u} = 0, \tag{37}$$

where $\mathrm{Re} = \bar{\rho}UL/\mu$ is the Reynolds number with $\bar{\rho}$ being the fluid density, U and L the characteristic velocity and length of the flow field, μ the viscosity; $\tilde{\mathrm{g}} = \bar{\rho}\mathrm{gL}^2/\mu\mathrm{U}$ is the dimensionless body force, which can be absorbed into the pressure term. This last parameter is the ratio of the pressure drop imposed over the channel, $\bar{\rho}gL$, to the characteristic viscous force, $\mu U/L$, (a Poiseuille number). The flow past the periodic nozzles can be simulated using finite element method (FEM), subjected to the no-slip boundary conditions.

In the MD simulation, the viscosity is unknown and it is also difficult to determine the viscosity from simulated results, since the viscosity-temperature relation cannot be derived from the molecular model. Also there are discontinuities in velocity and temperature across the wall surface. Some researchers determined the apparent viscosity by fitting the velocity profiles to the isothermal NS solution using no-slip boundary conditions for Poiseuille flow [28][29]. To compare FEM to MD results, here we use the Reynolds number as an adjustable parameter in the finite element simulation to match the flux of the MD simulation. The dimensionless body force (Poiseuille number) can be re-written as $\mathrm{Re}\,Lg/U^2$. The Reynolds number or Poiseuille number can be used as the similarity criterions to characterise the flows. That is, two flows with the same parameters (same Re and $\mathrm{Re}\,Lg/U^2$) and geometrically similar must be dynamically similar in classical fluid mechanics.

In the following sections, we present some MD results in two periodic nozzles and comparing them with NS solutions obtained by FEM. The validity of hydrodynamic similarity applying to nano channel flows is examined [42].

3.1 Flow through a smaller periodic nozzle

The computational domain of the smaller periodic nozzle is $-18.85 \le x < 18.85$, $-3.37 \le y < 3.37$, $-10.11 \le z \le 10.11$. This corresponds to about 7nm in width. Gravity (driving force) is set to 0.3 and is applied in the x-direction. The density of fluid, $\overline{\rho}$, is 0.8362 and the density of wall, $\overline{\rho}_w$, is $1.6\overline{\rho}$. There are 3800 molecules including 1100 wall molecules used in the simulation. The projection of the initial configuration of wall and fluid molecules is shown in Fig. 1.

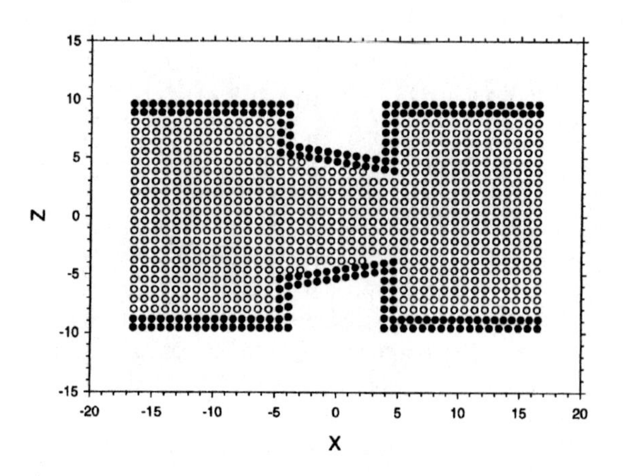

Figure 1. Initial configuration of wall and fluid molecules

The spring constants of wall molecules are: H_1 = 57, H_2 = 20, and H_3 = 50. The wall/fluid interaction is characterised by $\varepsilon_{wf} / \varepsilon = 3.5$. The wall temperature is kept at unity during the simulation. The integration time step is chosen to be 0.005. The whole domain is divided into 100×100 bins for collecting the local data.

The longitude velocity profiles on some typical flow sections through the passage of the channel are shown in Figs. 2 to 4. From these figures, it is seen that the nozzle has a strong influence on the flow behaviour. The fluid velocity in the central part of the channel accelerates as the fluid approaches the inlet and passes through the nozzle, and decelerates as it leaves the nozzle outlet. There are no vortices found behind the nozzle. Periodicity enforces the same velocity profiles on the inlet and outlet of the computational domain. For a given Lg/U^2, the Re number can be adjusted in the FEM simulation to obtain the same mean velocity to MD results in the entry of the flow domain. The positions of the wall surfaces near the entry boundary are at z = 8.188, i.e., L is about 16.38. When g = 0.3, the mean velocity is 0.8144 from the MD simulation. We adjust Re in the FEM to match the mean velocity of the MD and finally get the Reynolds number to be 8.144 and the Poiseuille number to be about 60.32. The results for the velocity profile obtained from the FEM are also plotted in Figs. 2 to 4 (solid lines).

In the region $-5.05 \leq x \leq 5.05$, as shown in Fig. 3, there is a significant slip velocity on the wall surface in the MD simulation. The discrepancy between two sets of results in this region is much more pronounced than that upstream and downstream of the nozzle, see Figs. 2 and 4. The wall slip in the MD simulation is closely related to the layered structure near the wall and the driving forces. Furthermore, NS simulation assumes a constant density throughout the flow field for isothermal incompressible fluid, which is far from the truth for the MD simulation, see Figs. 5 and 6. From these two figures, we can see that

1. There is a large density fluctuation near a solid wall. This is due to fluid/wall molecules interactions. The layered structure near a wall surface always exists.
2. The flow momentum would rearrange the molecules in the layers near a wall. The layered structure becomes more obvious at the inlet of the nozzle, since the flow momentum out the layers is reduced by the contraction of the nozzle. In the nozzle the layers are reduced somewhat and driven by the higher momentum of the flow. Hence a larger slip velocity on the wall surface is found there.
3. The contraction of the nozzle resists molecules with low momentum. Hence more molecules are staying in front of the nozzle than behind it and the density is not uniform along the flow direction, see curves for x = -6.2143 and 6.2143 in Fig. 5.

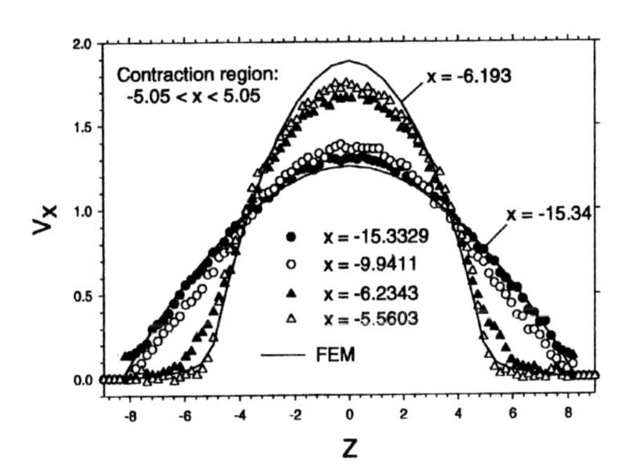

Figure 2. Velocity profile on a typical flow section upstream of the nozzle with g = 0.3

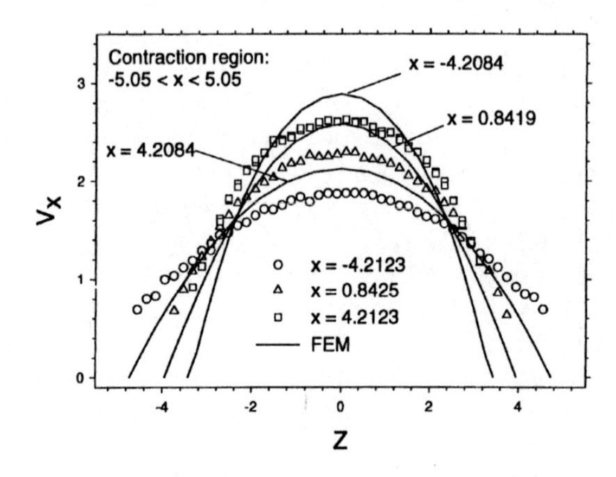

Figure 3. Velocity profile on a typical flow section in the nozzle with g = 0.3

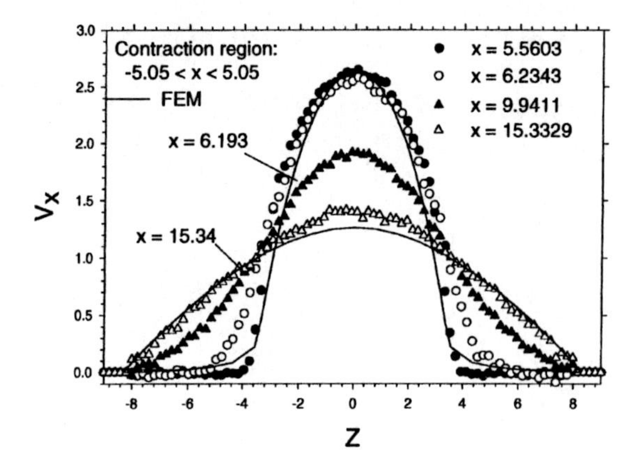

Figure 4. Velocity profile on a typical flow section downstream of the nozzle with g = 0.3

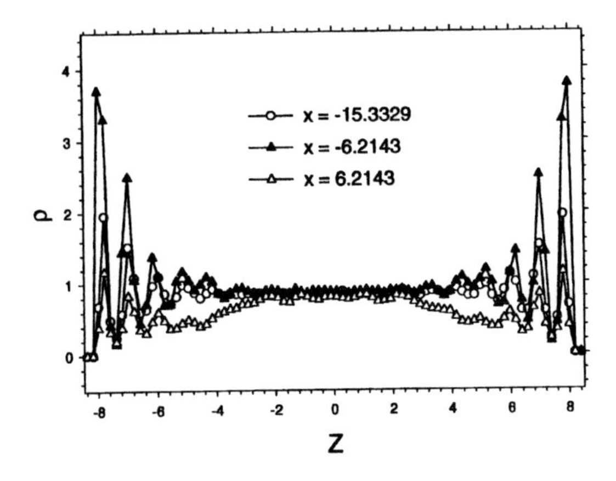

Figure 5. The density profile upstream and downstream of the nozzle

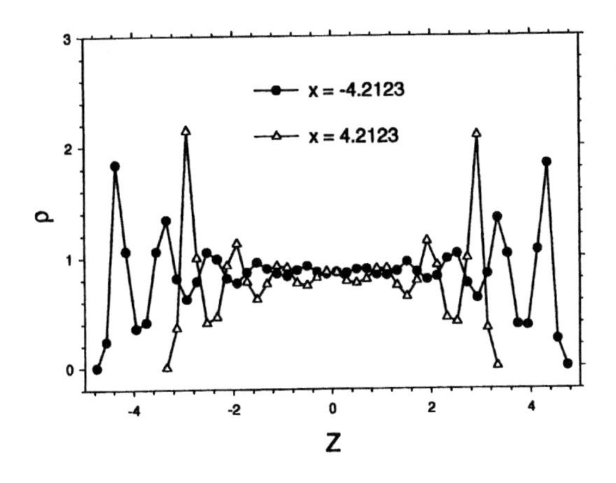

Figure 6. The density profile in the nozzle

The pressure distributions are shown in Figs. 7 and 8. There is a high degree of similarity in the density and pressure profiles, a consequence of the constitutive equations used to calculate the pressure, Eqs. (35) and (36); they are strong functions of the density and the second term in Eq. (35) dominates the stress. In the FEM simulation, it was assumed the density and temperature are uniform in the flow field. Hence there were no pressure and stress fluctuation. This is an obvious deviation from the NS hydrodynamics.

An appropriate comparison for pressure and stress between two simulations cannot be made. The constitutive equation, Eq. (35), can be highly non-Newtonian. In MD simulations, the flow behaviour is governed by intermolecular potentials, not by the constitutive equations for the stress tensor. It is still remarkable that the two simulations result in qualitatively comparable velocity field.

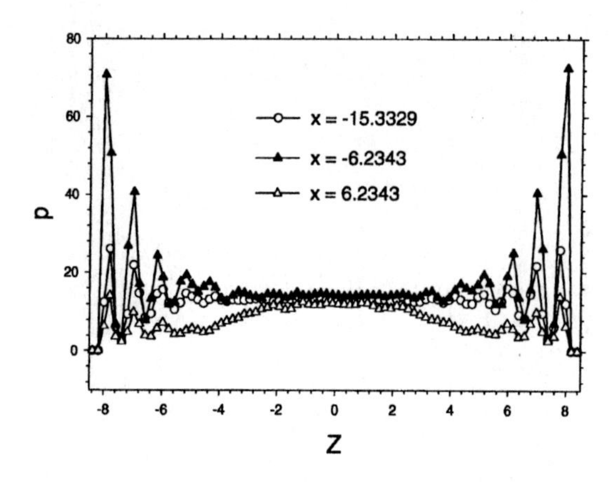

Figure 7. The pressure distribution upstream and downstream of the nozzle

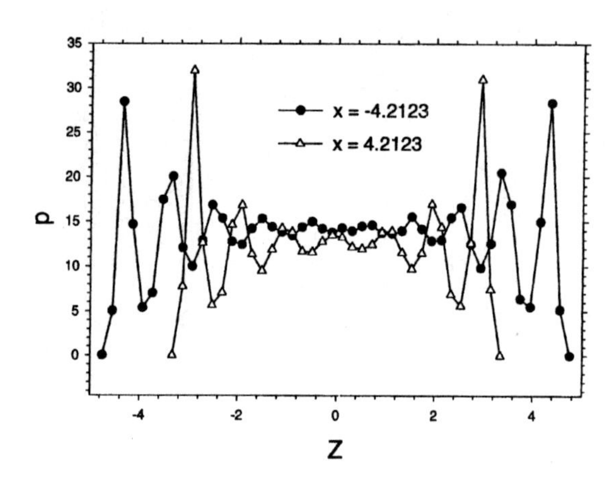

Figure 8. The pressure distribution in the nozzle

If the energy losses upstream and downstream of the nozzle are negligible comparing to those in the nozzle, the energy loss in the nozzle can be calculated as the decrease in potential energy through one periodic domain. MD and FEM simulations assume periodicity, hence the work done by gravity field would be converted into the internal energy and heat absorbed by the wall, which are just the loss of the mechanical energy. The loss of the mechanical energy is therefore

$$\Delta h_{loss} = \zeta \frac{\overline{v}^2}{2g}, \tag{38}$$

where Δh_{loss} is the energy loss, ζ is the energy loss coefficient and $\overline{v}$ is the mean fluid velocity in the channel. From the simulated data, we obtain the mean velocity near the inlet of the domain to be 0.8144, the energy loss to be 37.7. According to Eq. (38), the energy loss coefficient is 34.10 when $g = 0.3$.

3.2 Flow through larger periodic nozzle

To test the applicability of flow similarity in the MD simulation, we simulated the flow through a larger periodic nozzle, but maintaining geometric similarity. The flow domain is $-50.5481 \leq x < 50.5481$, $-2.5274 \leq y < 2.5274$ and $-26.9590 \leq z \leq 26.9590$ (18 nm in width). There are a total of 21090 molecules including 2768 wall molecules. The external field force is applied in the x direction with various strengths from 0.01 to 0.30. Other parameters are kept the same as those described in the previous subsection.

The flow patterns for $g = 0.05$, 0.10 and 0.20 are shown in Figs. 9 to 11. From these figures, we can see well-developed vortices behind the nozzle; their sizes increase rapidly with g. Since these vortices "block" the bulk flow, the flux is no longer proportional to the external field, for example, when $g = 0.02$, 0.05, 0.1, 0.2 and 0.30, the fluxes are 19.37, 48.05, 86.12, 139.2 and 180.0, respectively. A part of work done by the external forces is transferred into the kinetic energy of the vortex flow. The size of the nozzle is almost thrice larger than that considered previously. If the flows were governed by the Navier-Stokes equations, they would be similar as long as both have the same flow parameters (Reynolds and Poiseuille numbers), as is the case here. We find that when $g = 0.02$, the real Reynolds number is 8.55 and Poiseuille number is about 60.4. These numbers nearly match with those for the flow in the smaller nozzle with $g = 0.30$ (Re = 8.144, Poiseuille number 60.32). However, MD results for the two flows are certainly not similar. There is no vortex in the smaller nozzle simulation for $g = 0.30$, but vortices can be seen for the larger nozzle when $g = 0.02$. For the smaller nozzle, the space behind the nozzle is too small, about σ in width, for a vortex to form. The layered structure near the wall surface, about 3σ in width, resists any backward movement of the molecules. Hence there are no vortices developed there, even for $g = 0.30$. However, for the larger nozzle, the space behind the nozzle is about 15σ in width allowing a vortex flow to develop even for much lower driving forces.

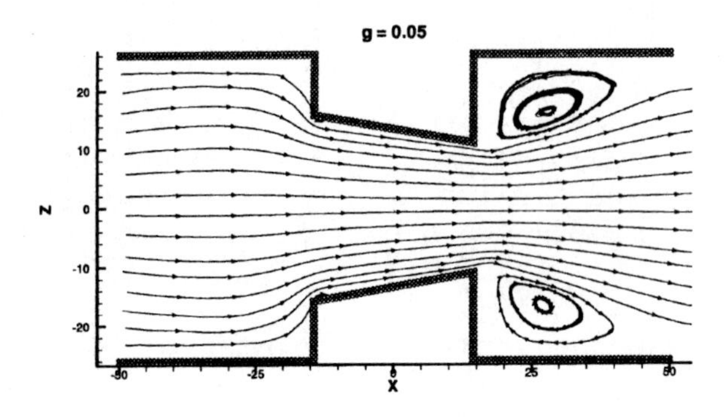

Figure 9. Streamlines with $g = 0.05$

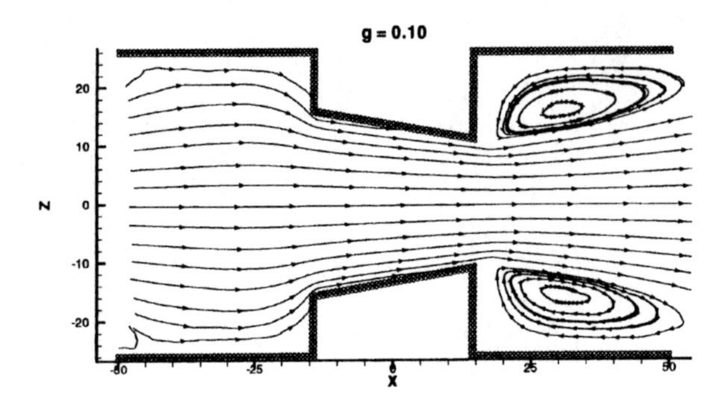

Figure 10. Streamlines with $g = 0.10$

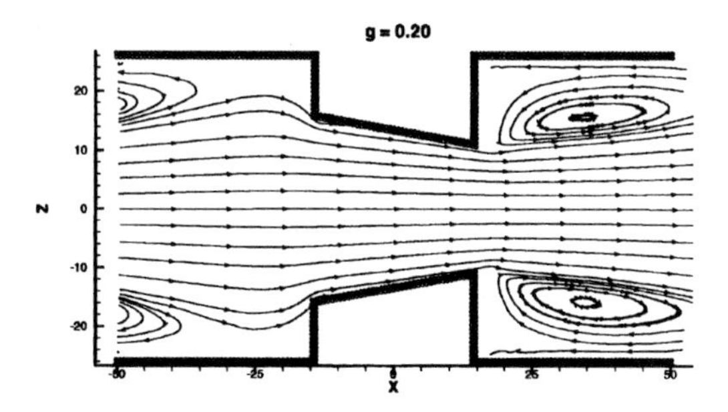

Figure 11. Streamlines with $g = 0.20$

The loss of similarity for MD simulation is not surprising. MD simulation is based on molecular motion and intermolecular interaction models. Similarity in the Poiseuille and Reynolds numbers do not contain any information on molecular sizes and interaction forces and therefore the above criterions are insufficient to guarantee the dynamic similarity at the molecular level. Another dimensionless number, Knudsen number defined as the ratio of the molecular size to the size of flow geometry, is required here. To keep the Knudsen number identical, the molecular size should be proportional to the flow dimension. If we require the Knudsen and the flow parameters be the same for two flows, the molecular unit of length, σ must necessarily vary. The similarity for molecular flows may be a subject for further investigation.

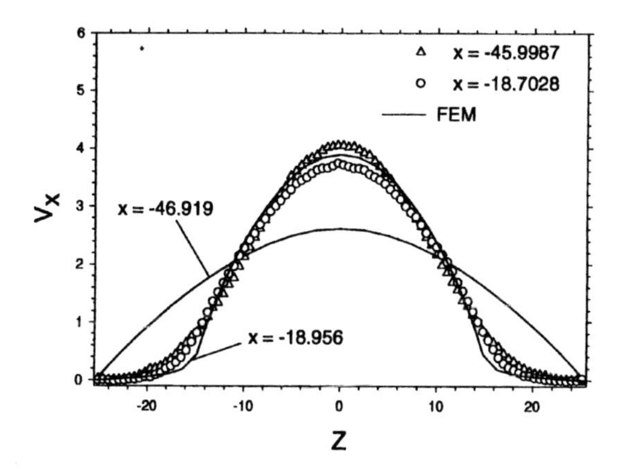

Figure 12. Velocity profile on some flow sections upstream of the nozzle with $g = 0.10$

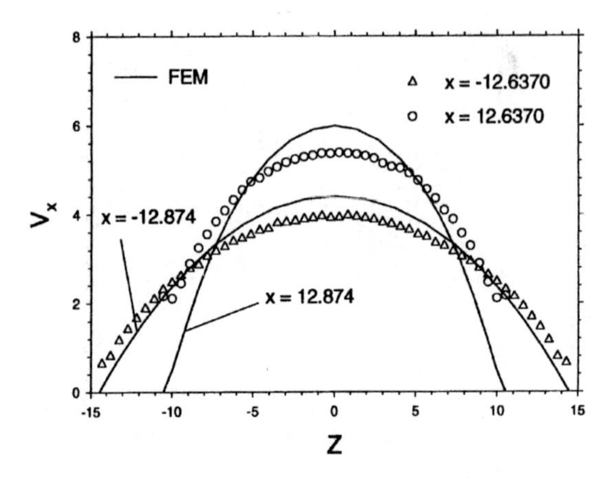

Figure 13. Velocity profile on some flow sections in the nozzle with $g = 0.10$

The deviation in the velocity profiles to FEM predictions become larger when the size of flow geometry increases. It was reported that in the Couette and Poiseuille flows for channels with width larger than 10σ, the velocity and stress fields are in complete agreement with Navier-Stokes solutions, using no-slip boundary conditions [21]. The present work indicates that the conclusion may not be valid in a complex flow.

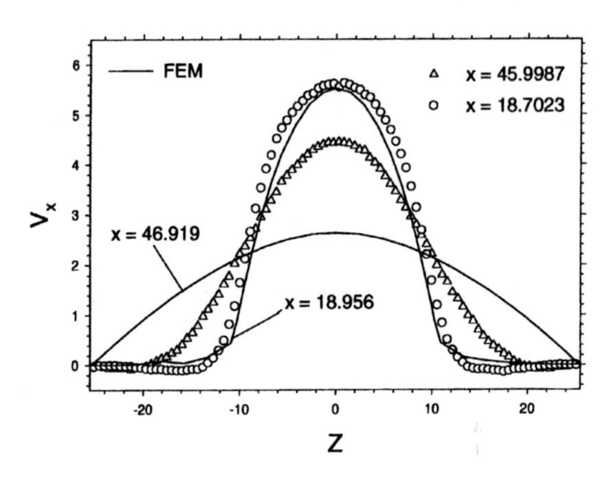

Figure 14. Velocity profile on some sections downstream of the nozzle with $g = 0.10$

The velocity profiles on some typical flow sections are plotted in Figs. 12 to 14 for the larger nozzle with g = 0.10. In these figures, the symbols denote the MD results and the solid lines those from FEM. Comparing these figures with Figs. 2 to 4 in the previous section, it is obvious that the FEM results agree better to the MD simulation for the flow through the smaller nozzle. This is in conflict with the conclusion drawn from simple flows mentioned above. The FEM cannot predict the vortex flow at such a low Reynolds number with no-slip boundary conditions. But in MD simulation, wall slip is inevitable when the driving force or molecules momentum out of the layers near the wall surface becomes relatively large. Molecules slipping out from the wall of the nozzle outlet carry larger momentum, drive molecules behind the nozzle to form vortex flow. If there is sufficient space, vortices will be developed. The flow pattern can then changed drastically from that predicted by FEM.

4. MICROCHANNEL FLOW OF BIO-MOLECULAR SUSPENSIONS

In this section, we present some results of DPD simulation of planar slit flow based on our recent research effort [67]. Firstly, we present some results for simple DPD fluids and then for bio-molecular suspensions.

We chose the parameters in Eqs. (9) to (11) basically according to [62]. Groot and Watten found that to satisfy the compressibility of water the coefficient of the conservation force should be

$$\alpha_{ij} = 75k_BT / \rho \tag{39}$$

and recommend $\sigma = 3$, with $\lambda = 0.5$ in the Verlet-type algorithm, Eqs. (28) to (31), when $\Delta t = 0.04$. The density is a freely chosen parameter. We chose $\rho = 4$. This leads to a unit period in the face-centred-cubic (FCC) lattice. If the density of wall, ρ_w, is equal to 4 as well, only two layers of wall particles arranged in a plane FCC are enough to calculate the interaction between wall and fluid particles. If the ratio of ρ_w / ρ is larger more layers are required. We used three layers of wall particles in the simulation. Like the molecular units used in MD, we set the energy unit as k_BT, i.e., $k_BT = 1$. Hence, from Eq. (39), $a_{ij} = a_{fluid} = 18.75$, if i and j both denote fluid particles or beads in molecular chains. For wall particles, we assumed $a_{wall} = 5.0$ and $a_{ij} = \sqrt{a_{fluid} * a_{wall}} = 9.6825$ when calculating the interaction between fluid and wall particles. The density distortion near the wall can be reduced in terms of softening the repulsion between wall and fluid particles. The mass of the particles, *m*, and the cutoff radius, r_c, were set to unity in the simulation involving DPD particles. As in MD simulation the field force was also assumed to applied to each fluid particle in the x-direction, $\mathbf{F}_e = g\boldsymbol{\delta}_x$.

4.1 Poiseuille flow for a simple DPD fluid

In the simulation of Poiseuille flow, a total of 11880 simple DPD particles are used: 10800 fluid particles are placed in the planar channels; 1080 wall particles at the FCC lattice sites of the three layers on each wall. The layers are parallel to the (x, y) plane and two inner layers are at z = ±15.25. The fluid domain is, in the x and y directions, $-30 \le x < 30$ and $-1.5 \le y < 1.5$. Periodic boundary conditions are applied to fluid boundaries in these two directions. A gravity of $g = 0.02$ is applied to each fluid particle providing the driving force. The time step is taken as 0.02. To measure the local properties of Poiseuille flow, the region is divided into 300 bins, of which 278 bins are located in the fluid domain. All local flow properties are obtained by averaging the sampled data in each bin over 10^4 time steps.

Applying the gravity field is equivalent to impose a pressure drop of $\rho g L_x$ on the channel (length L_x). The velocity development is shown in Fig. 15. The numbers indicated in the figure are the instant at which the sampling starts. The sampling lasts 200 time units. Form this figure, we can see that after about 2000 time units, the velocity profile is already fully developed. The agreement of the fully velocity profile at $t = 2500$ with the NS solution:

$$V_x = 8.639\left[1-\left(\frac{z}{15.25}\right)^2\right] \tag{40}$$

is excellent, as shown in Fig. 16. Here we took the half width of the channel, h_z, to be 15.25, which is just the distance from the centre plane of the channel to the inner layer of wall particles. We can further determine the apparent viscosity of DPD fluid from the solution of NS equations. Since

$$V_x(0) = \frac{\bar{\rho} g h_z^2}{2\mu} = 8.639, \tag{41}$$

where μ is the apparent viscosity of simple DPD fluids. We obtain $\mu = 1.077$.

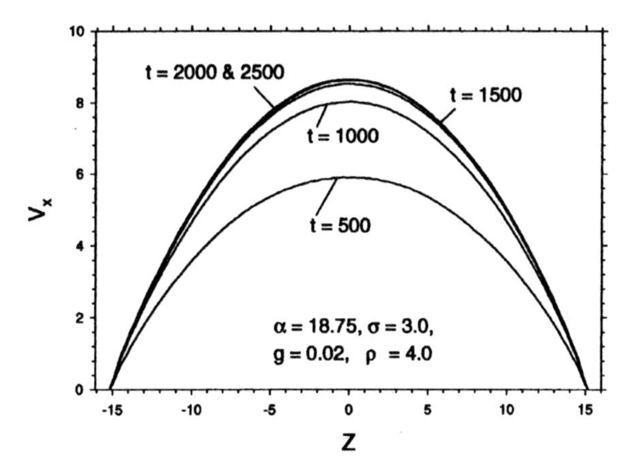

Figure 15. The development of velocity profiles in Poiseuille flow of a simple DPD fluid

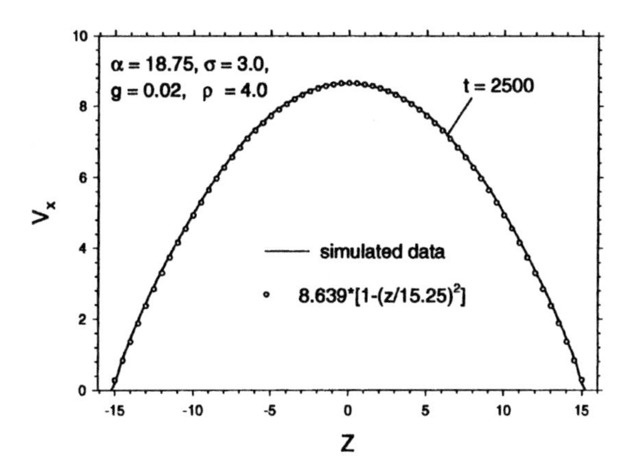

Figure 16. The fully developed velocity profile in Poiseuille flow: a comparison to N-S solution

In Fig. 17 the density and temperature profiles are shown. We can see that the density is almost uniform across the channel except for the region near the wall surfaces, where a fluctuation in density still exists. This is due to the soft repulsion between wall and fluid particles, and is not as severe as that predicted in MD simulations. The temperature is almost uniform across the channel, Fig. 17. The temperature drop near the wall is due to the low wall temperature ($T_w = 0$) and the thin layer is not hot enough to prevent the fluid particles from being cooled. This temperature drop decreases for increasing particle temperature in the thin layers near the walls, where the no-slip boundary condition is desired. The fluctuation may be further reduced if a lower repulsion strength between wall particles is used. The density fluctuation gives rise to pressure fluctuation in the region near the wall surfaces, as shown in Fig. 18. The analytical solution of the shear stress component in Poiseuille flow is $-\bar{\rho}gz$, valid for all, Newtonian or non-Newtonian fluids. An excellent comparison of the predicted shear stress by DPD simulation with the analytical solution is shown in Fig. 18 (except at the wall surfaces). In Fig. 19, we show the distributions of the first and second normal stress differences across the channel. It is seen that except in the region near the wall surfaces two normal stress differences are very small, near zero. Hence, we can conclude that the rheological properties of simple DPD fluids are Newtonian.

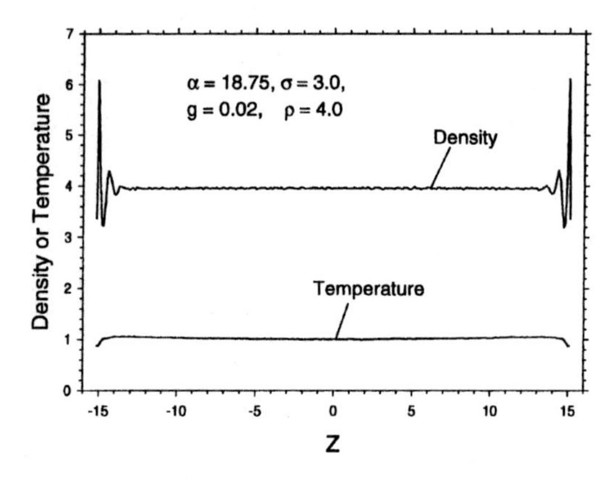

Figure 17. The density and temperature profiles of a simple DPD fluid in Poiseuille flow

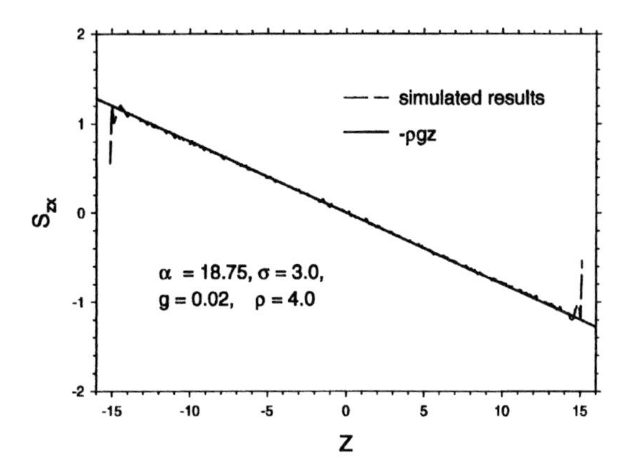

Figure 18. Shear stress distribution of a simple DPD fluid in Poiseuille flow

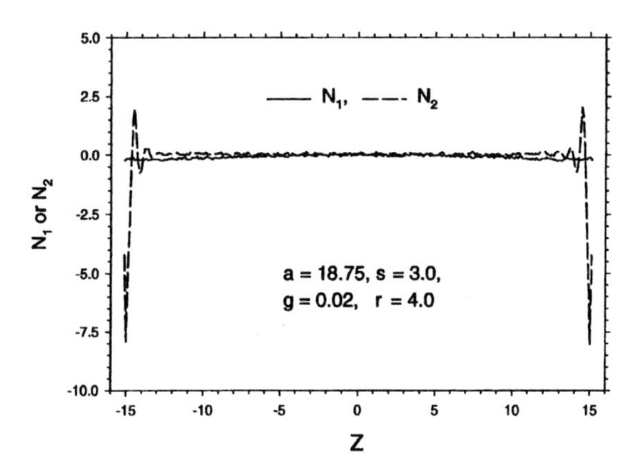

Figure 19. First and second normal stress difference distributions of a simple DPD fluid in Poiseuille flow

4.2 Flow of DNA suspension in a channel

The size of a bio-MEMS device is usually in the same order of magnitude as, or even smaller than that of a DNA molecule. The Knudsen number is $O(1)$. Hence flows of DNA suspensions in bio-MEMS devices cannot be modelled by the NS equations. On other hand, the flow domain size is too large to be handled by MD simulation. We believe that DPD is the best method to simulate this kind of flow problems.

In the simulation, we use simple DPD particles to model the suspending fluid and bead-spring chains to model DNA molecules. We prefer to use the FENE chain, rather than the worm-like chain model to model the DNA chains, because all of its rheological properties are available and it is simpler to implement. The FENE chain captures major non-linear mechanical properties of a flexible and finitely extendable macromolecule. Two kinds of FENE chains were used in simulations. The shorter chains consist of 29 segments with 30 beads and longer ones 59 segments with 60 beads. The parameters in the FENE spring, Eq. (18), are taken to be H = 6.0 and $r_m = 2.0$. The size of the fluid domain is $-30 \le x < 30$, $-3.5 \le y < 3.5$ and $-15.25 \le z \le 15.25$, so that $h_z = 15.25$.

We can estimate the coefficient of a bead, ζ, from the self-diffusion coefficient. In the present case, the self-diffusion coefficient of DPD particles, according to Marsh [45], is

$$D \approx \frac{36mk_BT}{\pi\rho\gamma}, \tag{42}$$

and the self-diffusion coefficient for a bead is

$$D = \frac{2k_BT}{\zeta} \tag{43}$$

Hence when $m=1$ and $\rho=4$, we obtain $\zeta \approx 0.7\gamma = 3.15$ for $\sigma = 3$ and $k_BT = 1$. From Eqs. (21), (22) and (23), we find $\lambda_H = 0.1313$, $\lambda_Q = 1.05$ and b = 24. The relaxation time of the FENE chain is calculated from Eq. (25): $\lambda_{fene} = 34.97$ when $N_b = 30$ and 140.01 when $N_b = 60$. The length of the chain segment is determined from Eq. (27): $L_c = 0.6433$. Hence, we can roughly estimate the contour length of molecules: 18.65 for the shorter chain and 37.95 for the longer chain. Hence, the Knudsen is 0.61 and 1.24 respectively.

The suspending fluid is modelled by simple DPD particles. The chain volume fraction can be determined to be the ratio of the number of beads to total number of particles. The total number of fluid particles used is 50400. For a dilute suspension with a volume fraction of $\phi = 0.00536$, we use 9 FENE chains, each of which has 30 beads, and 50130 simple particles for the suspending fluid. For a semi-concentrated suspension at volume fraction $\phi = 0.05357$, 90 FENE chains with 30 beads for each, and 47700 simple particles are used. And for a concentrated suspension at $\phi = 0.1071$, we use

90 FENE chains each with 60 beads and 45000 simple particles. The field force, g, is set to be 0.01. The local flow properties are obtained by averaging the sampled data in each bin over 20000 to 25000 time steps.

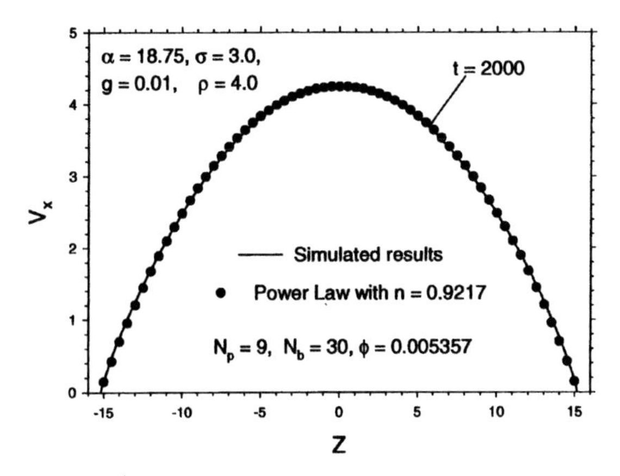

Figure 20. The velocity profile of DNA suspension in the slit at $\phi = 0.5357\%$. The solid line denotes the simulated results. The solid circles denote the velocity profile of a power-law fluid

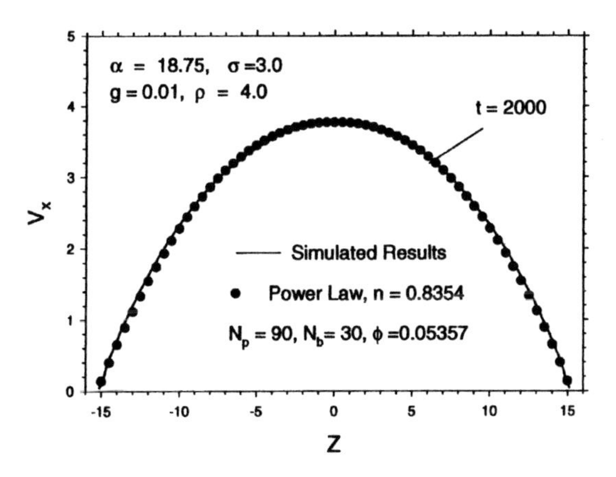

Figure 21. Velocity profile of DNA suspension flow through the slit at $\phi = 5.357\%$

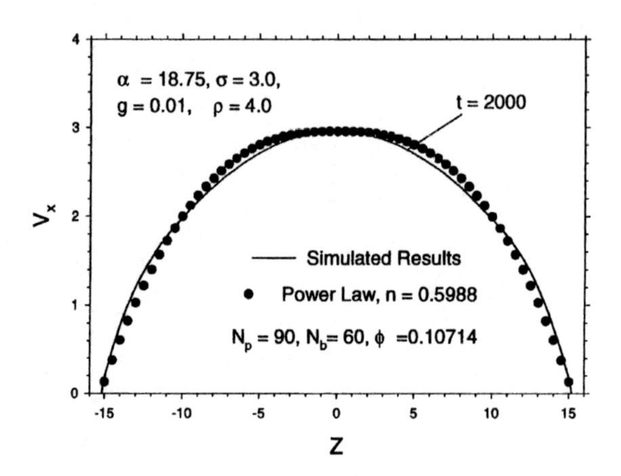

Figure 22. Velocity profile of DNA suspension flow through the slit at ϕ= 10.71%

The fully developed velocity profiles (after about t = 2000) at different volume fractions are shown in Figs. 20, 21 and 22. In these figures, the solid lines represent the simulated velocity profiles and solid circles denote the results calculated from a power-law distribution, which is

$$V_x = V_{\max}\left[1-\left(\frac{|z|}{h_z}\right)^{\frac{1}{n}+1}\right], \tag{44}$$

with

$$V_{\max} = \frac{nh_z^{\frac{1}{n}+1}}{n+1}\left(\frac{\overline{\rho}g}{m_p}\right)^{\frac{1}{n}} \tag{45}$$

being the maximum velocity determined from simulated data and n and m_p are the power-law index and consistency. We used numerical fitting to determine these parameters. The fitted values of n are 0.9217, 0.8354 and 0.5988, respectively, for $\phi = 0.5357\%$, 5.357% and 10.71%. The parameter of m_p can then be determined by n and $V_{\max}$ from Eq. (45). When n = 1, Eq. (44) reduces to the solution of NS equations, while n < 1, it describes a shear thinning non-Newtonian fluid. The solid circles in these figures are calculated from Eq. (44) with the fitted values of n. It is remarkable that the velocity profiles for DNA suspensions can be approximated by the power-law curves well, especially for those with lower volume fraction. The distribution of the shear stress component is shown in Fig. 24. From this figure we can see that, the simulated shear stress agrees with the predictions of continuum mechanics, obtained from a global balance of momentum. Figure 25 shows the profiles of the first, N_1, and second normal stress differences, N_2. Except for the region with large density fluctuation, N_1 is positive but N_2 is almost zero.

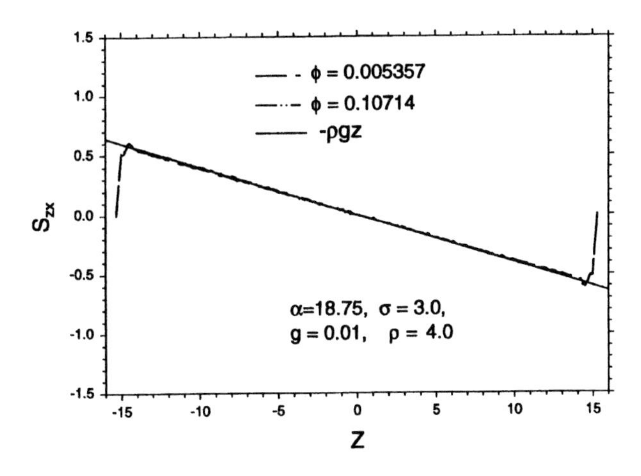

Figure 23. The shear stress distributions across the slit. The solid line denotes the prediction of continuum mechanics, $-\bar{\rho}gz$, the dotted and dashed lines denote the simulated results for $\phi = 0.5357\%$ and 10.71%, respectively

DPD simulation can provide detailed information on chain motion, such as molecular conformation, micro structure evolution and intramolecular tension, and so on. The conformation changes of some typical chains in the slit are shown in Figs. 25, 26 and 27 for suspensions with $\phi = 10.71\%$. Before the slit flow started, all molecules were at equilibrium and in relatively uniform length, Fig. 25. As the flow developed, the molecules moved, rotated and stretched. Figures 26 and 27 show the snapshot of molecular conformation at $t = 520$, when the flow was developing, and $t = 2000$, when flow was fully developed.

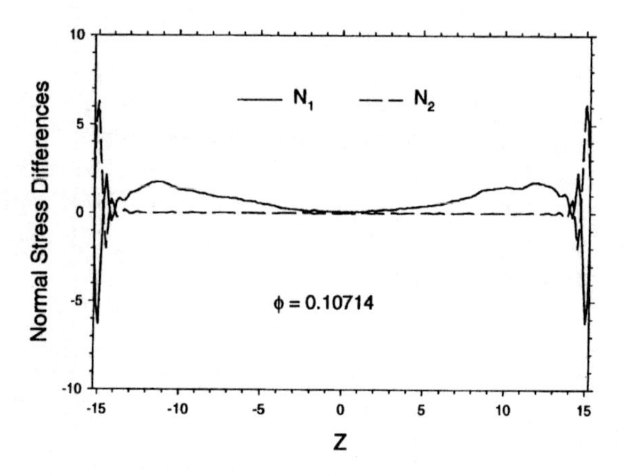

Figure 24. The first and second normal stress difference distribution, $\phi = 0.10714$

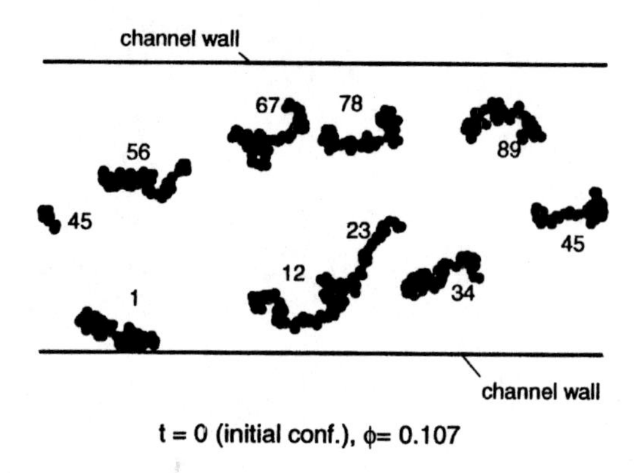

Figure 25. The conformation of some DNA molecules at equilibrium (no flow)

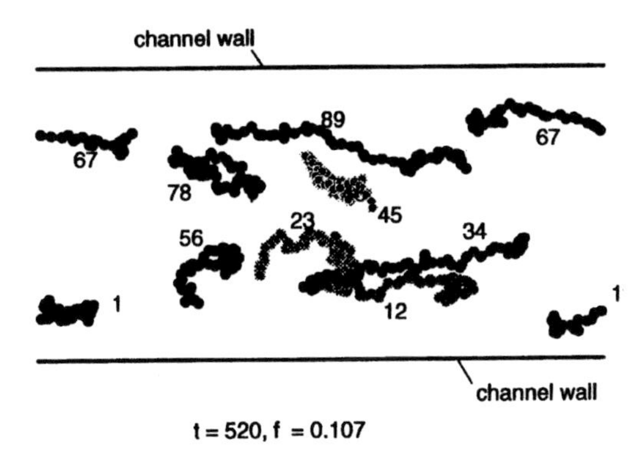

Figure 26. The conformation of some typical DNA molecules in developing slit flow, at t = 520

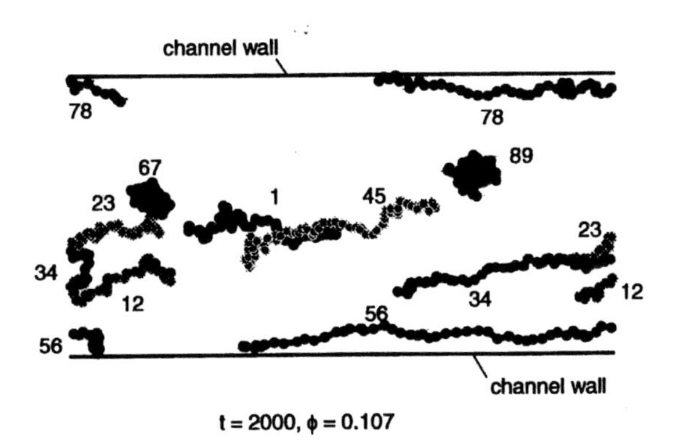

Figure 27. The conformation of some typical DNA molecules in fully developed channel flow, at t = 2000

From these two figures, we can see that molecules located near to the wall are stretched more, while coiled molecules are found in the central part of the channel. We also can see that more molecules are in the central part of the slit. This non-uniform concentration is due to the wall effect. It is known that there is a depletion layer near the wall surface. For large channels, the thickness of the depletion layer is about the equilibrium length of the molecules and for small channel, of width equal to or less than the equilibrium length, the concentration profile is non-uniform even at the mid-plane [70]. Figure 28 shows the concentration distributions of the suspension with $\phi = 5.357\%$ across the channel are plotted for $t = 500$, 1500 and 2500. It seems that the concentration is always non-uniform even in the central part of the channel.

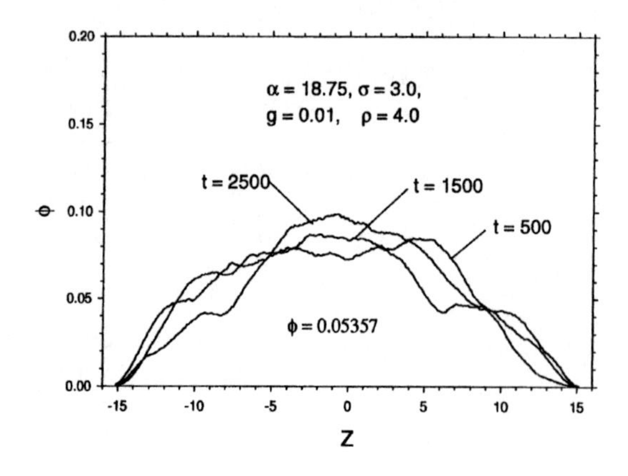

Figure 28. The concentration profiles for $\phi = 0.05357$ at various times

5. FINAL REMARKS

In this chapter, we described two numerical methods, molecular dynamics (MD) and dissipative particle dynamics (DPD), and employed these methods to simulate simple liquids and DNA suspension flows through nano and micro channels.

MD simulations were conducted in two flow domains which were similar in geometry but different in sizes. It was found that the similarity criterion often used in fluid mechanics is not sufficient for the molecular flow. The information about molecular size and intermolecular forces has to be taken into account.

MD results were compared to those obtained from a finite element analysis, assuming an isothermal, Newtonian and incompressible flow through two nozzles. There was a reasonable agreement between the two sets of results for the flow through a smaller nozzle. However, in the flow through the larger nozzle, a vortex flow was seen in the MD simulation, which was not predicted by the finite element analysis. This is in contrast to what we know in simple flows.

The main features of the MD simulation were that the fluid density is not uniform and wall slip is inevitable as long as driving force is strong. The momentum of the molecules near the surface can help vortex flow to develop in an area of tens of molecular sizes.

The insight gained from MD simulation is helpful to understand some of the micro channel flow phenomena. Firstly, the molecules in the layered structure near the wall surface have considerable momenta and driving forces. Secondly, the wall roughness is much larger than molecular size and provides enough space behind the peaks. Hence, molecules carrying large momenta slip over the peaks of the roughness and develop vortex flow. Thirdly, the ratio of surface-area to volume of micro channels is much larger than that of macro channels, and the flow behaviour near the surface would play an important role in micro channels. Some of the micro channel flow phenomena can be explained using slip flow pattern with vortices around roughness asperities. When the vortices are weak, slippage results in the reduction of the friction factor, but when vortices are strong the friction would be increased. The further developing vortices may give rise to the laminar-to-turbulence transition directly related to the molecule momentum in the layered structure, but not to the Reynolds number. This may explain why the critical Reynolds number for micro channels is less than that of macro channels.

DPD is a method for mesoscopic simulations and is suitable for dealing with complex fluids. We simulated the Poiseuille flow of simple DPD fluids and confirmed their Newtonian rheology. The DPD was applied successfully to simulate slit flow of DNA suspensions at three different concentrations, from dilute to concentrated. We found that the fully developed velocity profiles in slit flows can be fitted by the power-law equations, especially for suspensions at low concentration. The fitted power-law indices were less than 1.0 (shear-thinning) and decreased with concentrations.

DPD simulation can provide detailed information on molecular motion and conformation revolution. We showed that the molecules are severely stretched near the wall where the shear rate is maximal, and coiled molecules can be found in the central part of the slit, where the shear rate is low. When molecules are over stretched, they may break. Hence the degradation of DNA molecules may occur near the wall during delivery. The depletion layer near the wall makes the concentration across the channel highly non-uniform when channel is narrow.

How to implement the no-slip boundary conditions on the surface of solid walls is one of the central issues for DPD method. We use frozen wall particles plus a thin boundary layer to force the DPD particles to satisfy the boundary condition. This scheme works reasonably well but does not eliminate density fluctuations near the wall. The method may be refined further.

In DPD simulation, the units are not defined as explicitly as in MD simulation. Here we attempt to show the scaling in DPD is much larger than that in MD. For the model parameters used in the present simulations, $L_c = 0.6433[r_c]$, here we use $[\bullet]$ to denote the unit of $\bullet$, the contour length of a FENE chain is roughly equal to $L_c(N_b - 1)[r_c]$. Hence, $[r_c]$ can be determined by the contour length and number of beads. The unit of density can be defined to be the density of suspending fluids. If the length of a DNA molecule is 30μ m and we use 30 beads in a FENE chain, we obtain $[r_c] = 1.608 \times 10^{-4} cm$. The volume of one cell of FCC is $[r_c]^3 = 4.158 \times 10^{-12} cm^3$, for $\rho = 4$. If the suspending fluid is water with a density of $1g/cm^3$, the unit of density is $[\rho] = 0.25g/cm^3$. There are 4 particles in each cell and thus the mass of per particle is $[m] = \rho[\rho][r_c]^3/4 = 1.040 \times 10^{-12} g$. The mass of a bead in the chains can be determined by the number of beads in a chain and the mass of the bio-macromolecules. For simplicity, here we have assumed the mass of a bead to be identical to that of a DPD particle, i.e., one mass unit. We take the unit of energy to be $[\varepsilon] = k_B T$. If T = 298K, $[\varepsilon] = 4.115 \times 10^{-14} erg$, the unit of velocity is equal to $[V] = \sqrt{3[\varepsilon]/[m]} = 0.345 cm/s$ and the unit of time $[t] = [r_c]/[V] = 4.66 \times 10^{-4} s$. In the simulation, we take the time step to be 0.02, or in dimensional units $0.93 \times 10^{-5} s$. Hence we can see that the length scale and time scale in DPD is much larger than those in MD. In MD the length unit is the diameter of an argon atom, i.e., $3.4 \times 10^{-8} cm$ and one time step $\Delta t = 0.005$ is about $10^{-14} s$ [29]. The local flow properties for the suspension were obtained by averaging sampling data over 2×10^4 or 2.5×10^4 time steps, and this means that the results were the mean values over the time interval of 0.19s or 0.233s. Hence, these mean values can not be used as instantaneous flow properties.

The relaxation times for FENE chain with 30 beads is about 0.0163s (34.97[t]). If $N_b = 60$ and the contour length of the chain remains $30 \mu m$, the unit of time would be reduced to 0.79×10^{-4} and the relaxation time is about 0.011s (140[t]). The relaxation time for real DNA molecules is solvent-dependent. Shrewsbury et al [15] measured the relaxation time for λ-DNA to be 0.68 s for the buffered solvent with the viscosity of 15cP; it corresponds to 0.041 s in water at 25^{0}C. Babcock et al [71] obtained the relaxation time of DNA to be 6.3 s in a 60cP solution, i.e., 0.095s in water. Hence, the parameters used here for FENE chains are reasonable compared to actual parameters measured from DNA suspensions.

REFERENCES

[1] M. Gad-el-Hak, "The fluid mechanics of microdevices – The Freeman scholar lecture," J. Fluid Eng., Vol. **121**, 5 (1999).

[2] G.A. Bird, "Monte Carlo simulations of gas flows," Ann. Rev. Fluid Mech., Vol. **10**, 11 (1978).
[3] Qu Weilin, Gh Mohiuddin Mala and Li Dongqing, "Pressure-driven water in trapezoidal silicon," Int. J. Heat Mass Trans, Vol. **43**, 353 (2000).
[4] J. Harley, H. Bau, "Fluid flow in micro and submicron size channels," IEEE Trans,. **THO249-3**, 25 (1989).
[5] J. N. Pfahler. "Liquid Transport in Micro and Submicro Channels," Ph. D. thesis, Department of Mechanical Engineering and Applied Mechanics, University of Pennsylvania (1992).
[6] D. V. McAllister, M. G. Allen and M. R. Prausnitz, "Microfabricated microneedles for gene and drug delivery," Annu. Rev. Biomed. Eng., Vol. **2**, 289 (2000).
[7] K. Chun, G. Hashiguchi, H. Toshiyoshi, H. Fujita, "Fabrication of array of hollow microcapillaries used for injection of genetic materials into animal/plant cells," Jpn. J. Appl. Phys. Part 2, Vol. **38**, 279 (1999).
[8] J. Chen, K. D. Wise, "A multichannel neural probe for selective chemical delivery at the cellular level," IEEE Trans. Biomed. Eng. Vol. **44**, 760 (1997) .
[9] L. Lin, A. P. Pisano, "Silicon processed microneedles," IEEE J. Micromech. Syst., Vol. **8**, 78 (1999).
[10] J. D. Brazzle, S. Mohanty, A. B. Frazier, "Hollow metallicmicromachined needles with multiple output ports," Proc. SPIE Conf. Microfluidic Dev. Syst., Santa Clara, Vol. **3877**, 257 (1999).
[11] D. V. McAllister, F. Cros, S. P. David, L. M. Matta, M. R. Prausnitz, M. G. Allen, "Three-dimensional hollow microneedle and microtube arrays," Transducers 99, Int. Conf. Solid-State Sens. Actuators, 10th, Sendia, pp. 1098-1101. Tokyo: IEEE (1999).
[12] T. T. Perkins, D. E. Smith and S. Chu, "Single polymer dynamics in an elongational flow," Science, Vol. **276**, 2016 (1997).
[13] D. E. Smith and S. Chu, "Response of flexible polymers to sudden elongation flow," Science, Vol. **281**, 1335 (1998).
[14] D. E. Smith, H. P. Babcock and S. Chu, "Single-polymer dynamics in steady shear flow," Science, Vol. **283**, 1724 (1999).
[15] P. J. Shrewsbury, S. J. Muller, D. Liepmann, "Effect of flow on complex biological macromolecules in microfluidic devices," Biomed. Microdevices, Vol. **3**, 225 (2001).
[16] P. S. Doyle, E. S. G. Shaqfeh, A. P. Gast, "Dynamics simulation of freely-draining, flexible polymers in steady linear flows," J. Fluid Mech., Vol. **334**, 251 (1997).
[17] P. S. Doyle, E. S. G. Shaqfeh, "Dynamics simulation of freely-draining, flexible bead-rod chains: start-up of extensional flow," J. Non-Newtonian Fluid Mech., Vol. **76**, 43 (1998).
[18] P. S. Doyle, E. S. G. Shaqfeh, G. H. McKinley, S. H. Spiegelberg, "Relaxation of dilute polymer-solutions following extensional flow," J. Non-Newtonian Fluid Mech., Vol. **76**, 79 (1998).
[19] J. S. Hur, S. G. Shaqfeh, R. G. Larson, "Brownian dynamics simulation of single DNA molecules in shear flow," J. Rheol., Vol. **44**, 713(2000).
[20] D. J. Evans and G. P. Morriss, *Statistical Mechanics of Non-Equilibrium Liquids* (Academic Press, New York, 1990).
[21] J. Koplik and J. R. Banavar, "Continuum deduction from molecular hydrodynamics," Ann. Rev. Fluid Mech., Vol. **27**, 267 (1995).
[22] J. P. McDonald, *Theory of Simple Liquids*, 2nd ed. (Academic, New York, 1986).
[23] J. D. Weeks, D. Chandler and H.C. Andersen, J. Chem. Phys., Vol. **54**, 5237 (1971).
[24] U. Heinbuch and J. Fischer, "Liquid flow in pores: slip, no-slip, or multilayer sticking," Phys. Rev. A, Vol. **40**, 1144 (1989).

[25] J. H. Irving and J. G. Kirkwood, "The statistical mechanics of transport processes IV. The equation of hydrodynamics," J. Chem. Phys., Vol. **18**, 817 (1950).

[26] G. K. Batchelor, "The stress system in a suspension of force-free particles," J. Fluid Mech., Vol. **44**, 545 (1972).

[27] R. R. Huilgol and N. Phan-Thien, *Fluid Mechanics of Viscoelasticity: General Principles, Constitutive Modelling, Analytical and Numerical Techniques* (Elsevier, Amsterdam, 1997).

[28] B. D. Todd, D. J. Evans and P. J. Davis, "Pressure tensor for inhomogeneous fluids," Phys. Rev. E, Vol. **52**, 1627 (1995).

[29] D. C. Rapaport, *The Art of Molecular Dynamics Simulation* (Cambridge University Press, Cambridge, 1995).

[30] P. A. Thompson and M. O. Robbins, "Shear flow near solids: epitaxial order and flow boundary conditions," Phys. Rev. A, Vol. **41**, 6830 (1990).

[31] J. Koplik and J. Banavar, "Corner flow in the sliding plate problem," Phys. Fluids, Vol. **7**, 3118 (1995).

[32] J. Koplik and J. Banavar, "Reentrant corner flows of Newtonian and non-Newtonian fluids," J. Rheol., Vol. **41**, 787 (1997).

[33] A. Jabbarzadeh, J. D. Atkinson and R. I. Tanner, "Rheological properties of thin liquid films by molecular dynamics simulations," J. Non-Newt. Fluid. Mech., Vol. **69**, 169 (1997).

[34] A. Jabbarzadeh, J. D. Atkinson and R. I. Tanner, "Nanorheology of molecularly thin films of n-hexadecane in Couette shear flow by molecular dynamics simulations," J. Non-Newt. Fluid. Mech., Vol. **77**, 53 (1998).

[35] A. Jabbarzadeh, J. D. Atkinson and R. I. Tanner, "Effect of the wall roughness on slip and rheological properties of hexadecane in molecular dynamics simulation of Couette shear flow between two sinusoidal walls," Phys. Review E, Vol. **61**, 690 (2000).

[36] A. Jabbarzadeh, J. D. Atkinson and R. I. Tanner, "Lubrication processes near wall asperities: a molecular dynamics study," in: *Tribology Research: From Model Experiment to Industrial problem*, G. Dalmaz et al Ed., 779-785 (Elsevier Science, 2001).

[37] Y. Z. Hu, H. Wang, Y. Guo and L. Q. Zheng, "Simulation of lubricant rheology in thin lubrication, Part 1: simulation of Poiseuille flow," Wear, Vol. **196**, 243 (1996).

[38] Y. Z. Hu, H. Wang, Y. Guo, Z. J. Shen and L. Q. Zheng, "Simulation of lubricant rheology in thin lubrication, Part 2: simulation of Couette flow," Wear, Vol. **196**, 249 (1996).

[39] K. P. Travis, B. D. Todd and D. J. Evans, "Departure from Navier-Stokes hydrodynamics in confined liquids," Phys. Rev. E, Vol. **55**, 4288 (1997).

[40] K. P. Travis and K. E. Gubbins, "Poiseuille flow of Lennard-Jones fluid in narrow slit pores," J. Chem. Phys., Vol. **112**, 1984 (2000).

[41] P. J. Daivis, K. P. Travis and B. D. Todd, "A technique for the calculation of mass, energy, and momentum densities at planes in molecular dynamics simulations," J. Chem. Phys, Vol. **104**, 9651 (1996).

[42] X. J. Fan, N. Phan-Thien, Ng Teng Yong, Xu Diao, "Molecular dynamics simulation of a liquid in a nano channels," Phys. Fluids, Vol. **14**, 1146 (2002).

[43] P. J. Hoogerbrugge and J. M. V. A. Koelman, "Simulating microscopic hydrodynamic phenomena with dissipative particle dynamics," Europhys. Lett., Vol. **19**, 155 (1992).

[44] P. Espa ñ ol and P. Warren, "Statistical mechanics of dissipative particle dynamics," Europhys. Lett., Vol. **30**, 191 (1995).

[45] C. Marsh, "Theoretical Aspect of Dissipative Particle Dynamics," Ph. D. Thesis, University of Oxford (1998).

[46] R. B. Bird, C. F. Curtiss, R. C. Armstrong, O. Hasssager, *Dynamics of Polymeric Liquids*, Vol. 2: Kinetic Theory (John Wiley & Sons, 1987).
[47] Y. Kong, C. W. Manke, W. G. Madden, A. G. Schlijper, "Simulation of a confined polymer in solution using the dissipative particle dynamics method," Int. J. Thermophys. Vol. **15**, 1093 (1994).
[48] Y. Kong, C. W. Manke, W. G. Madden, A. G. Schlijper, "Effect of solvent quality on the conformation and relaxation of polymers via dissipative particle dynamics," J. Chem. Phys., Vol. **107**, 592 (1997).
[49] A. G. Schlijper, P. J. Hoogerbrugge, C. W. Manke, "Computer Simulation of dilute polymer solution with the dissipative particle dynamics method," J. Rheol., Vol. **39**, 567 (1995).
[50] A. G. Sclijper, C. W. Manke, W. G. Madden, Y. Kong, "Computer simulation of non-Newtonian fluid rheology," Int. J. Mod. Phys. C, Vol. **8**, 919 (1997).
[51] E. S. Boek, P. V. Coveney, H. N. W. Lekkerkerker, "Computer simulation for rheological phenomena in dense colloidal suspensions with dissipative particle dynamics," J. Phys. Cond. Matt., Vol. **8**, 9509 (1996).
[52] E. S. Boek, P. V. Coveney, H. N. W. Lekkerkerker, P. van der Schoot, "Simulating the rheology of dense colloidal suspensions using dissipative particle dynamics," Phys. Rev. E, Vol. **55**, 3124 (1997).
[53] J. M. V. A. Koelman, P. J. Hoogerbrugge, "Dynamic simulation of hard sphere suspensions under steady shear," Europhys. Lett., Vol. **21**, 363 (1993).
[54] P. V. Coveney, P. Espa ñ ol, "Dissipative particle dynamics for interacting multi-component systems," J. Phys. A. Maths and General, Vol. **30**, 779 (1997).
[55] P. V. Coveney, K. E. Novik, "Computer simulations of domain growth and phase separation in 2d binary fluids using dissipative particle dynamics," Phys. Rev. E, Vol. **54**, 5134 (1996).
[56] L. E. Novik, P. V. Coveney, "Using dissipative particle dynamics to model binary immiscible fluids," Int. J. Mod. Phys. C, Vol. **8**, 909 (1997).
[57] M. P. Allen, D. J. Tildesley, *Computer simulation of liquids* (Clarenden Press, Oxford, 1987).
[58] R. J. Sadus, *Molecular simulation of fluids, theory, algorithms and object-orientation* (Elsvier, Amsterdam, 1999).
[59] J. M. Haile, *Molecular dynamics simulation, elementary methods* (John Wiley & Sons, 1992).
[60] U. Frisch, B. Hasslachcher, Y. Pomeau, "Lattice-gas automata for the Navier-Stokes equations," Phys. Rev. Lett., Vol. **56**, 1505 (1986).
[61] M. J. Field, *A practical introduction to simulation of molecular system* (Cambridge University Press, New York, 1999).
[62] R. D. Groot, P. B. Warren, "Dissipative particle dynamics: bridging the gap between atomic and mesoscopic simulation," J. Chem. Phys., Vol. **107**, 4423 (1997).
[63] S. M. Willemsen, H. C. J. Hoefsloot, P. D. Iedema, "No-slip boundary condition in dissipative particle dynamics," Int. J. Mod. Phys. C, Vol. **11**, 881 (2000).
[64] M. Revenga, I. Zu' ñ iga, P. Esoa ñ ol, "Boundary models in DPD," Int. J. Mod. Phys. C, Vol. **9**, 1319 (1998).
[65] M. Revenga, I. Zu' ñ iga, P. Esoa ñ ol, "Boundary condition in dissipative particle dynamics," Computer Phys. Commun., Vol. **121-122**, 309 (1999).
[66] L. E. Wedgewood, D. N. Ostrov and R. B. Bird, "A finitely extensible bead-spring model for dilute polymer solution," J. Non-Newtonian Fluid Mech., Vol. **40**, 119 (1991).

[67] X. J. Fan, N. Phan-Thien, Ng Teng Yong, Xuhong Wu and Diao Xu, "A Simulation of Bio-macromolecular Suspension Flow through Microchannels," submitted to Phys. Fluids.

[68] C. Bustamante, J, F. Marko, E. D. Siggia and S. Smith, "Entropic elasticity of λ-phage DNA," Science, Vol. **265**, 1500 (1994).

[69] R. G. Larson, Hua Hu, D. E. Smith and S. Chu, "Brownian dynamics simulation of a DNA molecules in an extensional flow field," J. Rheol. Vol. **43**, 267 (1999).

[70] U. S. Argarwal, A. Dutta and R. A. Mashelkar, "Migration of macromolecules under flow: the physical origin and engineering implications," Chem. Engng. Sci., Vol. **49**, 1693 (1994).

[71] H. P. Babcock, D. E. Smith, J. S. Hur, E. S.G. Shaqfeh and S. Chu, "Relating the microscopic and Macroscopic response of a polymeric fluid in a shearing flow," Phys. Rev. Lett., Vol. **85**, 2018 (2000).

Chapter 9

TRANSPORT OF LIQUID IN RECTANGULAR MICROCHANNELS BY ELECTROOSMOTIC PUMPING

Chun Yang
School of Mechanical and Production Engineering, Nanyang Technological University, Singapore

Abstract: Rapid development in microfluidic systems has stimulated an interest in study of electroosmotic flow phenomenon in microchannels. In this paper, a theoretical framework for the description of such phenomenon is developed. Specifically, the characteristics of transient and steady state electroosmotic flows in a rectangular microchannel are analysed. Based on the Debye - Hückel approximation, an analytical solution to the 2D Poisson – Boltzmann equation governing the electrical double-layer field near the solid-liquid interface is presented. With implementation of the Green's function formula, an exact solution to the Navier-Stokes equation is found to describe the electroosmotic flow field. The effects of the zeta potential, electrolyte concentration, channel geometry and applied electrical field on the electroosmotic velocity distributions are discussed. The results of the evolution of the electroosmotic flow provide insights into the electroosmosis phenomenon.

Key words: Electroosmotic flow, BioMEMS pumping, Rectangular microchannel, Greens' function approach.

1. INTRODUCTION

During recent years a great deal of information has been generated on integrating Micro-Electro-Mechanical Systems (MEMS) technology into nucleic acid analysis, enzyme assays, and immunoassays. This leads to the development of the BioMEMS (or Biochip) technology, which represents the innovation of next generation's biotechnology [1–3]. The basic "unit

F.E.H. Tay (ed.), Microfluidics and BioMEMS Applications, 265–285.

process" operations in these systems are sample injection, mixing, chemical reaction, separation, and detection. Many of these fluidic manipulations are based on electrokinetic transport of materials exploiting the phenomena of electrophoresis and electroosmosis [4][5]. Specifically, the electroosmotic pumping, owing to its numerous advantages including ease of fabrication and control, no need for moving parts, high repeatability and reliability, and no noise, has been extensively used in the BioMEMS to transport liquid in microchannel networks [3–5].

Figure 1 illustrates the basic concept of the electroosmosis. In brief, most surfaces obtain an electrical charge when they are brought into contact with a polar medium. Due to electrostatic interaction, the surface charge, in turn, causes both counter-ions and co-ions in the liquid to be preferentially redistributed, resulting in the formation of the electrical double layer (EDL) [6]. Conceptually, the EDL can be divided into an inner compact layer and an outer diffuse layer. The inner layer, usually of several Angstroms, is immediately next to the charged surface and contains a layer of counter-ions that are strongly attracted to the surface and are immobile. In contrast, ions in the diffuse layer are less affected by the charged surface and hence are mobile. The thickness of the diffuse layer is dependent on the bulk ionic concentration and the electrical property of the liquid, usually ranging from several nanometres for high ionic concentration solutions up to the order of micrometers for ultra-filtered water. Within the diffuse layer, because of the presence of the EDL, the local net charge density is not zero. If an external electric field is applied in such a way that it is tangential to the EDL field, an electric body force is exerted on the ions in the diffuse layer of the EDL. The ions will move under the influence of the applied electric field, pulling the liquid with them and resulting in an electroosmotic flow. The liquid movement is carried through to the rest (beyond the EDL region) of the liquid in the channel by viscous forces. The driving forces for the electroosmotic flow in microchannels are interplay of the applied electric filed and the EDL. The electroosmotic flow, therefore, depends on the strength of the electric field and the EDL characteristics such as the Debye-Hückel length, the surface ζ-potential, and the net charge density [7].

In the literature, numerous studies have been reported on rigorous modelling of the electroosmotic flow in microchannels. Earlier works using analytical approaches for study of electroosmotic flows were limited to systems having simple geometry. Burgreen and Nakache [8] formulated a mathematical model for the electroosmotic flow in an ultrafine slit. Rice and Whitehead [9] studied the same problem in a narrow cylindrical capillary within the framework of the Debye-Hückel approximation. Levine et al. [10] extended the Rice and Whitehead's model to deal with the electrokinetic flow in cylindrical capillaries having high zeta-potentials. Koh and Anderson

[11] presented an analysis of the electroosmotic flow in charged microcapillaries of ellipse shape. On the other hand, numerical simulations were carried out to investigate of the electroosmotic flow in complex geometry of microchannel network. Results concerning the electroosmotic flow at the intersection of a cross microchannel during the chemical sample injection procedure were presented by Patankar and Hu [12]. Hu et al. [13] developed a numerical scheme to study the electroosmotic flow in intersecting channels of a T-shaped configuration. They showed the hydrodynamic effect is an important factor that influences fluid leakage out of a channel where the electric potential is left floating. It was also found that the flow at the T-section could be controlled by applying a potential at each reservoir connected to the two ends of a channel. With implementation of a finite-element formulation, Bianchi et al. [14] performed numerical computation of the electroosmotic flow at a T-junction of microscale dimensions. Their results indicated that relative zeta potentials and channel widths are two predominant parameters affecting the distribution of flow at the intersection.

Most of channels used in modern microfluidic devices and MEMS are made by micromachining technologies. The cross-section of these channels is close to a rectangular shape. As such, the characteristics of the electroosmotic flow in rectangular microchannels are studied in this paper. We use a two-dimensional model for the Poisson-Boltzmann equation to describe the EDL field. Such a two-dimensional model is necessary to include the "corner effect", which is expected to affect the electroosmotic flow in the rectangular microchannel [15][16]. Under the Debye-Hückel approximation, we solve the Poisson-Boltzmann equation analytically. With implementation of the Green's function approach, we obtain an exact solution to the equation of motion for the electroosmotic flow in rectangular microchannels. The advantages of our approach are that the solutions obtained to describe the electroosmotic flow field are analytical, and hence can provide much more insights than numerical results. The EDL field and the flow velocity distributions are studied as functions of zeta potential, liquid concentration, channel geometry and applied electrical field.

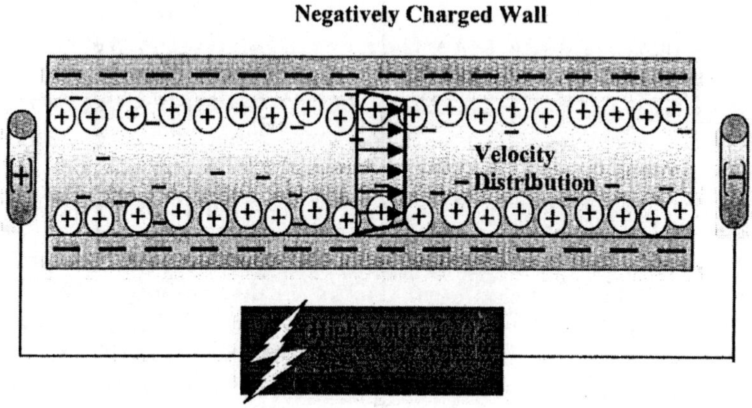

Negatively Charged Wall

Velocity Distribution

Figure 1. Illustration of electroosmotic flow in a ultra fine capillary

2. FORMULATION OF PROBLEM

2.1 Basic Transport Equations

In an excellent monograph on electrokinetic transport phenomena, Masliyah [17] has formulated basic transport equations for electrokinetic flows resulted from electroosmosis, electrophoresis, streaming potential, and sedimentation. For the electroosmotic flow of a Newtonian incompressible liquid fluid through arbitrarily shaped capillary structures, the basic governing equations are non-dimensionalized by using

$$\underline{r}' = \underline{r}/d_h \qquad t' = t/\left(d_h^2/\nu\right) \quad \underline{V}' = \underline{V}/V_0 \tag{1a}$$

$$p' = p/(\rho V_0^2) \quad \underline{E}' = \underline{E}/E_0 \qquad n_i' = n_i/n_0 \qquad \rho_e' = \rho_e/(zen_0) \tag{1b}$$

where r, t, V, p, E, n_i, and ρ_e are the position vector, time, local fluid velocity vector, pressure, local electric field vector, local number concentration of the type-i ion, and local net charge density, respectively. These quantities are normalized using d_h, $t^* = d_h^2/\nu$ (ν is the kinetic viscosity of the fluid), V_0, $\omega = \rho V_0^2$ (ρ is the density of the fluid), E_0, n_0, and zen_0 (z and e are the valence of ions of a symmetric electrolyte and elementary charge, respectively), which are the microchannel hydraulic diameter, characteristic time scale, characteristic electroosmotic flow velocity, characteristic kinetic energy of the electroosmotic flow, characteristic electric field strength, bulk ion number concentration, and characteristic electric charge, respectively. Then we can write down the continuity equation

$$\nabla' \cdot \underline{V}' = 0 \tag{2}$$

and the Navier-Stokes equation

$$\frac{\partial \underline{V}'}{\partial t'} + \mathrm{Re}_0 \, (\underline{V}' \cdot \nabla') \, \underline{V}' = -\mathrm{Re}_0 \, \nabla' p' + \underline{F}' + \nabla'^2 \, \underline{V}' \tag{3}$$

where the gradient and Laplacian operators are normalized with respective to d_h. The components of the time-dependent, non-dimensional velocity vector can be expressed as

$$\underline{V}' = u'\underline{i} + v'\underline{j} + w'\underline{k} \tag{4}$$

The normalized position vector is given by

$$\underline{r}' = x'\,\underline{i} + y'\underline{j} + z'\,\underline{k} \tag{5}$$

The Reynolds number is defined as

$$\mathrm{Re}_0 = V_0 d_h / \nu \tag{6}$$

The non-dimensional body force term on the right-hand side of Eq. (3) is the electric force exerting on the fluid resulted from the local electric field, $\underline{E}$, and is given by

$$\underline{F}' = d_h^2 \, z e n_0 \rho_e' E_0 \underline{E}' / (\mu V_0) \tag{7}$$

where μ is the dynamic viscosity of the fluid. It is such a body force that causes the liquid flow through the microchannel. Consider that the microchannel hydraulic diameter and the characteristic electroosmotic flow velocity respectively are of order 10 μm and V_0 = 1 mm/s, we can estimate Re_0 of order 10^{-2}. Furthermore, usually no external pressure gradient is applied. Hence the pressure gradient term on the right-hand side of Eq. (3), $\mathrm{Re}_0 \nabla' p'$ is not present, and also the inertia term, $\mathrm{Re}_0 (\underline{V}' \cdot \nabla') \underline{V}'$ is small enough to be neglected. Then, Eq. (3) reduces to the unsteady Stokes equation, which is valid for the creeping flow and is expressed as

$$\frac{\partial \underline{V}'}{\partial t'} = \underline{F}' + \nabla'^2 \underline{V}' \tag{8}$$

Clearly, it is through the body force term, $\underline{F}'$ that the Stokes equation is coupled with the governing equation for the EDL field. According to the theory of electrostatics, the electrical potential distribution, $\psi(\underline{r})$ in the EDL is described by the Poisson equation that takes the form:

$$\nabla'^2 \psi' = -(d_h \kappa)^2 \rho_e' \tag{9}$$

where the non-dimensional electric potential is defined as $\psi' = ze\psi / (k_b T)$ (k_b is the Boltzmann constant and T is the absolute temperature). κ is the Debye-Hückel parameter defined as

$$\kappa = (2 z^2 e^2 n_0 / \varepsilon_r \varepsilon_0 k_b T)^{1/2} \tag{10}$$

where ε_r is the dielectric constant of the fluid and ε_0 is permittivity of the vacuum. Physically, κ^{-1} represents the characteristic thickness of the EDL. The net electric charge density is associated with the number concentration of N ions through

$$\rho_e = e \sum_{i=1}^{N} z_i n_i \tag{11}$$

Obviously, a full formulation of governing equations should also include the mass transfer equation for each type of ions in the system. Strictly speaking, rigorous mathematical modelling of charge transport in the EDL region should take into account of unsteady effects. For a typical BioMEMS using a 10^{-1}-10^{-3} M electrolyte (the corresponding reciprocal Debye-Hückel parameter is of order $\kappa^{-1} \approx 1 - 10\ nm$) and a $d_h = 10$ μm microchannel, we have $d_h \kappa >> 1$, indicating the characteristic dimension of the electroosmotic flow, d_h is much larger than that of the EDL, κ^{-1}. Assuming reasonable values of the kinetic viscosity, $\nu \approx 10^{-6}$ m^2/s and the ion mass diffusion coefficient, $D_i = 10^{-10}$ m^2/s, we can readily show that the time scale related to electromigration in the EDL characterized by

$(\kappa^{-1})^2/D_i$ is of order 10^{-8}-10^{-6} sec, at least two orders smaller than the characteristic time associated with the evolution of the electroosmotic flow, which is of order 10^{-4} sec estimated from d_h^2/ν. This suggests a negligible transient effect of the EDL "relaxation". As such, in the microchannel electroosmotic flow, the ionic concentration distribution $n_i(\underline{r})$, generally described by the Nernst-Planck equation [17], takes the convective-diffusion form

$$Pe_0 \nabla' \cdot (\underline{U}_i' n_i') = \nabla'^2 n_i' \quad (12)$$

where $\underline{U}_i'$ is the normalized velocity of the type-i ion. The gradient and Laplacian operators are normalized with respective to κ^{-1}. Pe_0 is the Peclet numbers defined as

$$Pe_0 = V_0 \kappa^{-1}/D_i \quad (13)$$

Substituting given values of V_0 = 1 mm/s, $D_i = 10^{-10}$ m^2/s and κ^{-1}= 10 nm, we can estimate the Peclet numbers of order $Pe_0 = 10^{-1}$. For the microchannel electroosmotic flow, the ion velocity $\underline{U}_i$ can be decomposed into contributions from hydrodynamic velocity V and a velocity u_i due to electromigration (caused by the presence of the electric field). Then mathematically we can express

$$\underline{U}_i' = \underline{V}' + \underline{u}_i' \quad (14)$$

The velocity u_i is related to the electrostatic force exerting on the ions by the following equation

$$\underline{F}_{ei} = z_i e \underline{E} = f_i \underline{u}_i \quad (15)$$

where f_i is the hydrodynamic resistance coefficient, and it can be determined from the Stokes-Einstein equation [17]

$$f_i = k_b T/D_i \quad (16)$$

The local electric field, E can be decoupled as the summation of the applied electric filed and the EDL field, and thus E can be expressed as

$$\underline{E} = -\nabla\phi - \nabla\psi \quad (17)$$

where ϕ is the electric potential due to applied electric field, and it satisfies the following

$$\nabla^2 \phi = 0 \qquad \underline{E}_e = -(\partial \phi / \partial x)\, \underline{i} = E\, \underline{i} \tag{18}$$

where E is the strength of the applied electric field. Then we can readily show that the normalized velocity $\underline{u}_i'$ can be expressed by

$$\underline{u}_i' = -(\nabla' \phi' + \nabla' \psi')\, z_i / (z\, Pe_0) \tag{19}$$

Substituting Eqs. (14), (18) and (19) into Eq. (12) and making use of Eq. (2), we can show next

$$\nabla'^2 n_i' - Pe_0 \underline{V}' \cdot \nabla n_i' + \frac{\partial \phi'}{\partial x'} \frac{\partial n_i'}{\partial x'} + \nabla' \cdot [\frac{z_i}{z} n_i' \nabla' \psi')] = 0 \tag{20}$$

When the microchannel electroosmotic flow is fully-developed, the components of fluid velocity V satisfy $u = u(y, z)$ and $v = w = 0$ in terms of Cartesian coordinates. Equation (20) reduces to

$$\nabla'^2 n_i' + (\frac{\partial \phi'}{\partial x'} - Pe_0\, u') \frac{\partial n_i'}{\partial x'} + \nabla' \cdot (\frac{z_i}{z} n_i' \nabla' \psi') = 0 \tag{21}$$

Note that the ionic concentration is time-independent, implying no appreciable ionic concentration gradient is established along the flow direction, i.e., $\partial n_i' / \partial x' = 0$. Then Eq. (21) becomes

$$\nabla'^2 n_i' = \nabla' \cdot (-\frac{z_i}{z} n_i' \nabla' \psi') \tag{22}$$

We can readily solve Eq. (22), and obtain its solution, which is

$$n_i' = \exp\left(-\frac{z_i}{z} \psi'\right) \text{ or } n_i = n_o \exp(-\frac{z_i e \psi}{k_b T}) \tag{23}$$

This is the well-known Boltzmann distribution. Therefore, it can be concluded that for a microchannel electroosmotic flow in a fully developed state, the equilibrium Boltzmann distribution equation is still valid.

It can be seen that closure mathematical models of the electroosmotic flow through arbitrary shaped microchannels should include the unsteady Stokes equation, Eq. (8), the Poisson equation, Eq. (9), and the Boltzmann distribution, Eq. (23). In the next sections, we will apply these basic equations to the case of the electroosmotic flow through rectangular microchannels.

2.2 Electric Double Layer Field in Rectangular Microchannels

We consider a rectangular microchannel of width 2W, height H and length L as shown in Fig.2. Due to symmetry in the potential and velocity fields, the solution domain reduces to a half section of the channel.

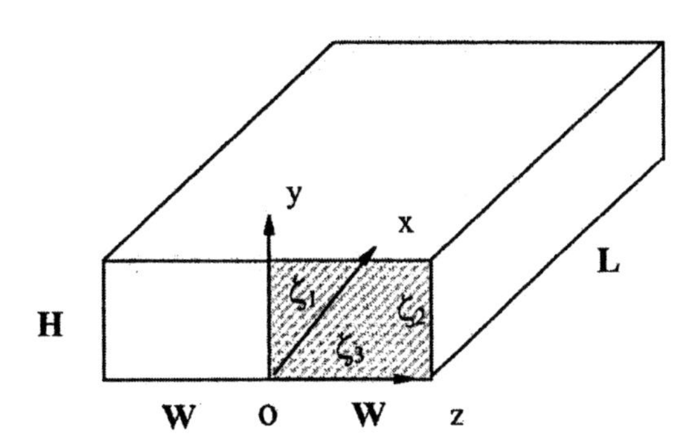

Figure 2. Geometry of a rectangular microchannel

In order to characterize the electroosmotic flow in a rectangular microchannel, we must evaluate distributions of electrical potentials and net charge density first. The Poisson equation, i.e., Eq. (9) governing the electric potential distribution in a rectangular microchannel is expressed in terms of Cartesian coordinates as

$$\frac{\partial^2 \psi'}{\partial y'^2} + \frac{\partial^2 \psi'}{\partial z'^2} = -(d_h \kappa)^2 \rho_e' \tag{24}$$

where the hydraulic diameter of the rectangular microchannel is defined as $d_h = 4\,HW/(H + 2W)$. Substituting the expression for the net charge density, Eq. (11) and the Boltzmann distribution, Eq. (23) into Eq. (24) leads to the well-known Poisson-Boltzmann equation that takes the form

$$\frac{\partial^2 \psi'}{\partial y'^2} + \frac{\partial^2 \psi'}{\partial z'^2} = -(d_h \kappa)^2 \sinh \psi' \tag{25}$$

Without loss generality, we consider channel walls having different zeta potentials as follows

$$y' = 0 \qquad \psi' = \zeta_3' \qquad y' = H/d_h \qquad \psi' = \zeta_1' \tag{26a}$$

$$z' = 0 \qquad z' = W/d_h \qquad \psi' = \zeta_2' \qquad \partial\psi'/\partial z' = 0 \tag{26b}$$

By definition, the zeta potential is a measurable potential at the shear plane, i.e. the boundary between the compact layer and the diffuse layer in the EDL theory [6]. Generally speaking, the non-linear 2D Poisson-Boltzmann (P-B) equation (24) can't be solved without using numerical method. However, for small values of the electrical potential (i.e., $|ze\psi| << \kappa_b T$, which physically means small electrical potential comparison with thermal energy of ions), the P-B equation can be linearized as

$$\frac{\partial^2 \psi'}{\partial y'^2} + \frac{\partial^2 \psi'}{\partial z'^2} = -(d_h \kappa)^2 \psi' \tag{27}$$

Using the separation of variables method, we can obtain an analytical solution to Eq. (27)

$$\psi'(y',z')$$

$$= 4\zeta_1' \sum_{m=1}^{\infty} \frac{(-1)^{m+1} \sinh[\sqrt{1+\frac{(2m-1)^2\pi^2 d_h^2}{4(d_h\kappa)^2 W^2}}(d_h\kappa)y']}{(2m-1)\pi \sinh[\sqrt{1+\frac{(2m-1)^2\pi^2 d_h^2}{4(d_h\kappa)^2 W^2}}(\kappa H)]} \cos[\frac{(2m-1)\pi d_h}{2W} z'] +$$

$$2\zeta_2' \sum_{n=1}^{\infty} \frac{[(-1)^{n+1}+1] \cosh[\sqrt{1+\frac{n^2\pi^2 d_h^2}{(d_h\kappa)^2 H^2}}(d_h\kappa)z']}{n\pi \cosh[\sqrt{1+\frac{n^2\pi^2 d_h^2}{(d_h\kappa)^2 H^2}}(\kappa W)]} \sin[\frac{n\pi d_h}{H} y'] -$$

$$4\zeta_3' \sum_{p=1}^{\infty} (-1)^{p+1} \cos[\frac{(2p-1)\pi d_h}{2W} z'] \times \left\{(2p-1)\pi \tanh[\sqrt{1+\frac{(2p-1)^2\pi^2 d_h^2}{4(d_h\kappa)^2 W^2}}(\kappa H)]\right\}^{-1} \times$$

$$\{-\tanh[\sqrt{1+\frac{(2p-1)^2\pi^2 d_h^2}{4(d_h\kappa)^2 W^2}}(\kappa H)] \cosh[\sqrt{1+\frac{(2p-1)^2\pi^2 d_h^2}{4(d_h\kappa)^2 W^2}}(d_h\kappa)y'] +$$

$$\sinh[\sqrt{1+\frac{(2p-1)^2\pi^2 d_h^2}{4(d_h\kappa)^2 W^2}}(d_h\kappa)y']\} \tag{28}$$

Once the electrical potential distribution is computed, the non-dimensional local net charge density can be calculated.

$$\rho_e'(y',z') = -2\sinh\psi'(y',z') \approx -2\psi' \tag{29}$$

This equation shows that the local charge density in the EDL region carries the opposite sign of the charged wall. For example, if the channel wall is negatively charged – this is the case for most of non-conductance materials in contact with aqueous solutions in neutral pH zones, the local net charge density is positive, indicating there exists more cations than anions in the EDL region. As such, when an external electric field is applied along the channel wall, the electroosmotic flow moves toward the cathode.

2.3 Electroosmotic Flow Field in Rectangular Microchannels

Consider a rectangular microchannel filled with an electrolyte initially in the still state. The electroosmotic flow in response to an applied electric field is governed by the unsteady Stokes equation, Eq. (8) which, under conditions of $u = u(y, z)$ and $v = w = 0$ in terms of Cartesian coordinates, is expressed as

$$\frac{\partial u'}{\partial t'} = d_h^2 zen_0 E_0 \rho_e' E'/(\mu V_0) + \frac{\partial^2 u'}{\partial y'^2} + \frac{\partial^2 u'}{\partial z'^2} \tag{30}$$

Substituting $\rho_e'(y', z')$ by Eq. (29) and defining a dimensionless parameter $\Theta = zen_0 E_0 d_h^2 E'/\mu V_0$ (Θ represents the ratio of electrokinetic force to viscous force), we, therefore, can rewrite Eq. (30) as

$$\frac{\partial u'}{\partial t'} = \frac{\partial^2 u'}{\partial y'^2} + \frac{\partial^2 u'}{\partial z'^2} + \Theta \psi' \tag{31}$$

The initial, no-slip and symmetric conditions are applied

$$t' = 0 \quad u' = 0 \tag{32a}$$

$$y' = 0 \quad y' = H/d_h \qquad u' = 0 \tag{32b}$$

$$z' = 0 \quad \partial u'/\partial z' = 0 \qquad z' = W/d_h \qquad u' = 0 \tag{32c}$$

By using the Green's function formulation, the solution of Eq. (31) subjecting to the above boundary conditions is obtained as

$$\begin{aligned} &u'(y', z', t') \\ &= -\int_0^{t'} d\tau \int_{Y=0}^{H/d_h} \int_{Z=0}^{W/d_h} G(y', z', t' \mid Y, Z, \tau)[\Theta \psi'(Y, Z)]\, dY dZ \end{aligned} \tag{33}$$

Here $G(y', z', t' \mid Y, Z, \tau)$ is the Green's function, which can be found by using the separation of variable method [18]. The expression for $G(y', z', t' \mid Y, Z, \tau)$ is given as

$$G(y',z',t' \mid Y,Z,\tau)$$
$$= \frac{4d_h^2}{WH} \sum_{m=1}^{\infty} \sum_{n=1}^{\infty} \exp\left\{ -\pi^2 d_h^2 [\frac{n^2}{H^2} + \frac{(2m-1)^2}{4W^2}](t'-\tau) \right\} \tag{34}$$
$$\times \sin[\frac{n\pi d_h}{H} y'] \sin[\frac{n\pi d_h}{H} Y] \cos[\frac{(2m-1)\pi d_h}{2W} z'] \cos[\frac{(2m-1)\pi d_h}{2W} Z]$$

Substituting Eq. (34) into Eq. (33) and rearranging it, we can obtain the non-dimensional fluid velocity profile in the microchannel as follows:

$$u'(y',z',t') = -\frac{4\Theta}{\pi^2 HW} \times$$
$$\sum_{m=1}^{\infty} \sum_{n=1}^{\infty} \frac{1 - \exp\left\{ -\pi^2 d_h^2 [\frac{n^2}{H^2} + \frac{(2m-1)^2}{4W^2}] t' \right\}}{\frac{n^2}{H^2} + \frac{(2m-1)^2}{4W^2}} \sin[\frac{n\pi d_h}{H} y'] \cos[\frac{(2m-1)\pi d_h}{2W} z']$$
$$\times \int_{Y=0}^{H/d_h} \int_{Z=0}^{W/d_h} \sin\left[\frac{n\pi d_h}{H} Y\right] \cos\left[\frac{(2m-1)\pi d_h}{2W} Z\right] \psi'(Y,Z) dY dZ$$

(35)

Case I: For a small time elapsed, i.e., $t' << 1$, $e^{-\beta^2 t'} \approx 1 - \beta^2 t'$, then the velocity distribution reduces to

$$u'(y',z',t' << 1)$$
$$= -\frac{4\Theta d_h^2}{HW} t' \sum_{m=1}^{\infty} \sum_{n=1}^{\infty} \sin[\frac{n\pi d_h}{H} y'] \cos[\frac{(2m-1)\pi d_h}{2W} z']$$
$$\times \int_{Y=0}^{H/d_h} \int_{Z=0}^{W/d_h} \sin[\frac{n\pi d_h}{H} Y] \cos[\frac{(2m-1)\pi d_h}{2W} Z] \psi'(Y,Z)\, dY\, dZ$$

(36)

indicating that the velocity monotonically increases with time at initial moment.

Case II: For a large time, i.e., $t' \to \infty$, $e^{-\beta^2 t'} \approx 0$, then the velocity distribution becomes

$$u'(y',z',t' \to \infty)$$

$$= -\frac{4\Theta}{\pi^2 H W} \sum_{m=1}^{\infty} \sum_{n=1}^{\infty} \frac{1}{\frac{n^2}{H^2} + \frac{(2m-1)^2}{4W^2}} \sin[\frac{n\pi\, d_h}{H} y'] \cos[\frac{(2m-1)\pi\, d_h}{2\, W} z'] \quad (37)$$

$$\times \int_{Y=0}^{H/d_h} \int_{Z=0}^{W/d_h} \sin[\frac{n\pi\, d_h}{H} Y] \cos[\frac{(2m-1)\pi\, d_h}{2\, W} Z]\, \psi'(Y,Z)\, dY\, dZ$$

which is the formulation for the electroosmotic velocity distribution under the steady-state situation.

From Eq. (35), we can show the mathematical expression for the average velocity of the electroosmotic flow is

$$u'_{ave}(t') = \frac{8\Theta}{\pi^4 d_h^2}$$

$$\times \sum_{m=1}^{\infty} \sum_{n=1}^{\infty} \frac{1 - \exp\left\{-\pi^2 d_h^2 [\frac{n^2}{H^2} + \frac{(2m-1)^2}{4W^2}]\, t'\right\}}{n(2m-1)[\frac{n^2}{H^2} + \frac{(2m-1)^2}{4W^2}]} \{[1-(-1)^n](-1)^m\}$$

$$\times \int_{Y=0}^{H/d_h} \int_{Z=0}^{W/d_h} \sin[\frac{n\pi d_h}{H} Y] \cos[\frac{(2m-1)\pi d_h}{2W} Z]\, \psi'(Y,Z)\, dY\, dZ \quad (38)$$

Thus the non-dimensional volume flow rate through the rectangular microchannel, defined by $Q'_v = Q_v / 2HWV_0$, is given by

$$Q'_v = u'_{ave} \quad (39)$$

3. RESULTS AND DISCUSSION

Case studies were carried out for a *NaCl* electrolyte at a typical room temperature, T = 298 K, at which the properties of the electrolyte are

$\varepsilon = 80$, $\rho_f = 998$ kg/m^3, and $\mu_f = 0.9 \times 10^{-3}$ kg/ms. Since the zeta potential is strongly dependent on the surface material and electrolyte concentration. It is quite common that for a microchannel surface (e.g., silicon wafer) in NaCl electrolyte of concentration 10^{-4}, 10^{-5}, and 10^{-6} M, the corresponding values of the zeta potential may be chosen as 25, 50, and 75 mV, respectively. Bear in mind that the linear approximation used for solving the P-B equation is valid only when the zeta potential is small (usually less than 25 mV is suggested). However, if one is interested in the flow behaviour of a highly charged system where only the outer region of the diffuse double layer is important, the linear approximation can give a good prediction even when the zeta potential is over 100 mV [6].

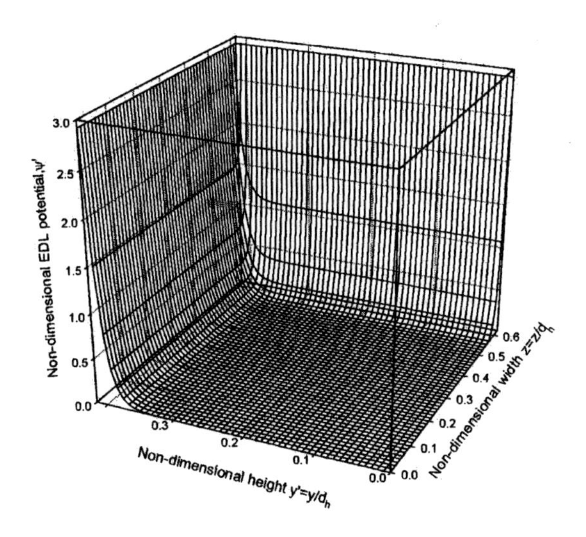

Figure 3. Non-dimensional electrical potential profile in a quarter section of a rectangular microchannel with $\kappa d_h = 78.5$, $H{:}2W = 2{:}3$, and $\zeta_1 = \zeta_2 = \zeta_3 = 75$ mV

The non-dimensional electrical potential distribution across a quarter section of the rectangular channel is shown in Fig. 3. As illustrated in Fig. 3, the potential field drops off sharply very close to the wall. The region where the net charge density is not zero is limited to a small region close to the channel surface. Furthermore, it can be seen a corner effect on the electrical potential profile. This is an important feature of the 2D P-B equation, which is expected to affect the flow field in such a rectangular microchannel.

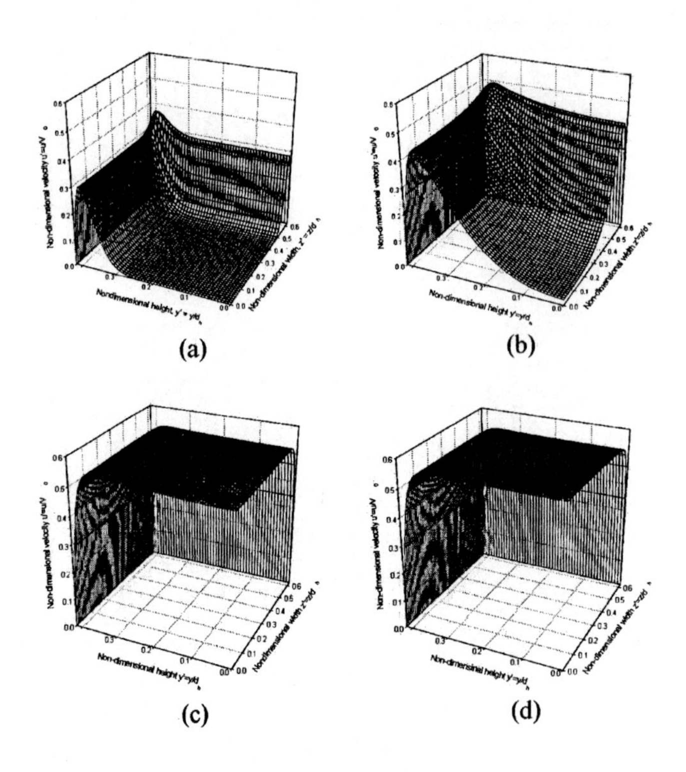

Figure 4. Evolution of electroosmotic non-dimensional velocity field in a quarter section of a rectangular microchannel with κd_h = 78.5, $H{:}2W$ = 2:3, E = 100 V/cm, and $\zeta_1 = \zeta_2 = \zeta_3$ = 75 mV. (a) t = 1 μs; (b) t = 10 μs; (c) t = 100 μs; (d) t = 10 ms

Figure 4 shows the evolution of electroosmotic velocity field in a quarter of the rectangular channel of dimensions $H \times 2W = 20 \times 30\ \mu m$. Within elapsed time of 1 μs, the velocity develops rapidly from zero to reach a maximum velocity inside the EDL region and drops gradually as approaching the centre of the channel. This phenomenon reveals a unique characteristic of the electroosmotic flow. Because the flow is driven by the electrical forces resulted from the action of an external electrical field and the EDL field. Such driving forces exist only inside the EDL region where the electrical net charge is present. The flow in this region may be viewed as "active". In contrast, the flow outside the EDL region may be considered as "passive" flow caused by viscous forces. Also an important feature observed from Fig. 4 is the "corner effect", which is clearly demonstrated during the course of the evolution of the electroosmotic flow. Moreover, it is noted that the flow velocity within the EDL reaches almost the steady-state value in a very short time period, while for the region away from the EDL, the velocity is almost zero. This suggests that within the EDL the viscous forces are much smaller than the electrical forces at initial moments. Then the viscous drug increases as the flow develops, and finally is balanced with the electrical forces when the flow reaches the steady state. At the steady-state situation, the velocity profile exhibits a plateau, resembling a plug flow. Under parameters studied, it takes of the order of 10^{-3} sec for the electroosmotic flow to reach a fully developed state. In addition, there is a correlation between the velocity distribution and the EDL potential profile, which can be reflected from Fig. 5. The velocity distribution has the highest gradient where the EDL potential profile also hits its highest gradient. For instance, the region close to the wall where the velocity sharply reduces to zero is the region where the EDL potential increases from approximately zero to the zeta potential.

The electroosmotic flow, as explained earlier, is the result of a tangentially applied electric field on a channel with the presence of the EDL. The electric field strength therefore has a significant influence on the electroosmotic flow across the microchannel. As shown in Fig. 6, the velocity increases with an increase in the electric field strength, indicating a linear relationship between applied voltage and velocity.

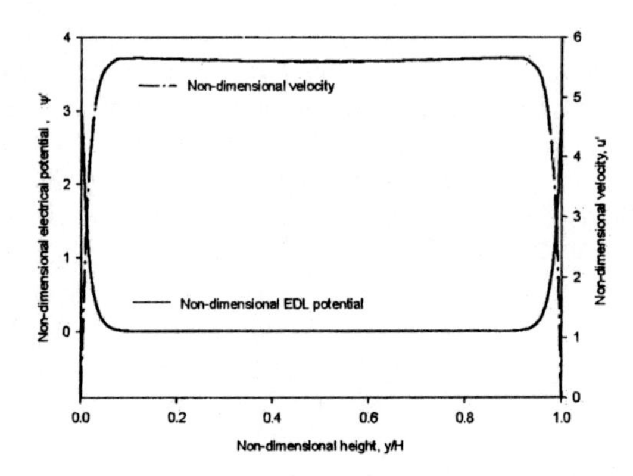

Figure 5. Non-dimensional velocity and electrical potential profiles along the height of a microchannel with $\kappa d_h = 78.5$, $H{:}2W = 2{:}3$, $\zeta_1 = \zeta_2 = \zeta_3 = 75$ mV, and $E = 100$ V/cm. In this Figure, both $y/H = 0.0$ and $y/H = 1.0$ represent the channel wall

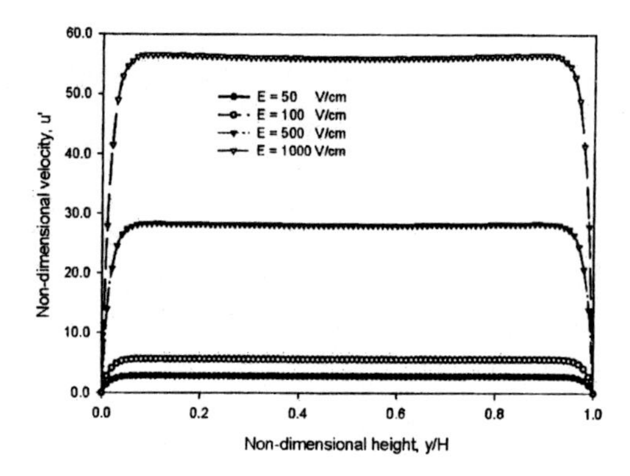

Figure 6. Variation of non-dimensional velocity distributions along the height of a microchannel with $\kappa d_h = 78.5$, $H{:}2W = 2{:}3$, and $\zeta_1 = \zeta_2 = \zeta_3 = 75$ mV and for four different electrical fields

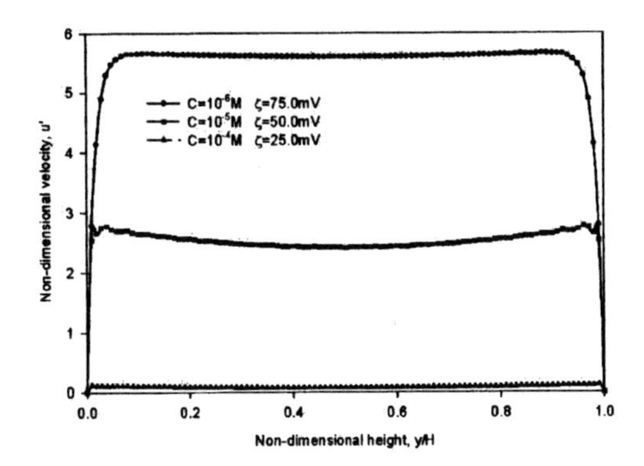

Figure 7. Non-dimensional velocity profiles along the height of a microchannel with $H{:}2W$ = 2:3, and E = 100 V/cm for three different combinations of concentrations and zeta potentials

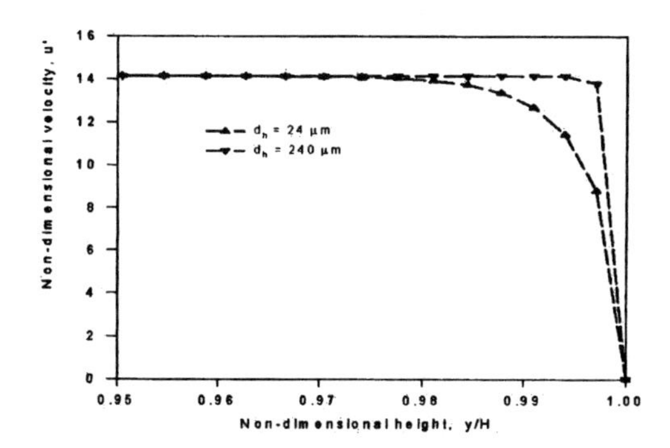

Figure 8. Non-dimensional velocity profiles along the height of a microchannel with $H{:}2W$ = 2:3, and $\zeta_1 = \zeta_2 = \zeta_3$ = 75 mV and E = 200 V/cm for two different hydraulic diameters

As the electrolyte concentration and zeta potential are concerned, it is found from Fig. 7 that the electroosmotic flow for the case of low concentration and high zeta potential is stronger than that for the case of the high concentration and low zeta potential. This is due to the fact that the mean electroosmotic velocity is proportional to the zeta potential and the EDL field is compressed for the high concentration.

The hydraulic diameter d_h affecting the electroosmotic velocity is shown in Fig. 8, in which only a very small fraction of the velocity profiles near the channel wall is displayed to see the differences clearly. It is noted that although variation of d_h does cause a change in the velocity profiles near the channel wall, the maximum electroosmotic velocity is independent of the channel size d_h. This is an important feature of the electroosmotic flow. For design purpose, whenever large pumping flow rates are desired, relatively larger diameter channels would seem to be a better choice.

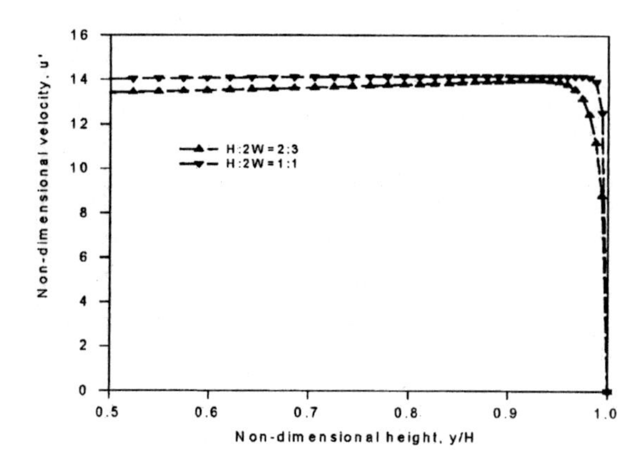

Figure 9. Non-dimensional velocity profiles along the height of a microchannel with $d_h = 24$ μm, $\zeta_1 = \zeta_2 = \zeta_3 = 75$ mV, and $E = 200$ V/cm for two different ratios of height to width

Due to the impact of the channel geometry on the EDL filed, the geometric aspect ratio is another factor influencing the electroosmotic flow. When keeping a fixed hydrodynamic diameter and other physicochemical parameters, it is shown in Fig. 9 that the electroosmotic flow velocity decreases as the aspect ratio approaches 1:1 (for a square channel). This is due to the large role that corner effects have on the development of the EDL field and the velocity profiles in square channels.

REFERENCES

[1] A. Manz, N. Graber, and H. M. Widmer, "Miniaturized total chemical analysis systems: A novel concept for chemical sensing," Sensors Actuators B, Vol. **1**, 244 (1990).

[2] van den A. Berg, and T. S. J. Lammerink, "Micro total analysis systems: Microfluidic aspects, integration Concepts and applications," Topics Curr. Chem., Vol. **194**, 21 (1998).

[3] D. J. Harrison, K. Fluri, K. Seiler, Z. Fan, C. S. Effenhauser, and A. Manz, "Micromachining a miniaturized capillary electrophoresis-based chemical analysis system on a chip," Science, Vol. **261**, 895 (1993).

[4] A. Manz, C. S. Effenhauser, N. Burggraf, D. J. Harrison, K. Seiler, and K. Fluri, "Electroosmotic pumping and electrophoretic separations for miniaturized chemical analysis systems," J. Micromech. Microeng., Vol. **4**, 257 (1994).

[5] C. T. Culbertson, R. S. Ramsey, and J. M. Ramsey, "Electroosmotically induced hydraulic pumping on microchips: Differential ion transport," Anal. Chem., Vol. **72**, 2285 (2000).

[6] R. J. Hunter, *Zeta Potential in Colloid Science: Principles and Applications* (Academic Press, New York, 1981).

[7] G. M. Male, C. Yang, and D. Li "Electrical double layer potential distribution in a rectangular microchannel," Colloids Surf. A, Vol. **135**, 109 (1998).

[8] D. Burgreen, and F. R. Nakache, "Electrokinetic flow in ultrafine capillary slits", J. Phys. Chem., Vol. **68**, 1084 (1964).

[9] C. L. Rice, and R. Whitehead, "Electrokinetic flow in narrow cylindrical capillaries," J. Phys. Chem., Vol. **69**, 4017 (1965).

[10] S. Levine, J. R. Marriott, G. Neale, and N. Epstein, "Theory of electrokinetic flow in fine cylindrical capillaries at high zeta potentials," J. Colloid Interface Sci., Vol. **52**, 136 (1975).

[11] W.-H. Koh, and J. L. Anderson, "Electroosmosis and electrolyte conductance in charged microcapillaries," AICHE J., Vol. **21**, 1176 (1975).

[12] N. A. Patankar, and H. H., Hu, "Numerical simulation of electroosmotic flow," Anal. Chem., Vol. **70**, 1870 (1998).

[13] L. Hu, J. Harrison, and J. H. Masliyah, "Numerical model of electrokinetic flow for capillary electrophoresis," J. Colloid Interface Sci., Vol. **215**, 300 (1999).

[14] F. Bianchi, R. Ferrigno, and H. H. Girault, "Finite element simulation of an electroosmotic-driven flow division at a T-junction of microscale dimensions," Anal. Chem., Vol. **72**, 1987 (2000).

[15] C. Yang, and D. Li, "Electrokinetic effects on pressure-driven liquid flows in rectangular microchannels," J. Colloid Interface Sci., Vol. **194**, 95 (1997).

[16] C. Yang, D. Li, and J. H. Masliyah, "Modeling forced liquid convection in rectangular microchannels with electrokinetic effects," Int. J. Heat Mass Transfer, Vol. **41**, 4229 (1998).

[17] J. H. Masliyah, *Electrokinetic Transport Phenomena, AOSTRA Technical Publication Series No. 12* (AOSTRA, Edmonton, 1994).

[18] J. V. Beck, *Heat Conduction Using Green's Functions* (Hemisphere, London, 1992).

Chapter 10

A DEVELOPMENT OF SLIP MODEL AND SLIP-CORRECTED REYNOLDS EQUATION FOR GAS LUBRICATION IN MAGNETIC STORAGE DEVICE

Eddie Yin-Kwee Ng, Ningyu Liu, and Xiaohai Mao
Nanyang Technological University, Singapore

Abstract: Reynolds equation with slip boundary conditions is widely applied for gas lubricant films in magnetic recording storage system. With the decrease of the clearance between the slider and the disk surface, conventional slip models become inaccurate. By implementing the direct simulation Monte Carlo (DSMC) method to the slider-bearing problem, it is found that the slip velocity is affected not only by shear stress, but also by density itself. In the ultra-thin film lubrication environment, the slip velocity has a linear relationship with the ratio of shear stress to density near the wall. Through this study, a new slip velocity model named stress-density ratio (SDR) model that is particularly suitable to the flows with large Knudsen numbers is developed. Based on this novel model, a slip-corrected Reynolds equation is also derived and solved numerically. The SDR model is confirmed to offer a better applicability than the previous slip-flow models for modern computer magnetic storage device, where the Knudsen number of slider air bearing is in the range of unity or larger.

Key words: SDR model, DSMC, Reynolds equation, slip coefficient, film lubrication

1. INTRODUCTION

One of the successful applications of gas bearing lubrication is the flying head for computer magnetic storage device. In a modern Winchester-type disk drive, the air between the read/write head and the surface of the spinning platter head forms a so-called slider air bearing to support the head

F.E.H. Tay (ed.), Microfluidics and BioMEMS Applications, 287–306.

float above the disk [1][2]. The small spacing between the slider and the disk can increase the intensity of the magnetic signal and thereby permit an increase in the areal density. With the rapid development of the modern computer magnetic storage device, there is an increasing demand for smaller, more compact disk storage that requires minimising the film thickness. In recent year, the minimum spacing under the slider could reach the order of 50 nm or less (Fig. 1). In such a narrow spacing, the gas flow has its special governing law because it performs in the scale of one molecular mean free path.

Figure 1. High-capacity, high-performance portable storage

Due to the extremely small sizes, gas flow in small passages becomes rarefied. Knudsen number, defined as the ratio of the molecular mean free path λ to the minimum spacing *h*, represents the level of rarefaction effect. The rarefied gas flows are typically classified into four flow regimes according to their Knudsen numbers [2][3]:

- $Kn \leq 10^{-2}$ continuum flow,
- $10^{-2} < Kn < 0.1$ slip flow,
- $0.1 \leq Kn < 10$ transition flow,
- $Kn \geq 10$ molecular flow.

The continuum flow is governed by the Navier-Stokes equation and the free molecular flow is governed by the collisionless Boltzmann equation.

In slip flow and transition flow regimes, the flow is too rarefied for Navier-Stokes equation based analysis, but is not rarefied enough to apply the collisionless Boltzmann equation. The numerical results predicted by the continuum Reynolds equation without slip corrections are in considerable errors [4][5].

In the slider air bearings, the gas flows usually fell into the slip and transition regimes. Traditionally, there are two main methods in attend to model slider air bearing problems. One method, the direct simulation of Monte Carlo method (DSMC) introduced by Bird [6], is an efficient numerical algorithm used for gas flow simulation in either slip or transition regime by capturing its important physical features. Wagner [7] has proved that the DSMC solution is an equivalent numerical solution of Boltzmann equation. Nevertheless, DSMC method has some considerable shortcomings. The unavoidable statistical error, and the fact that the error declines only with the square root of the sample size, are the main disadvantages in this method. Therefore, DSMC is applicable only to problems that involve larger perturbations or smaller sample size. For lower speed flow, it would be an expensive method because it requires a huge CPU time in samplings. When the fluid become denser, the number of real molecules increases. With a fix number of simulated molecules, the ratio of the number of real molecules to simulated molecules increases, and could lead to a statistics scatter. On the other hand, by increasing the number of simulated molecules would make the DSMC method a costly tool to use.

The second method is to solve the macroscopic hydrodynamics equations with the slip boundary condition, the so-called slip-corrected Reynolds equation [8], or to solve a molecular generalised lubrication equation [9]. Many attempts have been made to develop the slip correction models since the discovery of the "velocity slip" effect near a moving wall in nineteenth century. Hitherto, the well-known models used to predict the pressure distribution in air slider bearing include the first-order slip model of Schaaf and Sherman [10]; the second-order model of Hsia and Domoto [11]; the 1.5-order slip of Mitsuya [12] and the Beskok's model [13]. The slip-corrected Reynolds equations associated with these models have the advantages of simplicity in handling and compatibility with computer programs. However, they are limited to a narrow K_n range, and have the accuracy constraints. Even those models that are being claimed by many authors as promising approaches, they still resulted in some deviations over the transition flow regime [12][13]. The best approach applied in the transition regime today is still the expensive DSMC method. Unfortunately, in a modern magnetic storage device, the length scale of microchannel flow under the slider is the order of magnitude of the mean free path of gas molecular, and is within the vicinity of the so-called transition flow regime. Therefore, to develop a slip model suitable to transition regime is meaningful.

In current study, an attempt is made to develop an improved slip model suitable especially to transition regime. A simple model called the stress-density ratio (SDR) model, with a universal slip coefficient is proposed for ultra thin gas lubricant film. By means of DSMC method [14], from the results of numerical experimentation, the slip coefficient is investigated and determined. Based on this novel model, a new slip-corrected Reynolds equation is finally derived and solved numerically. The results confirm that the SDR model and the SDR slip-corrected Reynolds equation perform better than the previous models for the flow in the inclined slider air bearings with a high Knudsen number.

2. THEORETICAL ANALYSIS

According to Maxwell [15], a rarefied gas at the surface of a solid body is satisfied with the conditions of:

$$u_s - G(\frac{du}{dy} - \frac{3}{2}\frac{\mu}{\rho T}\frac{d^2T}{dxdy}) - \frac{3}{4}\frac{\mu}{\rho T}\frac{dT}{dx} = 0 \tag{1}$$

where

$$G = \frac{1}{2}\mu(2\pi)^{\frac{1}{2}}(p\rho)^{-\frac{1}{2}}(\frac{2}{\sigma} - 1) \tag{2}$$

If there is no inequality of temperature, and the fluid is perfect gas, we can derive the slip velocity as:

$$u_s = (\frac{2}{\sigma} - 1)\frac{\tau}{\rho\sqrt{(2RT/\pi)}} \tag{3}$$

where σ is an accommodation coefficient, ρ the density near the wall, p the pressure near the bottom wall, T the absolute temperature near the bottom wall, which is a constant in isothermal flow, μ the coefficient of viscosity, and τ the shear stress. The gas constant R = 208.12 J / (kg K) for argon, and R = 287 J / (kg K) for air. The x is in streamwise direction and y is in the bottom wall-normal direction.

Maxwell wrote the slip velocity, Eq. (3) as:

$$u_s = G\frac{\partial u}{\partial y} \tag{4}$$

The coefficient G was introduced with the term of Gleitungs-coefficient, or coefficient of slipping. Then Schaaf and Sherman [10] proposed the similar slip velocity:

$$u_s = (\frac{2}{\sigma} - 1)\lambda\frac{\partial u}{\partial y} \tag{5}$$

This slip boundary condition is the so-called first-order slip-flow model.

When the dimension of the characterised spacing of flow field becomes the same order of magnitude as one mean free path or less, the mean free path of a molecule would have undefined physical meaning, and one has to revert to Eq. (3). Based on Eq. (3), a new proposed slip coefficient is thus defined as:

$$C_s = \left(\frac{2}{\sigma} - 1\right)\frac{1}{\sqrt{(2RT/\pi)}} \tag{6}$$

The slip boundary condition can then be re-written as:

$$u_s = C_s \frac{\tau_w}{\rho} \tag{7}$$

This novel slip flow velocity model is termed as the stress-density ratio model. Because Eq. (6) was derived from the Maxwell's qualitative expression of boundary conditions used in Eq. (1), the slip coefficient in the SDR model can not be determined using Eq. (6), and thus an empirical mean is required. In the current work, we use DSMC method [14] to determine the slip coefficient in SDR model.

3. DSMC RESULTS AND SLIP COEFFICIENT

In the DSMC algorithm, the flow is simulated as a collection of discrete particles. Each particle has a position, velocity and energy etc, and represents a number of real molecules having the same properties. Based on the Boltzmann equation, the particle evolution is divided into two independent phases within a time interval: movement phase and collision phase. Intermolecular collisions are performed according to the theory of probability. The flow geometrical domain is divided into many cells. The dimension of each cell is smaller than one molecular mean free path. The collision pairs of particles are selected by probability method and all collision events involve two particles so that the computational time can be reduced. After collision, each particle changes its velocity and direction. The macroscopic physical quantities are obtained using statistics method by sampling the microscopic quantities of all the particles in every computational cell.

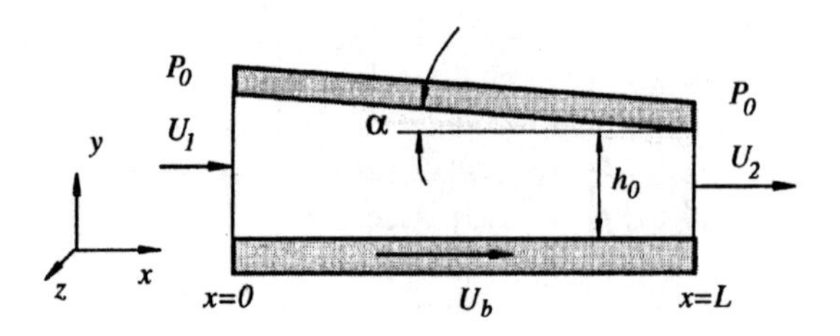

Figure 2. The geometry of slider bearing

Consider a microchannel formed by a stationary inclined wall above a horizontal wall. The inclined wall has a pitch angle of α = 0.01 rad and the horizontal wall moves with a velocity U_b in x direction, as shown in Fig. 2.

In the simulations, argon is chosen as a working fluid because it is a simple monatomic molecule gas with no internal degrees of freedom. In the inlet and outlet boundaries of the channel, there is a constant ambient temperature of T_0 = 273 K, and a pressure of p_a = 1 atm. (101,325 Pascal). At the ambient conditions, the number density (the number of molecules per unit volume, which depends on the temperature and pressure) of argon is $n_0 = p_a/(k\,T_0) = 2.6883\times10^{25}\text{m}^{-3}$ (here the Boltzmann constant $k = 1.380622\times10^{-23}$ J/K). The molecular mass is $m = 6.63\times10^{-26}$ kg; the mean free path is 62 nm; the coefficient of viscosity is 2.117×10^{-5}, and the viscosity-temperature exponent is $\omega = 0.81$. The molecule is represented using the variable hard sphere (VHS) model with diameter of $d = 4.17\times10^{-10}$ m. The accommodation coefficient, σ, for simplicity, is set as unity so that the wall is assumed as full accommodation and a diffuse reflection model is used in the DSMC solver.

The computational domain is divided into 2,000 trapezoid cells. In x direction, there are 200 uniformly spaced cells. In y direction, at each specified x location, the channel height is divided into ten uniform cells. Therefore, the size of the cell can be represented as:

$$\Delta x = L/200 = 25nm \tag{8}$$

$$\Delta y = [h_0 + (L - x)tg\alpha]/10 \tag{9}$$

The time step used here is Δt = $1.5\text{x}10^{-11}$ s, which is far less than the mean collision time for a particle. With this time step, the particle will need more than four time steps to pass through a single cell.

In all cases, a unified number of 56,000 particles are set initially. Every system arrives its equilibrium with a different particle number in the computational domain. For example, in the case of K_n =1.24 with U_b = 50 m/s, the particle number starts with around 56,000 particles, and it equilibrates to a steady state number of around 70,000 particles after 0.5 μs. The sampling function begins at 1.0 μs and lasts about 2.0 μs. With the fixed ambient condition, the minimum number of particles in each cell is more than 30. This range of particle number used here could ensure the statistical accuracy of the DSMC.

The pressure of the first and last cells are monitored and adjusted by changing the inflow and outflow velocities so as to fix the pressure at $x=0$ and $x=L$ to the ambient pressure.

The shear stress value is produced by the definition in DSMC, which is:

$$\tau_{xy} = -n_0(\overline{mu_1u_2}) \tag{10}$$

A bar over a quantity denotes the average value over all molecules in the sample. u_1 and u_2 are the x- and y- components of molecular velocity.

The results of u, τ_{xy}, ρ on bottom wall (y = 0) are computed by first-order extrapolation.

In the present DSMC simulation, the slip velocity on bottom wall is calculated by:

$$u_s = u\big|_{y=0} - U_b \tag{11}$$

Finally, the slip coefficient can be obtained from DSMC results:

$$C_s = \frac{u_s\rho}{\tau_{xy}} \tag{12}$$

A series of study cases are performed using the DSMC method to study the slip coefficient.

3.1 The slip coefficients with different boundary velocities at a specified Kn and bearing number

The Knudsen number is defined as:

$$K_n = \frac{\lambda}{h_0} \tag{13}$$

where h_0 is a minimum spacing of channel, and h_0 = 50 nm in this case. Therefore the Knudsen number is Kn = 1.24.

The bearing number is defined as:

$$\Lambda = 6U_b \mu L / (h_0^2 p_a) \tag{14}$$

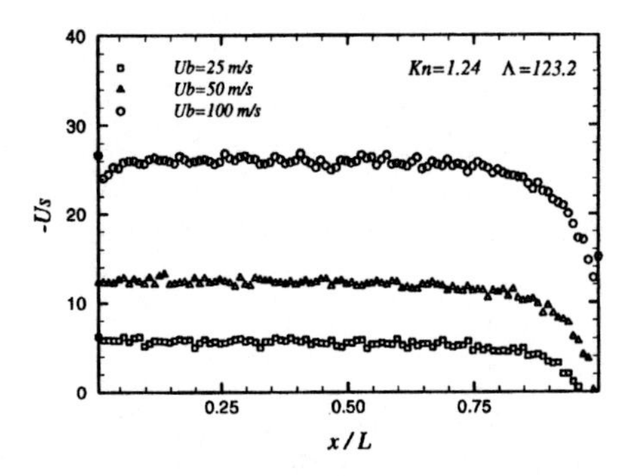

Figure 3. The slip velocities [ms^{-1}] on bottom wall at various boundary velocities

In this case, the effect of boundary driven velocity, U_b, on slip coefficient C_s is investigated. For dispelling the effect of bearing number, the value of Λ is fixed at 123.2. When boundary velocity is changed, the channel length should also be altered accordingly to keep a constant bearing number. The boundary velocities being considered here include U_b = 25 m/s, 50 m/s, 100 m/s, and the respective channel length are L=1.0×10^{-5} m, 5.0×10^{-6} m, and 2.5×10^{-6} m.

The DSMC results confirmed that no direct relationship could be established between slip velocity as shown in Fig. 3 and shear stress distribution near bottom wall as given in Fig. 4. According to Eq. (7), the density would play an important role, therefore a figure of stress-density ratio is drawn for comparison. Figures 3 and 5 indicate that the slip velocity has exactly the same trend as this ratio. There exists only a constant coefficient between the slip velocity and the ratio of shear stress to density as presented in Fig. 6. This constant, named slip coefficient and determined by Eq. (12) has a value of about 0.047 s/m.

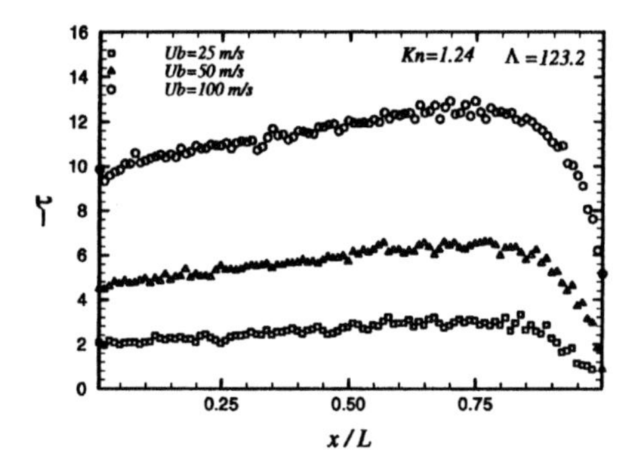

Figure 4. Shear stress [KNm^{-2}] near bottom wall at various boundary velocities

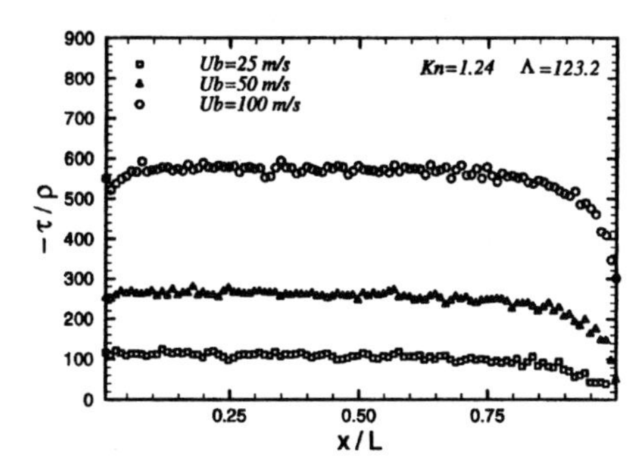

Figure 5. The stress-density ratio [m^2s^{-2}] near bottom wall at various boundary velocities

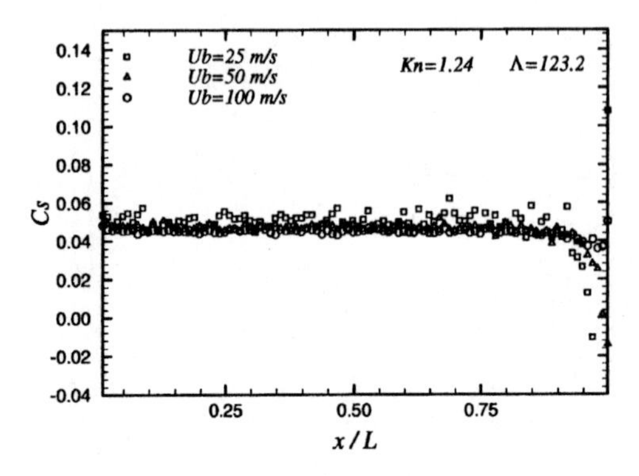

Figure 6. Slip coefficients [sm^{-1}] for bottom wall at various boundary velocities, $\Lambda = 123.16$

When the bearing number is decreased from $\Lambda = 123.16$ to $\Lambda = 61.6$, similar trend and results with same order of slip coefficient can be observed as shown in Fig. 7.

3.2 The slip coefficients for different Knudsen numbers at a specified U_b and bearing number

This case investigates the effect of Knudsen number on the slip coefficient. The bottom wall velocity is fixed as $U_b = 50$ m/s.

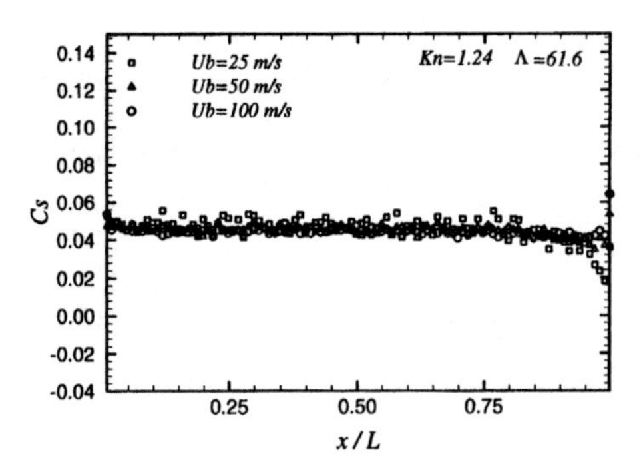

Figure 7. Slip coefficients [sm^{-1}] for bottom wall at various boundary velocities, $\Lambda = 61.6$

Generally, in a modern slider air bearing in the compact hard disk, the K_n number is usually above 1. Therefore, the current analysis begins by choosing the K_n value from 1 to 10.

When K_n number is changed, the minimum spacing of channel, h_0, is also varied with respect to Eq. (13). In order to keep the bearing number a constant value, the channel length should then be adjusted accordingly with respect to Eq. (14). When K_n number being considered here include K_n = 1.24, 2.5, 5.0 and 10.0, then the respective minimum channel height h_0 = 50 nm, 24.8 nm, 12.4 nm and 6.2 nm, and the respective channel length are L = 5.0 μm, 1.23 μm, 0.3075 μm and 0.07688 μm.

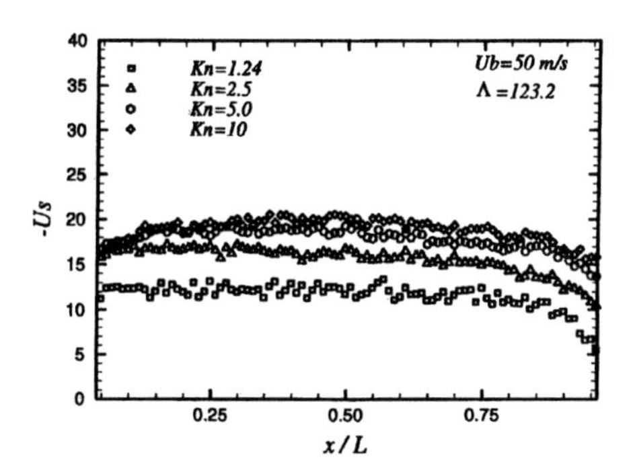

Figure 8. Slip velocities [ms^{-1}] on bottom wall in transition regime

Figure 8 shows the streamwise distribution of slip velocity on bottom wall. The slip velocity distribution on bottom wall has the similar tendency as the ratio of shear stress to density (Fig. 9). When comparing Fig. 8 with Fig. 9 for different K_n values, it is obvious that there exists a slip coefficient between the slip velocity and the ratio of shear stress to density. The results of the study on the slip coefficient are included in Fig. 10. Throughout the range from K_n = 1 to K_n = 10, the coefficient again appears well behaved in similar trend as Figs. 6 and 7. The results confirm again that this coefficient can be considered as a constant value of about C_s = 0.047 s/m. The more rarefied the gas, the more stable would be the value of slip coefficient. Therefore, the proposed SDR model in Eq. (7) shows its validity and applicability and yet remains simple with a constant slip coefficient when deriving a new set of Reynolds equation.

More calculations are further performed to verify the application of the new model, and investigate various parameters that may impact the slip coefficient. For example, the effects of film thickness ratio on the slip coefficient are also calculated by the DSMC method. The results suggest that for different film thickness ratios, the slip coefficients show less difference, and are thus excluded here.

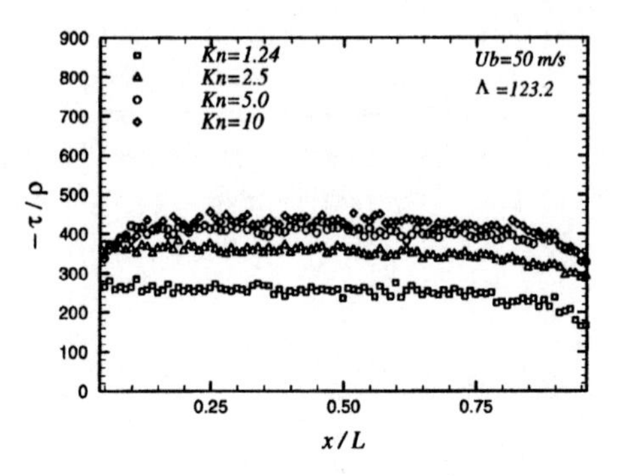

Figure 9. The stress-density ratio [m^2s^{-2}] near bottom wall in transition regime

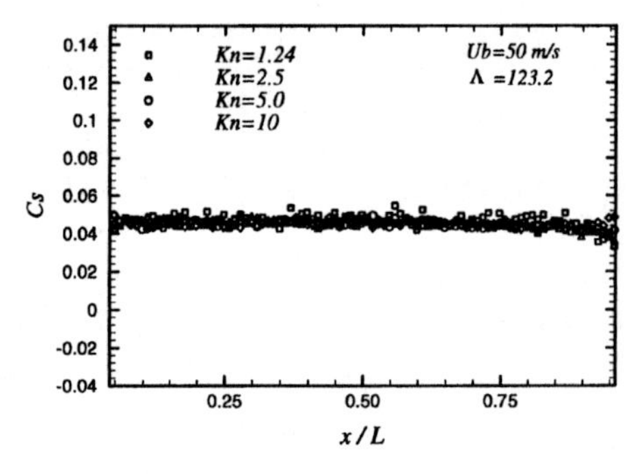

Figure 10. Slip coefficients [sm^{-1}] in transition regime, $K_n > 1$

4. DERIVATION OF SDR MODEL FOR THE MODIFIED REYNOLDS EQUATION

The mass conservation equation is given as:

$$\int_0^h \frac{\partial}{\partial x}(\rho u)dy + \int_0^h \frac{\partial}{\partial y}(\rho v)dy + \int_0^h \frac{\partial}{\partial z}(\rho w)dy + \int_0^h \frac{\partial \rho}{\partial t}dy = 0 \tag{15}$$

Because of the following expression:

$$\int_0^h \frac{\partial f(x,y,z)}{\partial x} dy = -\left[f(x,h,z) - f(x,0,z)\right]\frac{\partial h}{\partial x} + \frac{\partial}{\partial x}\left(\int_0^h f(x,y,z)dy\right) \tag{16}$$

the first term in Eq. (15) becomes:

$$\int_0^h \frac{\partial}{\partial x}(\rho u)dy = (\rho u)_{y=0}\frac{\partial h}{\partial x} + \frac{\partial}{\partial x}\left(\rho\int_0^h udy\right) \tag{17}$$

and the mass conservation equation is rewritten as:

$$(\rho u)_{y=0}\frac{\partial h}{\partial x} + \frac{\partial}{\partial x}\left(\rho\int_0^h udy\right) + (\rho v)_{y=0}\frac{\partial h}{\partial y} + \frac{\partial}{\partial y}\left(\rho\int_0^h vdy\right) + (\rho w)_{y=h} - (\rho w)_{y=0} + h\frac{\partial \rho}{\partial t} = 0 \tag{18}$$

Assuming that the flow is only driven by shear stress, then there are:

$$(w)_{y=h} = 0.0 \tag{19}$$

and

$$(w)_{y=0} = (u)_{y=0}\frac{\partial h}{\partial x} + (v)_{y=0}\frac{\partial h}{\partial y} \tag{20}$$

By substituting Eqs. (19) and (20) into (18), we have:

$$\frac{\partial}{\partial x}\left(\rho\int_0^h udy\right) + \frac{\partial}{\partial y}\left(\rho\int_0^h vdy\right) + h\frac{\partial \rho}{\partial t} = 0 \tag{21}$$

Using the new slip boundary condition as given in Eq. (7), the velocity distribution and the boundary condition are obtained as:

$$u = -\frac{1}{2\mu}\frac{\partial p}{\partial x}\left(C_s vh + yh - y^2\right) + U_b\left(1 - \frac{y + C_s v}{h + 2C_s v}\right) \tag{22}$$

$$u_{y=0} = U_b + C_s \nu \left.\frac{\partial u}{\partial y}\right|_{y=0} \qquad u_{y=h} = -C_s \nu \left.\frac{\partial u}{\partial y}\right|_{y=h} \tag{23}$$

where ν is kinematic viscosity, and $\nu = \mu / \rho$. Then, one obtains:

$$\int_0^h u dy = -\frac{h^3}{12\mu}(1 + \frac{6C_s \nu}{h})\frac{\partial p}{\partial x} + \frac{U_b h}{2} \tag{24}$$

In the analogy with x direction, but without the wall driven velocity in spanwise direction, we have:

$$\int_0^h v dy = -\frac{h^3}{12\mu}(1 + \frac{6C_s \nu}{h})\frac{\partial p}{\partial z} \tag{25}$$

With the geometrical stability assumption of $\partial h / \partial t = 0$, Eq. (21) becomes:

$$\begin{aligned} &\frac{\partial}{\partial x}\left[(\rho h^3 + 6\rho h^2 C_s \nu)\frac{\partial p}{12\mu \partial x}\right] + \\ &\frac{\partial}{\partial z}\left[(\rho h^3 + 6\rho h^2 C_s \nu)\frac{\partial p}{12\mu \partial z}\right] = \frac{U_b}{2}\frac{\partial(\rho h)}{\partial x} + \frac{\partial(\rho h)}{\partial t} \end{aligned} \tag{26}$$

Finally, by introducing the non-dimensional parameters as follows:

$$P = \frac{p}{p_a} \qquad H = \frac{h}{h_0} \qquad X = \frac{x}{L} \qquad Z = \frac{z}{L} \qquad T = \frac{U_b}{L} t \tag{27}$$

and for an isothermal process, which is a widely used assumption in the majority of gas bearing applications,

$$p = \rho RT \tag{28}$$

one then obtains:

$$\begin{aligned} &\frac{\partial}{\partial X}\left\{\left(PH^3 + 6C_s \mu \frac{RT}{h_0} H^2\right)\frac{\partial P}{\partial X}\right\} + \\ &\frac{\partial}{\partial Z}\left\{\left(PH^3 + 6C_s \mu \frac{RT}{h_0} H^2\right)\frac{\partial P}{\partial Z}\right\} = \Lambda \frac{\partial(PH)}{\partial X} + 2\Lambda \frac{\partial(PH)}{\partial t} \end{aligned} \tag{29}$$

This is the three dimensional Reynolds equation for an ultra thin gas bearing lubrication using the newly proposed SDR model. For steady state operating conditions, $\partial(PH)/\partial t = 0$, Eq. (29) has the form of:

$$\frac{\partial}{\partial X}\left\{\left(PH^3 + 6C_s\mu\frac{RT}{h_0}H^2\right)\frac{\partial P}{\partial X}\right\} + \frac{\partial}{\partial Z}\left\{\left(PH^3 + 6C_s\mu\frac{RT}{h_0}H^2\right)\frac{\partial P}{\partial Z}\right\} = \Lambda\frac{\partial(PH)}{\partial X} \quad (30)$$

Obviously, the two ndimensional form of the modified Reynolds equation is:

$$\frac{\partial}{\partial X}\left\{\left(PH^3 + 6C_s\mu\frac{RT}{h_0}H^2\right)\frac{\partial P}{\partial X}\right\} = \Lambda\frac{\partial(PH)}{\partial X} \quad (31)$$

If: $$\psi = C_s\mu\frac{\sqrt{2}\pi d^2}{m} \quad (32)$$

Then: $$\frac{\partial}{\partial X}\left\{\left(PH^3 + 6\psi K_n PH^2\right)\frac{\partial P}{\partial X}\right\} = \Lambda\frac{\partial(PH)}{\partial X} \quad (33)$$

where the VHS molecular model assumption with diameter d, molecular mass m and coefficient of viscosity μ is applied.

The flow rate of the plane Poiseuille flow for the newly proposed slip approximation can further be obtained by substituting Eq. (22) into the below expression:

$$Q_r = -\frac{\int_0^h u dy}{h^3/(2\mu D)\cdot(\partial p/\partial x)} = \frac{D}{6} + \left(\frac{2}{\sigma} - 1\right)\frac{\sqrt{\pi}}{2}\psi \quad (34)$$

where the non-dimensional parameter D is an inverse Knudsen number defined as :

$$D = \sqrt{\pi}/(2K_n) \quad (35)$$

Similarly, the Poiseuille flow rates for first-order, 1.5-order, second-order slip approximations and for the continuum flow can be expressed as:

1st-order: $$Q_1 = \frac{D}{6} + \frac{\sqrt{\pi}}{2}\left(\frac{2}{\sigma} - 1\right) \tag{36}$$

1.5-order: $$Q_{1.5} = \frac{D}{6} + \frac{\sqrt{\pi}}{2}\left(\frac{2}{\sigma} - 1\right) + \frac{\pi}{9D} \tag{37}$$

2nd-order: $$Q_2 = \frac{D}{6} + \frac{\sqrt{\pi}}{2} + \frac{\pi}{4D} \tag{38}$$

Continuum: $$Q_c = \frac{D}{6} \tag{39}$$

Figure 11 compares the flow rates of Poiseuille case for SDR model and those of the previous slip flow models. The flow rate from the Boltzmann equation [12] is included here for better comparison. In transition regime, where the inverse Knudsen number varies from D = 8.86 to D = 0.0886, the first-order slip-flow approximation underestimates the flow rate from the Boltzmann equation, while second-order slip flow and 1.5-order slip flow approximations overestimate it. Using the SDR model, the flow rate produces a good approximation that is closer to the flow rate from the linear Boltzmann solution.

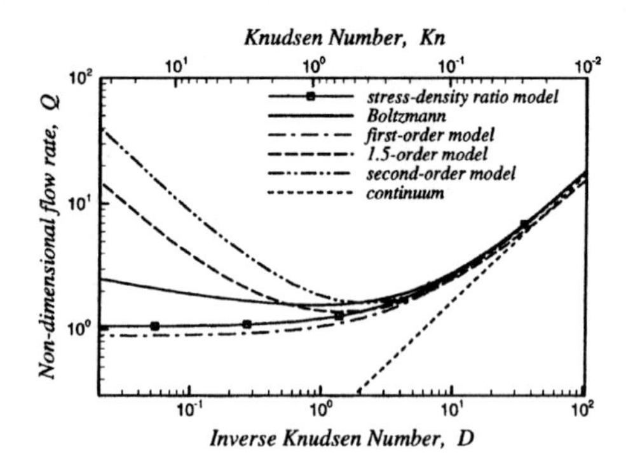

Figure 11. Comparison of non-dimensional flow rate of plane Poiseuille flow

The Reynolds equation is solved using a fourth order Runge-Kutta method. The pressure profiles obtained from the stress-density ratio slip-corrected Reynolds equation and the previous Reynolds equation are shown in Fig. 12, in which the K_n number is 1.24 with the boundary wall velocity and bearing number of 25 m/s and 61.6, respectively. The current DSMC results, those of Alexander et al. [4] and Reynolds equation results with the first-order slip and the second-order slip models are also included in Fig. 12. The Reynolds equation results using SDR model are in good agreement with the DSMC computation. The differences between the DSMC prediction and the conventional approximations using previous models are significant.

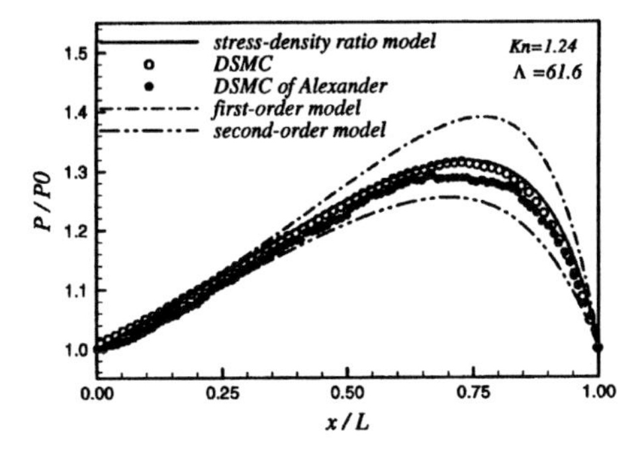

Figure 12. Comparison of slider bearing pressure profiles for $K_n = 1.24$, $\Lambda = 61.6$ and $U_b = 25m/s$

When the boundary wall velocity or bearing number is increased, a similar conclusion is obtained. For example, the results in Fig. 13 with boundary wall velocity of 50 m/s show the similar trend to Fig. 12.

The load carrying capacity is defined as

$$W = \frac{1}{L}\int_0^L \left(\frac{P(x) - P_a}{P_a}\right)dx \tag{40}$$

The relationship between the load carrying capacity W and inverse Knudsen number D for the bearing number, $\Lambda = 10$, is shown in Fig. 14. Obviously, in transition regime, especially for the Knudsen value ranging from 1 to 10, the stress-density ratio slip-corrected Reynolds equation result produces an improved approximation that is closer to the linear Boltzmann equation outcome. The results of the linear Boltzmann equation adopted here for comparison are taken from the Fukui and Kaneko [9] and Mitsuya [12].

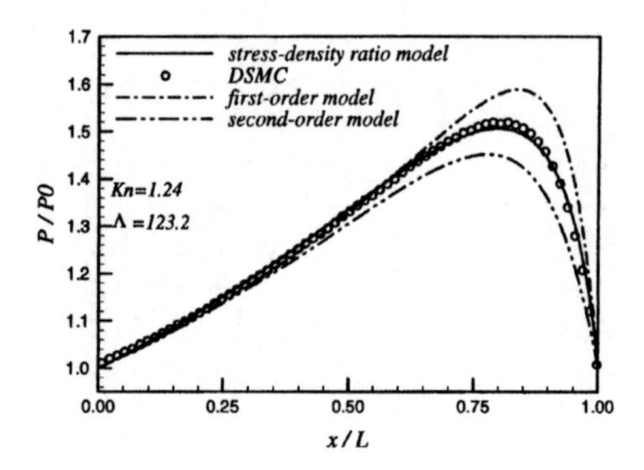

Figure 13. Comparison of slider bearing pressure profiles for $K_n = 1.24$, $\Lambda = 123.2$ and $U_b = 50$ m/s

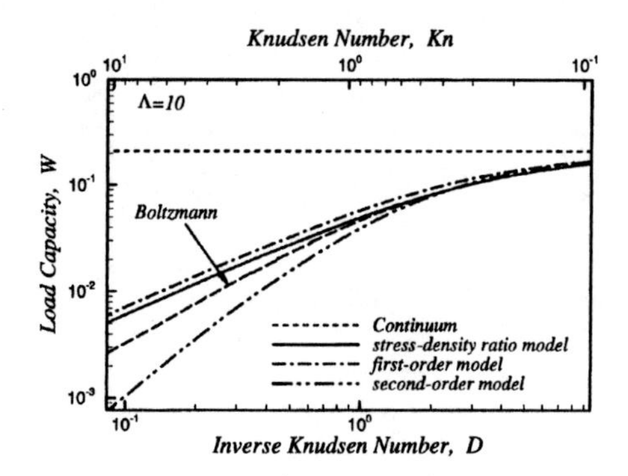

Figure 14. Comparison of non-dimensional load capacity vs. inverse Knudsen number

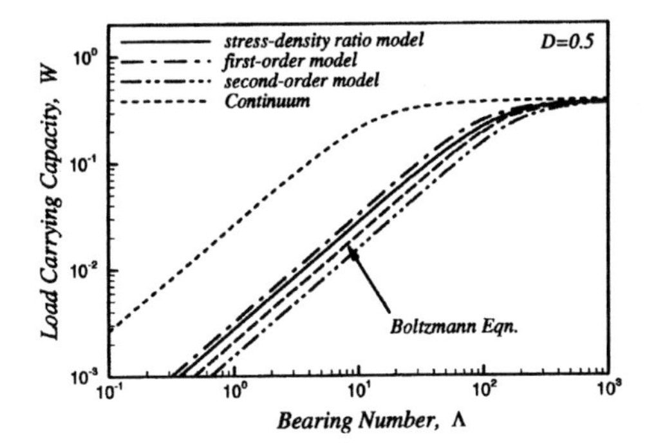

Figure 15. Comparison of non-dimensional load capacity vs. bearing number

The relationship between the load carrying capacity W and bearing number Λ for the mid-range of transition regime ($D = 0.5$) is also presented in Fig. 15. In the whole range of bearing number, the stress-density ratio slip-corrected Reynolds equation produces a better load carrying capacity prediction when compared with other models. The SDR model considers not only the effect of shear stress, but also the explicit density effect on slip velocity. The results indicated that the slip-corrected Reynolds equation with SDR model performs better both in slip and transition regimes than previous models.

5. CONCLUSION

DSMC method is a power numerical method in investigating the gas flows in either slip or transition regimes. By taking this advantage, it is applied to the gas lubrication problem in a slider air bearing. From the DSMC results, the relationship between the slip velocity and other physical properties is unveiled. The study shows that in slip and transition regime, the slip velocity is affected not only by shear stress near the wall, but also by density itself. An improved slip model named SDR model, particularly suitable to the regime where the conventional slip models fail to describe the slippage phenomena accurately. This new model has the following features:

- bases on the reasonable physical background, rather than the simple mathematical approximations.
- has simple expression with a uniform constant slip coefficient, so it can be easily applied in Reynolds equation for gas lubrication.

- is applicable to the commonly encountered velocity range of spinning disk,
- is suitable to the medium range of bearing number,
- is applicable to regime where the traditional models fail,

Nevertheless, the new model has some limitations. It is a model that works better for the transition regime with $K_n > 1$. The numerical results of the new slip-corrected Reynolds equation confirm that the SDR model can provide a useful applicability to the modern computer magnetic storage device. Meanwhile, the physical interpretation of this new model remains to be explored.

REFERENCES

[1] B. Bhushan, *Tribology and Mechanics of Magnetic Storage Devices* (Springer-Verlag, New York, 1990).

[2] W. A. Gross, L. A. Matsch, V. Castelli, A. Eshel, J. H. Vohr, and M. Wildmann, *Fluid Film Lubrication* (Wiley, New York, 1980).

[3] A. Z. Szeri, *Fluid Film Lubrication: Theory and Design* (Cambridge University Press, 1998).

[4] F. J. Alexander, A. L. Garcia, and B. J. Alder, "Direct simulation Monte Carlo for thin-film bearings," Phys. Fluids A Vol. **6**, 3854 (1994).

[5] C. S. Chen, S. M. Lee, and J. D. Sheu, "Numerical analysis of gas flow in microchannels," Numerical Heat Transfer A Vol. **33**, 749 (1998).

[6] G. A. Bird, *Molecular Gas Dynamics and The Direct Simulation of Gas Flows* (Clarendon, Oxford, 1994).

[7] W. Wagner, "A convergence proof for Bird's direct simulation Monte Carlo method for the Boltzmann equation," Journal of Stat. Phys., Vol. **66**, 1011 (1992).

[8] A. Burgdorfer, "The influence of the molecular mean free path on the performance of hydrodynamic gas lubricated bearing," ASME Journal of Basic Engineering, Vol. **81**, 94 (1959).

[9] S. Fukui, S. and R. Kaneko, "Analysis of ultra-thin gas film lubrication based on linearized Boltzmann equation," ASME Journal of Tribology, Vol. **110**, 253 (1988).

[10] S. A. Schaaf, S.A. and F. S. Sherman, "Skin friction in slip flow," Journal of Aeronautical Sciences, Vol. **21**, 85 (1953).

[11] Y. T. Hsia and G. A. Domoto, "An experimental investigation of molecular rarefaction effects in gas lubricated bearings at ultra-low clearance," ASME Journal of Tribology, Vol. **105**, 120 (1983).

[12] Y. Mitsuya, "Modified Reynolds equation for ultra-thin film gas lubrication using 1.5-order slip-flow model and considering surface accommodation coefficient," ASME Journal of Tribology, Vol. **115**, 289 (1993).

[13] A. Beskok, A. and G. E. Karniadakis, "A model for flows in channels, pipes, and ducts at micro and nano scales," Microscale Thermophysical Engineering, Vol. **3**, 43 (1999).

[14] N. Liu and E. Y.-K. Ng, "The posture effects of a slider air bearing on its performance with a direct simulation Monte Carlo method," J. Micromech. and Microeng., Vol. **11**, 463 (2001).

[15] W. D. Niven, *The Scientific Papers Of James Clerk Maxwell, Chap. XCIII.* (Dover Publications Inc., New York, 1965).

Chapter 11

SHORT NOTES ON PARTICLE IMAGE VELOCIMETRY FOR MICRO/NANO FLUIDIC MEASUREMENTS

C. Y. Lim and Francis E. H. Tay
Institute of Materials Research & Engineering, Singapore

Abstract: Particle image velocimetry (PIV) has emerged as an effective and mature tool for flow visualization and measurement technique for its non-intrusive nature. The basic working principle of a PIV system is deriving the flow velocity from particles' motion in an interrogation window by determining the average displacement within a short period. This approach only yields velocity accuracy up to the first order and inherently assumes that particles travel freely while faithfully following the flow. We briefly discuss the application of PIV in micro/nano fluidic measurements by addressing several important issues, such as the limit of particle size, the limit of diffraction of light, and the instrumentation. Lastly, we include some research directions that we plan to carry out in future.

Key words: particle image velocimetry, micro/nano fluidic measurements, diffraction limit, particle size, epi-fluorescence, CCD camera, fast Fourier transform, cross correlation

1. INTRODUCTION

The ability of making micron-scaled devices integrated with electronic circuits (Micro-Electro-Mechanical Systems, MEMS) frees up plenty of room at the bottom for research. Things are made smaller to achieve better response time, materials/waste saving and cost effectiveness. These devices, including microfluidic components and biotech integrated chips-on-glass, are inherently involved in low-Reynolds-number flows, in which inertial forces are negligible and

F.E.H. Tay (ed.), Microfluidics and BioMEMS Applications, 307–329.

viscous and other forces dominate. Some interesting properties of these flows are that turbulence is nonexistent by nature and hence that diffusion plays a vital role in mixing and that the onset of slip of liquid at this reduced dimensions has been reported, bringing much concern about the applicability of the typical no-slip condition at solid wall. Numerical simulations and experimental investigations show change on viscosity in reduced dimensions within molecular scales. These research opportunities lead to a need for a versatile flow visualization and measurement technique that can characterize flow behaviors at submicron or even nanoscales.

Particle image velocimetry (PIV) has emerged as an effective tool for flow visualization and measurement technique for its non-intrusive nature. With a constantly pulsed light source (normally by laser), an image-recording device captures the displacements of particles within short time intervals, allowing the determination of the flow velocity vector field. Two key performance indicators of PIV system are its dynamic velocity range (DVR) and dynamic spatial range (DSR) [1]. DVR of a measurement, or the ratio of maximum measurable velocity to minimum resolvable velocity, is primarily dependent on the interval of two successive images that are produced by the light source; this is also implicitly reliant on the resolution of the recording device. By considering only a smaller time interval, DVR can be improved considerably by having a shorter pulse interval under unchanged experimental conditions, which will ultimately reach the bottleneck due to recording device's frame speed and sensitivity to light. On the other hand, the ability to resolve detectable particle displacement over dimension of interest defines the DSR. As the dimension goes down, detectability of displacement should follow the same scale to keep the dynamic spatial range constant, which is eventually limited by the resolution of the imaging device. We briefly discuss a PIV system tailored for this application in the following section. We then look at some critical concerns that arise in conducting PIV measurements as the dimensions reduce to micro/nano scales, which include arisen issues due to the size of seeding particle and the required instrumentation associated some theoretical background. Lastly, we discuss some potential research interests in related applications.

2. BACKGROUND OF MPIV

A basic PIV system produces particle images with a thin sheet of light often provided by a laser beam and a camera for image recording. These images are then correlated using fast Fourier transform (FFT) and are usually treated with other post-processing techniques before a good quality measurement can be obtained. Optical microscope is often required for magnifying the tiny region of interest and projecting it to the CCD array. Vibration isolation is essential in singling out environmental noise from Brownian motions of particles.

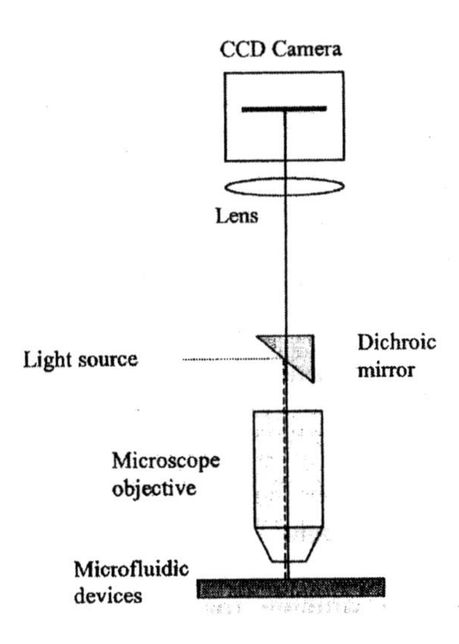

Figure 1. A typical μPIV set up, including the illumination, CCD recording device, a microscope, and microfluidic devices

Good quality images for PIV can be produced by appropriate choice of equipment in system integration. Pulsed lasers gain their popularity as the light source for their ability to deliver thin light sheet with precise duration and interval to freeze particles in motion without using a mechanical shutter. A high-speed high resolution CCD camera is desired to yield micron-order resolution, and an objective nano-positioner to enable focusing on nanometre order. An epi-fluorescence microscope is usually required to zoom in as well as to yield inelastic light scattering for viewing tiny particles (nanometre range). PIV measurements require a synchronized system that integrates all devices, starting from triggering laser pulses for illumination and CCD camera frame grabbing, digital data transfer and storage for offline data analysis. Lastly, the post data processing can be done using mature commercial software, which is essentially an auto- or cross-correlation algorithm locating the peak and deriving the velocities of particles by statistical means within interrogation windows. Digital approach is preferred and used extensively due to occurrence of high-resolution digital cameras; optical means together with photographic-film camera was widely used in early days in performing correlations for displacement peaks. Figure 1 is a schematic diagram depicting a typical μPIV system.

2.1 Working Principle

The basic working principle of a PIV system is deriving the flow velocity from particles' motion in an interrogation window by determining the displacement Δx within a short period Δt as in the following equation:

$$v = \frac{d\mathbf{x}}{dt} \approx \frac{\Delta \mathbf{x}}{\Delta t} \quad (1)$$

This approach inherently assumes that particles travel freely while faithfully following the flow. Note that due to its simplified nature, the velocity obtained is only accurate up to the first order; no knowledge on curved motion can be retrieved (see Fig. 2). In this case, particle size gives an upper limit below which they will follow the flow faithfully, while Brownian motion and light scattering impose the lower limit on the size suitable for seeding. PIV is in overall a statistical means for measuring flow velocities, i.e. particles within an interrogation window with density N_p yield only one velocity vector. Hence, flow imaging and image capturing are important in conducting experiments to produce accurate and quality results; this is particularly true for micro/nanoflows with extremely small dimensions and velocities. With the above system, resolution (which is the size of the first interrogation window) of order of a micron can be achieved in principle, while further improvement in resolution may be achieved by minor change of hardware as well as innovative design of experiment. Local measurements such as concentration (obtained from different fluorescence levels of dyed particles) and velocity can also be obtained, which can be used to further derive other flow properties such as vorticities, strain rate, Reynolds fluxes, etc.

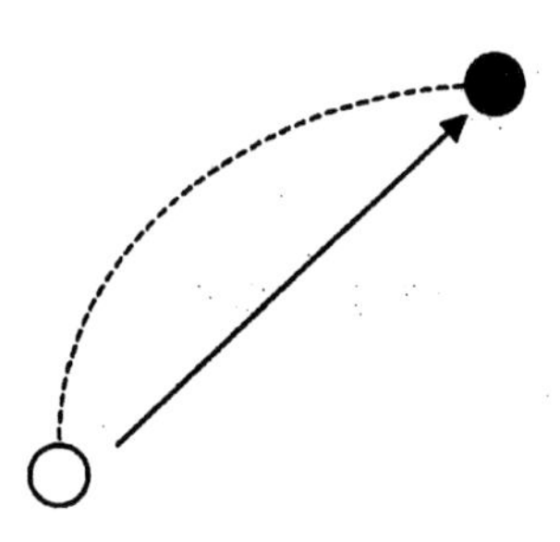

Figure 2. The first order velocity accuracy in PIV system, in which non-linear motion is essentially overcome by having small time interval Δt Auto- and Cross-Correlations

Early PIV [2] used primarily autocorrelation technique in locating displacement peak due to technological availability and historical reasons (particle streaking), which involved essentially doubly exposed images. This effectively increases the image density within an interrogation spot and enhances the correlation peaks. However, not only this method has a narrow range of DVR and hence limited to low-speed measurements, it imposes directional ambiguity for two displacement peaks occur by each side of the self-correlated peak, as shown in Fig. 3(a). Figure 3(b) briefly shows how an auto-correlation can be carried out with a frame of image. Image shifting using a rotating mirror solves the directional ambiguity and broadens the DVR, but further complicates the operations and setting up of the apparatus.

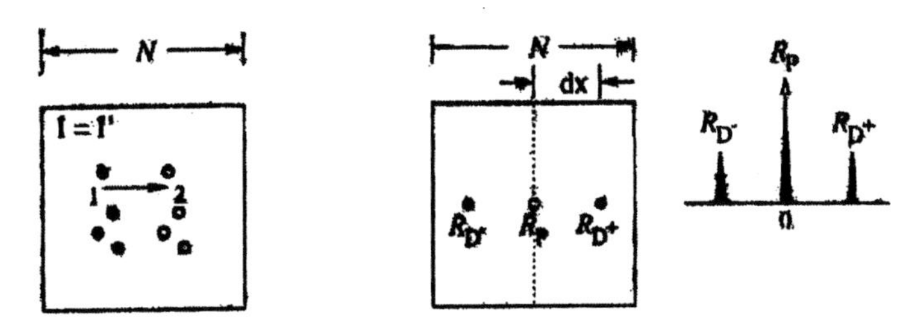

Figure 3(a). Autocorrelation by doubly exposed frames, yielding a strong self-correlation peak and two identical displacement peaks that cause directional ambiguity

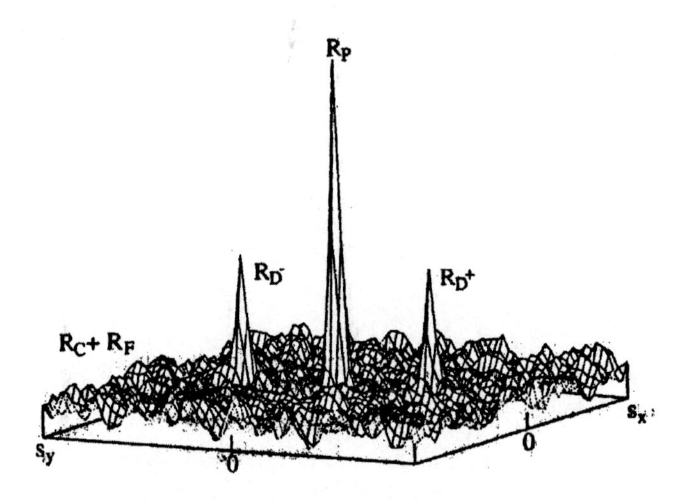

Figure 3(b). Intensity peaks of a self- (R_S) and two displacement-correlations (R_D) in symmetrical arrangement, as well as other much smaller peaks that are composed from convolution R_C and background noise R_F

On the other hand, cross-correlating two consequent images eliminates the directional ambiguity as the sequence of images is known. This approach, often used together with digital cameras, becomes very popular among its practitioners due to its clear directionality and simplicity. Figures 4(a) and 4(b) show how a cross-correlation of two images can be carried out and the resulting peaks after cross-correlation. By the correlation theorem, it can be shown that Fourier transform of the cross-correlation of two functions is equivalent to a complex conjugate multiplication of their individual Fourier transforms:

$$\hat{R}_{II} \Leftrightarrow \hat{I} \cdot \hat{I}'^{*} \tag{2}$$

where R_{II} is the cross-correlation, I and I' intensity values of images and superscript ^ and * denote the Fourier transform and the complex conjugate respectively. Hence, two real-to-complex, two-dimensional (2-D) Fourier transforms plus one complex-to-real, 2-D inverse Fourier transform are needed for each cross-correlation. Figure 5 summarizes the implementation of cross-correlation using Fourier transform and its inverse. In practice, the Fourier transform is efficiently implemented for discrete data using the fast Fourier transform or FFT which reduces the computation from $O(N^2)$ operations to $O(N\log_2 N)$ operations [3]. This effectively reduces the normal cross-correlation procedure of two two-dimensional images $O(N^4)$ operations to only $O(N^2 log_2 N)$ operations, as shown in Fig. 5.

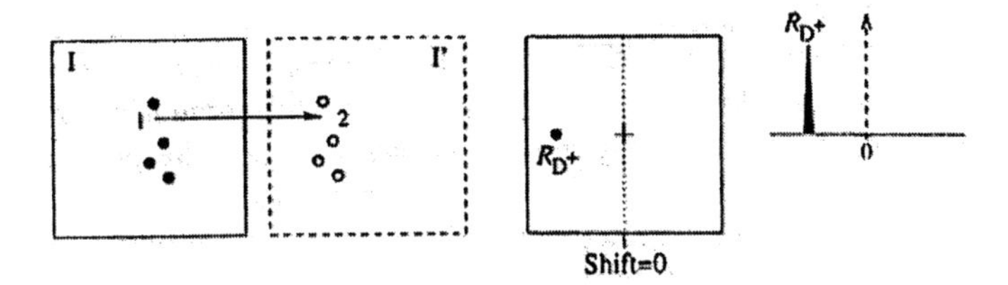

Figure 4(a). Cross-correlation by two different frames, giving only one clear displacement peak eliminating directional ambiguity encountered in auto-correlation

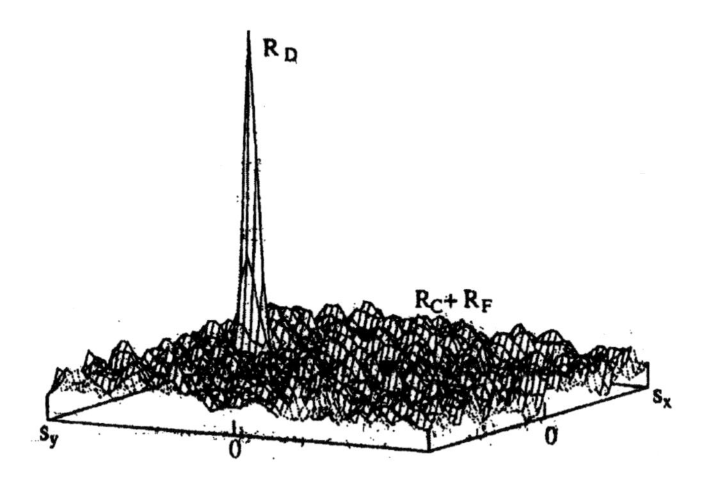

Figure4(b). A typical cross-correlation function, including a strong displacement peak R_S and others due to convolution R_C and background fluctuation R_F

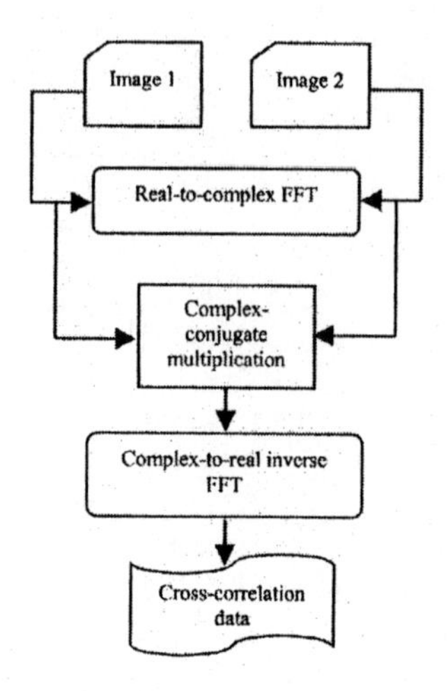

Figure 5. Two 2-D real-to-complex FFT, followed by complex-conjugate multiplication and then a complex-to real inverse FFT in computing a cross-correlation digitally

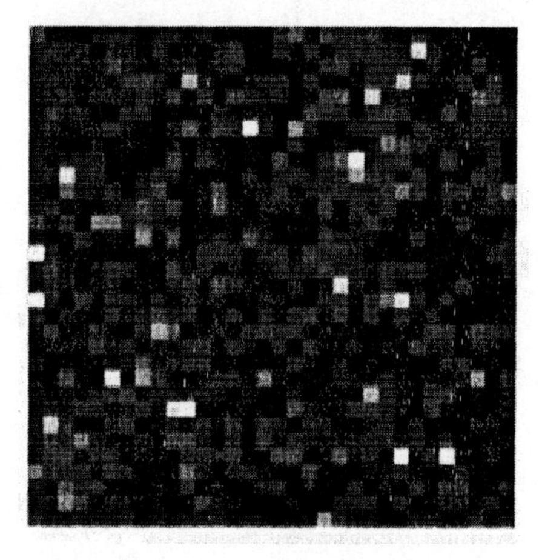

Figure 6(a). A fraction (32X32 pixels) of PIV recording taken in experiments for wake vortex flow behind a transport aircraft, which was originally imaged with digital resolution of 1280X1024 pixels

Figure 6(b). A fraction (32X32 pixels) of PIV recording taken in experiments for microchannel flow with digital resolution of 1100X930 pixels

Figures 6(a) and 6(b) give some idea of how those tracer (seeding) particles will appear on a digital CCD image, which are extracted from macroscopic wake vortex flow [4] and micro-channel flow respectively [5]. These particles appear to be squares on a checkerboard rather than of their normal shapes and that they seem to have higher noise ratio for micro-channel flow under high magnification. By presuming that particle images are Gaussian distributions, centroids of particles and therefore their locations can be determined within one tenth of their sizes, which in turn improves the resolution considerably [5].

3. SEEDING PARTICLES

Apart from fabrication of tiny channels with various complex geometries, which is not directly related to the measurement system, the size of the particle stands out as the primary obstacle for tiny length-scale measurements, as the size of those particles must be negligible compared to the area of interest. This is to ensure negligible disturbance of flow by the seeded particles, and to maintain the faithfulness of particles in tracking the flow. Conventional PIV measurements involve mostly particles of 1μm size, which brings in no significant drawbacks discussed above. Sharp [6] reported clogging of microcapillaries by 1μm particles, in which further reduction of the size of laden particles was restricted by their image detectability due to light refraction at round capillary wall.

A corresponding size for the seeding particles used in visualizing microfluidics is often much smaller than the flow dimensions. At these scales, normal elastic scattering of light will not be sufficient in delivering good images. The submicron requirement in seeding particles in modern PIV measurements calls for inelastic light scattering such as fluorescence technique in rendering sufficient intensity of light. Little investigation has been carried out on the direct scale-down of seeding particles. Adrian *et al.* [7] concluded that at least ten particle images were required to yield sufficient signal to noise ratio for good quality cross correlations via Direct Simulation of Monte Carlo (DSMC). Note also bare measurements without additional optical equipment are almost impossible. Modern PIV system often incorporates a microscope when used in measurements of micro/nanofluidics, which only requires minor optical alignment work [8]. In ref. 5 for example, fluorescent particles of size 200 nm were used in 300x30 μm^2 channel, which is a typical example of roughly 150:1 scale-down on seeding particles. More details on the instrumentation will be discussed in the following section.

3.1 Limit of Particle Size

In addition to the direct scale-down of the particle size, we need to consider the constraint of size imposed by the contraction of area of interest in the context of fluid mechanics. Assuming that the particles follow the flow 'faithfully' so that the velocity difference of particle and flow is small compared to the flow velocity, from Newton's second law

$$|\vec{v}-\vec{u}|(\vec{v}-\vec{u})=\left[\frac{2\rho_p d_p}{3\rho C_D}\vec{a}\right] \quad (3)$$

where $\vec{v}$, $\vec{u}$, $\vec{a}$, ρ, and d are respectively particle velocity, flow velocity, particle acceleration, density (flow) and particle diameter, subscript p denotes particle, and the drag coefficient

$$C_D=\frac{24}{\mathrm{Re}} \quad (4)$$

according to Stokes' law for flow past a sphere, $\mathrm{Re}=\vec{u}d_p/\nu$ is the Reynolds number, ν is the kinematic viscosity. Equations (1) and (2) ultimately gives 0.06 μm at condition $|\vec{v}-\vec{u}|\leq 0.0001|\vec{u}|$ with a time scale of $1\mu s$. This limit renders a guideline when determining size of particle for μPIV measurements, before further consideration of other factors such as effect of Brownian motion, sufficiency of scattering light, and magnification and diffraction limit.

Brownian motion of particles due to diffusion has to be considered when submicron particles are used. By definition, Brownian motion is the irregular movement which small particles of microscopic size carry out due to osmotic pressure when suspended in a liquid, which defines the root mean square of the diffusion distance Δ [9]:

$$\sqrt{\overline{\Delta^2}} = \sqrt{2D}\sqrt{\tau} \tag{5}$$

where τ is the time scale of diffusion and D is the coefficient of diffusivity, given by

$$D = \frac{kT}{3\pi\eta d_p} \tag{6}$$

in which k is the Boltzmann constant, T is the absolute temperature, η is the coefficient of dynamic viscosity, and d_p is the diameter of suspended particles. Santiago *et al.* [8] estimated the relative error due to Brownian motion by displacement of particle Δx following faithfully the flow with velocity u within time τ ($\Delta x = u\tau$) to be

$$\varepsilon_B = \frac{\sqrt{\overline{\Delta^2}}}{\Delta x} = \frac{1}{u}\sqrt{\frac{2D}{\tau}} \tag{7}$$

In their work, the error was reduced by ensemble averaging, which improves the diffusive uncertainty to $\varepsilon_B/\sqrt{N}$, where N is the total independent samples constituted by eight instantaneous realizations of an interrogation spot with five particle images on average. This ensemble averaging reduces effectively the unbiased Brownian uncertainty, and for the example of $N = 40$, the error can be improved by more than six times.

3.2 Limit of Light Diffraction

The image of a distant point source through a circular aperture does not appear as a point, but forms a circular diffraction pattern known as Airy disk on the image plane, as illustrated in Fig. 7(a). If the intensity distribution of the point response function of the lens and the light distribution of the particle by objective are approximated by Gaussian functions, then the image intensity, according to Adrian and Yao [10], will also be Gaussian with the effective diameter given by the following equation:

$$d_e = \sqrt{M^2 d_p^2 + d_s^2} \tag{8}$$

where M is the magnification of lens, d_p is the particle diameter, and d_s is the diameter of point response function of a diffraction-limited lens measured at the first dark ring of the airy disk intensity distribution:

$$d_s = 2.44(M+1) f_{\#} \lambda \tag{9}$$

in which $f_{\#}$ is the focal length divided by the aperture diameter and λ is the wavelength of light. Figure 7(b) shows the close fitting of a Gaussian approximation in representing Airy function for mathematical simplification. Note that if the particle size goes below the wavelength of light, the effective diameter captured will be nominated by the diffraction. Consider a 0.06 μm particle that contributes to only 0.8% of the effective image for 100X and numerical aperture (*N.A.*) of 1.4, which clearly depicts the dominance of diffraction at small diameters. *N.A.* is a direct indication of the ability of lens in gathering light and resolving objects. A direct approximated relationship between numerical aperture and focal number is given as below [11]:

$$NA = \frac{1}{2 f_{\#}} \tag{10}$$

It is clear from the above example that reducing the size of the particle will have insignificant effect on the effective image captured, which in strict sense can only be improved by using better lens (higher *N.A.*) or shorter wavelength of light. Hence, the appropriate size of particle is compromised in between Brownian motion uncertainty and the faithfulness of particle in following the flow from fluid mechanics point of view; also volume occupation of particles within an interrogation spot should be an indication of whether the existence of suspended particles are disturbing the flow. Meinhart *et al.* [5] used 0.05% occupation volume of 200 nm particles for their 13.6X0.9X1.8 μm^3 interrogation window, which was a slightly conservative value compared to conventional flows (0.07% for 1x1x1 mm^3 with 1 μm particles).

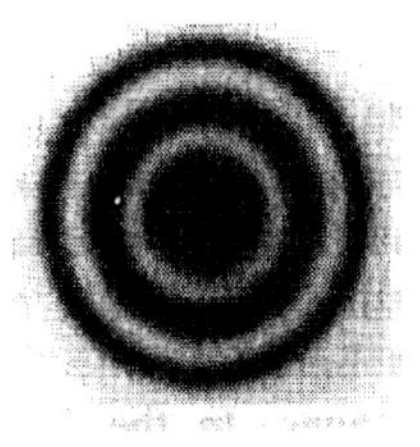

Figure 7(a). A typical Airy disk from light passing through small circular aperture

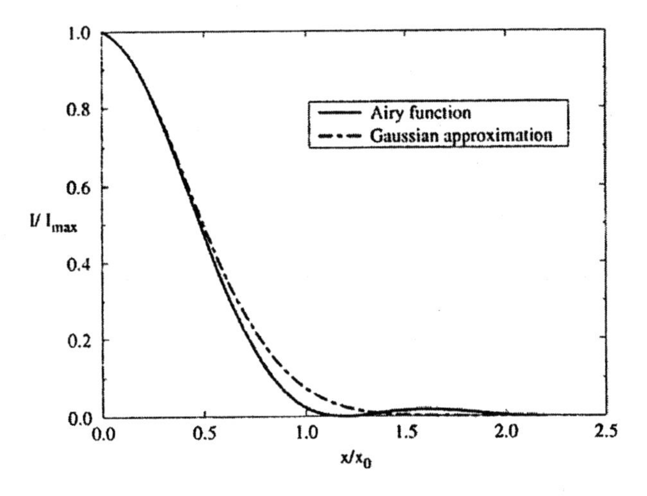

Figure 7(b). A fitting of Gaussian function to approximate Airy function, which simplify extensively mathematical equations involved in optics theory

4. INSTRUMENTATION

Tiny particles in submicron range often have the problem of yielding adequate scattered light for image capturing, as the scattering power is proportional to the fourth power of their diameters [12]. Together with optical and technological limitations on instrumentations, the size of the particle suitable for μPIV applications is confined to a narrow notch, within which yields sufficient light for their images to be captured (often by a CCD camera) while tracking the flow faithfully in strict sense of fluid mechanics. Fluorescent particles are becoming a popular alternative solution as they fluoresce instead of having light scattered elastically ($\propto d_p^4$). This inelastic scattering of light requires an excitation ray of certain range of wavelength in order to emit light of longer wavelength (Stokes' shift). With diameters of 300 nm or less (below visible light wavelength), the light intensity is still very limited by the size; relatively strong pulse of laser is required to achieve acceptable intensity that is good enough for image capturing and processing. This imposes another problem, as the light reflected by the background surfaces often acts as high background noise that will deteriorate the signal to noise ratio of particle images. In the following sections, we briefly describe the instrumentation used in modern flow visualization system for miniaturization, which consist of an image recording system, an optical microscope, and delivery systems.

4.1 Image Recording

The basic working principle of PIV lies on recording images of particle displacements, which show the fluidic motion within interrogation spots after performing data processing. As a result, good quality images are therefore essential for high quality measurements. Photographic films are preferred for its much higher spatial resolution compared to CCD sensors, and are frequently used in turbulence and other measurements featuring high velocity and spatial resolution. However, photographic film technique often has a disadvantage in data processing as analogue data need to be converted to digital data manually for auto or cross correlations using a computer. On the other hand, CCD sensors convert photons received into electric signals, and these digitised data can be fed directly to a computer for further processing. Note that CCD array is strictly an analogue device by itself, but the associated technology in transfer and storage and its heavy dependence on computer make it a digital device. As high-resolution CCD cameras are commercially available at affordable prices now, CCD cameras are becoming very popular in modern PIV measurements. As the scarcity of light is the major factor in getting good images of particles for μPIV application, the CCD array is normally nitrogen or air-cooled to eliminate dark current due to thermal effect. Large dynamic range of a CCD array is also preferred, i.e. a 12 bit range will have 4096 grey levels in resolving background noise and real signals. It carries 16 times more information than that of an 8 bit range by providing higher resolution on the grey levels and therefore is more beneficial in quantitative analysis. In addition, an elegant alternative is to employ an epi-fluorescence microscope, which is discussed in the following section. Another way to improve the signal to noise ratio is through binning, in which several adjacent pixels are combined during readout to yield greater signal. A 2X2 binning will improve four times the signal to noise ratio by providing four times more electron with constant read noise while shrinking four times the measurement area. A side advantage of binning is to increase the frame rate with the reduction of number of pixels to be readout.

The size of a single CCD pixel, which is also the smallest resolvable length in terms of instrumentation, has an important role in defining the DVR and DSR of the device for arbitrary flow measurements, which in turn limits the maximum flow velocity and physical area of interest. However, the image of an effective particle projected on CCD sensors is usually an order larger than a pixel size, this makes the pixel size not a premium consideration factor in improving experimental resolution. Take our previous particle for instance, the effective diameter under 100X with *N.A*=1.4 is roughly 50 μm, while a typical CCD pixel size is about 6~7 μm. In this case, about ten pixels are required for an image of a particle; this implies that further refining the size of pixel does not improve the resolution effectively. On the other hand, since the flow area is zoomed in under magnifications, larger array of CCD sensors gives more allowance on the physical area of interest, or better resolution (in terms of vector counts) with the same physical area provided the particle size is larger than 3~4 pixels [13]. The spatial resolution is usually defined as the first interrogation window of the image, which is often associated with an optimum seeding density. Depending on experiments, optimum number of particles within an interrogation window ranges from 5 to 20. With larger array of CCD sensors, the measurement resolution can be increased accordingly because either the flow domain can be better resolved spatially with more pixels within the same size of interrogation window or the flow field being measured can be extended with respect to the increase of CCD sensors. Larger array of CCD sensors also tends to have less error in capturing and locating particle images that fall in between sensors. A projected image on CCD with 13μm (corresponding to a minimum of 2 pixels in resolving an particle image) forms the lower limit of which tracer particle size can be reduced taking into consideration of magnification and diffraction.

Normal CCD camera that has a 30 frames per second can be used in conjunction with the frame straddling technique, which arranges two closely delayed laser pulses at the change of two successive frames (frame straddling) [14], as illustrated in Fig. 8. Two back-to-back images can be obtained within 200 ns with current CCD camera's technology, which is sufficient to cover almost all the velocity range in fluidic measurements. Such a high temporal resolution is impracticable with a continuous wave laser and a mechanical shutter economically. While short temporal delay of laser pulses is desired to improve correlation peak intensities by eliminating the out of plane movements and in-plane losses (loss of particles within two correlation windows), it is a rigid parameter whose change is associated with other factors such as the fluid velocity, optical magnification, and interrogation window. Note also that time resolved studies are impossible using this technique.

We mentioned earlier that background reflection constitutes a huge portion of noise, and suggested that a cooled, high dynamic range CCD camera to be used. A direct quick means is to place a filter in front of the CCD camera that will block out the excitation light reflected by the background. We will look at another way to overcome this problem, taking advantage of the Stokes' shift of the fluorescence in the following section. We devote a paragraph of basic quantum physics to depict what Rayleigh (elastic) scattering, Raman scattering, fluorescence (inelastic scattering), and phosphorescence are in the following.

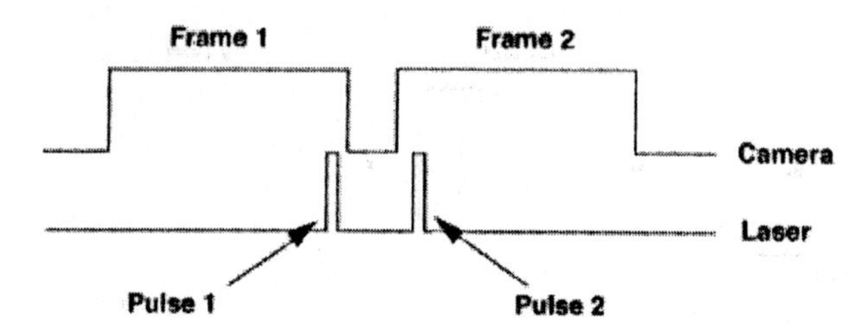

Figure 8. Frame straddling used in PIV measurements to provide a much faster frame rate where two laser pulses are shot at the changing of frames. Note that these pictures taken back-to-back cannot render time-resolved behaviors at the rated frame rate

Rayleigh scattering is the absorption and re-radiation of light by an object, which does not involve electron transition and energy loss, and therefore causes no shift in wavelength. When a molecule in the base state S_0 is exposed to light (energy higher than the threshold), the kinetic energy of the electrons in the molecule is increased, shifting the molecule into the excited state S_1 with a higher energy level as shown in Fig. 9(a). The excited molecule is unstable, so the high-energy electrons will undergo some radiation-less transition to a lower excited state within a short period of time, from which they go back to the base state radiating the excessive energy in the form of light. This is called a Stokes' shift. The associated energy loss during the radiation-less transition is dissipated via vibration and heat, and the emission after Stokes shift therefore has lower frequency/longer wavelength than that of the excitation in accordance to the energy equation:

$$E = hf = \frac{hc}{\lambda} \tag{11}$$

where E is energy of emission, h is Planck's constant 6.626×10^{-34} Js, c is the speed of light, and λ is the wavelength of radiation. Further radiation-less transition to lower energy level leads to phosphorescence, which has longer life span than fluorescence. Figure 9(b) shows the excitation and emission spectra of 0.06μm dyed-polystyrene beads in DI water with 2% solid, where a 30nm Stokes' shift was reported. Note that the sharp excitation spike in the emission spectrum that is caused by Rayleigh scattering of suspended particles.

4.2 Epi-Fluorescence Microscopy

A typical epi-fluorescence microscope comprises a filter cube capable of allowing only the emission of the fluorescence from the particles to pass through the barrier filter and at the same time reflecting any other rays having shorter wavelength by a dichroic mirror. Inverted microscope has recently been used for this purpose in microchannel measurements [5], together with a pulsed Nd:YAG laser system capable of delivering 5 ns pulses with 500 ns delay. The additional component that a μPIV system requires compared to its macroscopic counterpart is essentially the microscope. By incorporating the epi-fluorescence function to the microscope, not only it helps magnify the test section to the desired size, it plays a vital role in coordinating the excitation and emission light to yield high signal to noise ratio in recorded images.

Volume illumination rather than sheet illumination is applied in microfluidic measurements because of the difficulty in delivering a comparatively thin laser sheet corresponding to the flow measurement area. For instance, a typical micro-channel measurement requires 1~10μm laser sheet properly aligned to the area of interest for a typical out of plane resolution of similar scale, and this is of great difficulty if not impossible in both producing and aligning such a thin sheet of laser, and only volume illumination remains practical [15]. As a result, the out of plane resolution is defined by the measurement depth of the system. The measurement depth is twice the distance from centre of the object plane at which a particle can be located such that its maximum image intensity is an arbitrarily specified fraction of its maximum in-focus intensity, beyond which the particle image intensity will only be perceived as noise.

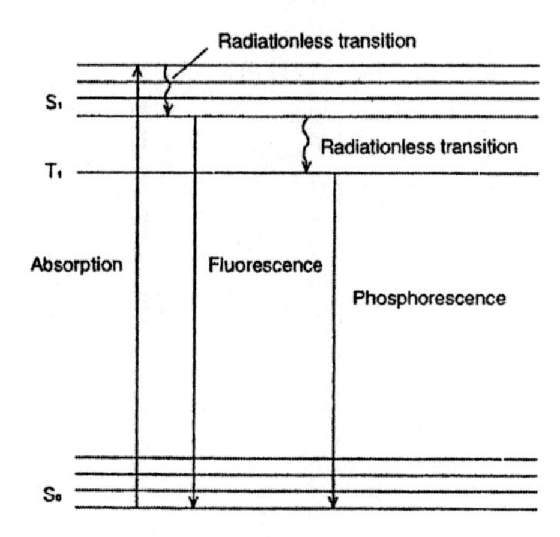

Figure 9(a). The principle of fluorescent emission involving absorption and conversion of photon energy into electron energy, in which energy loss due to vibration and heat imposes a shift of energy into longer wavelength

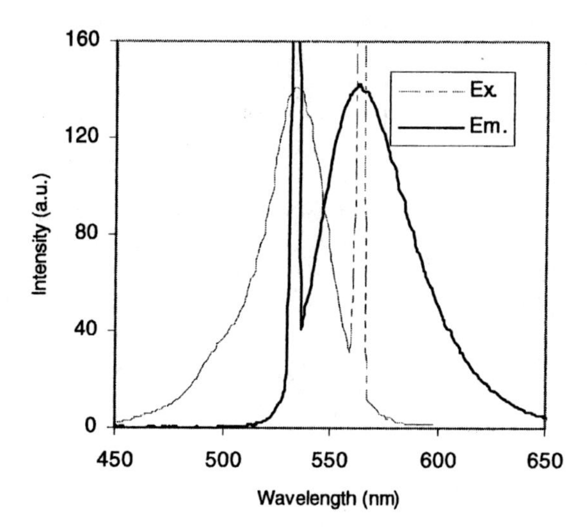

Figure 9(b). Excitation and emission spectra of *Bangs Labs Inc.*'s 0.06μm dyed polystyrene microbeads captured using *Shimadzu*'s RF-5301PC spectrophotometer, in which the Rayleigh scattering of suspended particles are clearly shown in both spectra

4.3 Delivery Systems

Delivering liquid flow at desired experimental conditions is a mandatory procedure for any microfluidic experiment to be carried out. Precision syringe pumps have been popular in delivering constant flow rate of typically 200 μl/hr [5]. A 'blow-down' of a highly charged pressure vessel provides consistent flows for a considerable period as the pressure deteriorates, which generates high velocity of flow for high Re applications [6]. Another approach worth mentioning is the electrokinetically driven flow by Paul *et al.* [16], which produces plug-like flow profile by applying a high-voltage dc electric field along the capillary between platinum electrodes. Other innovative practices include vertical adjustment of manometers' positions that are connected to ends of microchannels and by gravity (slanting the microchannel).

The microchannel may be fabricated from silicon wafer, etched from glass, or even off-the-shelf flexible fused silica capillaries, depending upon the research motives. Take the capillary tubing that is commercially available for example: the tiny tubing is usually coated with polyimide for protection, and ferrules are used at ends to enhance better grip between the tubing and other adapters connecting to conventional tubes (1/8 or ¼ inch in diameter). In the case of a larger interaction area, such as a mixing field for micromixer, only glass etching and anodic bonding can be feasible at laboratory scale. Nanoports that can be adhered on glass by *Upchurch Scientific* provide reliable fluid connections for lab-on-a chip devices.

5. POTENTIAL INTERESTS OF RESEARCH

The research opportunities are many-folded with this enabling instrumentation. The slip and viscosity measurements can be obtained directly from the velocity field obtained by PIV, in contrast to effort found in the literature that mainly concentrate on measurement of flow rate for slip investigation and measurement of the induced force by SFA (Surface Force Apparatus) for viscosity derivation. Whole-field visualization of a mixing junction [17] yields a better insight of fluidic interaction and vortices formation at low Reynolds number. Another interesting phenomenon under this low Reynolds number condition is the mixing of two fluids by merely diffusion, which is insignificant compared to that induced by turbulence at macroscopic dimensions. An investigation on diffusion of particles within two fluids through a mixing channel has been reported in the literature [18]. The ability in visualizing fluidic flow in reduced dimensions renders better understanding in fundamentals in micro/nano fluid dynamics. Fundamental fluid dynamics at submicron scales can be better understood by carrying out field measurements over obstructing objects and over complex geometries. We briefly discuss some research areas that we identify as interesting research scopes below.

5.1 Miniaturization Effects

At much reduced length scales, it is generally agreed that the static properties of fluid will remain unchanged. However, increasing interests have been focused on the dynamic properties, such as the dynamic viscosity of the fluid. Pfahler *et al.* [19] reported a decrease in apparent viscosity by measuring the flow rate without considering slip effect, while other groups concluded an increased apparent viscosity while losing fluidity (becoming solid-like) at monolayers above the wall by surface force apparatus (SFA) [20]. An associated phenomenon in this reduced length scale is the onset of slip. Bocquet and Barrat [21] reported no slip above ten monolayers, while Churaev *et al.* [22] discovered remarkable slippage of liquids over lyophobic solid surfaces. Here we propose to study both viscosity and slip of liquid at small dimensions, looking at the facts that both of them are related to shear on surfaces and that both are constituted by velocity, of which our proposed PIV system is capable of measuring.

5.2 Fundamentals and Fluid Dynamics

Fundamentals of fluid dynamics for micro/nano scales can be studied using the proposed PIV system, which renders a direct flow measurements compared to other indirect measurements, such as shear force, bulk flow rates, etc:

- Flow past an electrically/magnetically held still μsphere – a particle is to be held still electrically/magnetically in a microchannel against a stream of fluid flowing past; where we intend to study Stokes' law and Navier-Stokes equation for submicron spheres.
- Two dimensional micro/nanofluidics – visualizing flow past a circular cylinder matrix in Hele-Shaw cell at submicron level can be meaningful in studying wake and vortex formation in the strong presence of viscosity.
- Vortex formation in low-Reynolds-number flows – turbulence is inherently non-existent for $Re \ll 1$ as any perturbation will be damped out by the dominating viscous force; however, Brody [23]

reported vortex formation near an air/fluid interface. We intend to study these interfaces by forcing a bubble through a microchannel; the vortices formed may be a good source for mixing in low-Reynolds-number flow.

5.3 Other Potential Research Opportunities

With this PIV system, characterizations of microfluidic devices can also be done. For example, a crossed junction can be fabricated to characterize the hydrodynamic focusing of main flow by 2 side flows [17]; diffusion of fluorescent particles to a buffer can be investigated along a mixing microchannel by combining velocity and concentration measurements; in both cases fluid dynamics can be studied at submicron scales, which may eventually decrease down to nano scales. The *in situ* hydrodynamics in a carbon nanotube can now be studied, in which movement of bubble in a 100 nm nanotube was reported [24]. This will be the benchmark of our ultimate goal of achieving a nanoscale PIV system.

6. CONCLUDING REMARKS

PIV is a mature technique for macroscopic flow measurements since 1980s. Extending its application to micro/nanoscales is straightforward if a few issues are well taken care of. These include primarily finding a suitable size of fluorescent particles according to optics theories and fluid mechanics, and secondarily, integrating proper instrumentations to yield good quality images within desired range of velocity and spatial resolution for post-data processing. Subsequently, several methods in providing wide range of flow velocities are identified and discussed. We infer that μPIV will serve as an enabling tool in modern flow visualization and measurement technique in the foundational and developmental exploration in this miniaturization era. Its enhanced features extend current research capabilities and provide more flexibility in maneuvering experiments. This provides a better insight of fluid dynamics at reduced dimensions, and makes itself an essential tool in characterizing and helping in design of microfluidic in addition to the viscosity deviation and slip onset. In particular, studies on flow past a nanosphere and micro/nanofluidics under electric and magnetic fields are also made possible. While μPIV remains an enabling equipment for micro/nanosystems, it serves as a stepping-stone in designing other enabling instrumentations which would better complement micro/nanosystems by understanding the prerequisites in characterizing micro/nanofluidic behaviors and measurements.

REFERENCES

[1] R. J. Adrian, "Dynamic ranges of velocity and spatial resolution of particle image velocimetry," Measurement Sci. Tech., Vol. **8**, 1393 (1997).

[2] R. J. Adrian, T. Asanuma, D. F. G. Durao, F. Durst, J. H. Whitelaw, "Laser anemometry in fluid mechanics-III, Selected papers from the Third international symposium on applications of laser anemometry of fluid mechanics," Ladoan, Lisbon, Portugal (1988).

[3] E. O. Brigham, *The fast Fourier transform* (Prentice-Hall, New Jersey, 1974).

[4] J. Kompenhans, L. Dieterle, H. Vollmers, R. Stuff, G. Schneider, T. Dewhirst, M. Raffel, C. Kähler, J. C. Monnier, K. Pengel, "Aircraft wake vortex investigations by means of particle image velocimetry" measurement technique and analysis methods," In Proceedings of 3rd International Workshop on PIV, Santa Barbara, USA (1999).

[5] C. D. Meinhart, S. T. Wereley, J. G. Santiago, "PIV measurements of a microchannel flow," Experiments fluids, Vol. **27**, 414 (1999).

[6] K. V. Sharp, "Experimental investigation of liquid and particle-laden flows in microtubes," Ph. D. Thesis, University of Illinois at Urbana-Champaign (2001).

[7] R. D. Keane and R. J. Adrian, "Theory of cross-correlation analysis of PIV images," Appl. Scientific Res., Vol. **49**, 191-215 (1992).

[8] J. G. Santiago, S. T. Wereley, C. D. Meinhart, D. J. Beebe, R. J. Adrian, "A particle image velocimetry system for microfluidics," Experiments Fluids, Vol. **25**, 316 (1998).

[9] A. Einstein, *The Brownian Movement* (Methuen & Co., London, 1926).

[10] R. J. Adrian and C. S. Yao, "Development of pulsed laser velocimetry for measurement of fluid flow," in Proceedings, 8th Biennial Symposium on Turbulence, G. Patterson and J. L. Zakin, Eds., 170-186 (1984).

[11] M. Born and E. Wolf, *Principles of optics* (Pergamon Press, Oxford, 1991).

[12] R. J. Adrian and C. S. Yao, "Pulsed laser technique application to liquid and gaseous flows and the scattering power of seed materials," Appl. Optics, Vol. **24**, 44 (1985).

[13] A. K. Prasad, R. J. Adrian, C. C. Landreth, P. W. Offutt, "Effect of resolution on the speed and accuracy of particle image velocimetry interrogation," Experiments Fluids, Vol. **13**, 105 (1992).

[14] M. Raffel, C. Willert, J. Kompenhans, *Particle Image Velocimetry: A Practical Guide* (Sringer, Berlin, 1998).

[15] C. D. Meinhart, S. T. Wereley, M. H. B. Gray, "Volume illumination for two-dimensional particle image velocimetry," Measurement Sci. Tech., Vol. **11**, 809 (2000).

[16] P. H. Paul, M. G. Garguilo, D. J. Rakestraw, "Imaging of pressure- and electrokinetically driven flows through open capillaries," Anal. Chem., Vol. **70**, 2459 (1998).

[17] J. B. Knight, A. Vishwanath, J. P. Brody, R. H. Austin, "Hydrodynamic Focusing on a Silicon Chip: Mixing Nanoliters in Microseconds," Phys. Rev. Lett., Vol. **80**, 3863 (1998).

[18] J. P. Brody and P. Yager, "Diffusion-Based Extraction in a Microfabricated Device," Sensors Actuators, Vol. **58**, 13 (1997).

[19] J. Pfahler, J. Harley, H. Bau, "Gas and Liquid Flow in Small Channels," Micromechanical Sensors, Actuators, and Systems, ASME DSC Vol. **32**, 49 (1991).

[20] J. Van Alsten and S. Granick, "Molecular Tribometry of Ultrathin Liquid Flims," Phys. Rev. Lett., Vol. **61**, 2570 (1988).

[21] L. Bocquet and J.-L. Barrat, "Hydrodynamic Boundary Conditions and Correlation Functions of Confined Fluids," Phys. Rev. Lett., Vol. **70**, 2726 (1993).

[22] N. V. Churaev, V. D. Sobolev, A. N. Somov, "Slippage of Liquids over Lyophobic Solid Surfaces," Journal of Colloid and Interface Science," Vol. **97**, 574 (1984).

[23] J. P. Brody, "Fluid and Cell Transport through a Microfabricated Flow Chamber," Ph. D. Thesis, Princeton University (1994).

[24] Y. Gogotsi, J. A. Libera, A. G. Yazicioglu, and C. M. Megaridis, "*In-situ* Fluid Experiments in Carbon Nanotubes," Materials Research Society Symposium Proceedings, Vol. **633**, A7.4.1-6 (2001).

INDEX

Printed in the United States
23406LVS00001B/149